1991 Year Book of Developmental Biology

Joel M. Schindler

CRC PRESS, INC.
Boca Raton, Florida

Year Book is a federally registered trademark of Year Book Medical Publishers, Inc. and is used pursuant to a license from Year Book Medical Publishers, Inc.

This book represents information obtained from authentic and highly regarded sources. Reprinted material is quoted with permission, and sources are indicated. A wide variety of references is listed. Every reasonable effort has been made to give reliable data and information, but the author and the publisher cannot assume responsibility for the validity of all materials or for the consequences of their use.

Direct all inquiries to CRC Press, Inc., 2000 Corporate Blvd., N.W., Boca Raton, Florida, 33431.

©1991 by CRC Press, Inc.

International Standard Book Number 0-8493-3308-3
International Standard Serial Number 1042-8607

Printed in the United States

THE EDITOR

Joel Schindler received his B.Sc. degree in biology from the Hebrew University in Jerusalem, Israel, in 1973 and his M.Sc. degree in biochemistry from the same institution in 1975. The following year, he returned to the United States with his doctoral mentor, Professor Maurice Sussman, to complete his doctoral studies. He was awarded his Ph.D. from the University of Pittsburgh in 1978.

From 1978 to 1981, Dr. Schindler was a postdoctoral research fellow at the Roche Institute of Molecular Biology in Nutley, New Jersey. During this time period, Dr. Schindler's research efforts focused on the study to investigate changes in gene expression during the peri-implantation period of mouse development. In addition, he was involved in a series of studies aimed at unraveling the mechanism of action of retinoids in inducing murine embryonal carcinoma cell differentiation.

Following his tenure at Roche, Dr. Schindler became an Assistant and, subsequently, an Associate Professor in the Department of Anatomy and Cell Biology at the University of Cincinnati College of Medicine in Cincinnati, Ohio. In addition, he was a member of the Graduate Program in Developmental Biology at the Institute for Developmental Research, Children's Hospital Research Foundation, Cincinnati. Dr. Schindler participated in several team-taught courses to both graduate and medical students and was primarily responsible for the areas of cell differentiation and early embryo development. His research efforts remained focused on the regulation of gene expression during cell differentiation and specifically included defining the role of polyamines in regulating the differentiation of both murine and human embryonal carcinoma cells.

In 1987, Dr. Schindler accepted a position in the Genetics and Teratology Branch at the National Institute of Child Health and Human Development (NICHD), in Bethesda, Maryland. His current responsibilities include developing and overseeing NICHD-supported projects in the areas of basic developmental genetics and early embryo development. His unique position allows Dr. Schindler to closely monitor current progress and publications in the field of developmental biology.

Dr. Schindler has received several fellowships and awards; has been a Visiting Fellow at Macquarie University, New South Wales, Australia; and has served as both an editorial and grant reviewer for numerous journals and funding institutions. He is a member of the Society for Developmental Biology, Sigma Xi, the American Society of Cell Biology, the American Association for the Advancement of Science, and the New York Academy of Sciences. He is the author or coauthor of scientific reports in numerous journals, books, and symposia volumes, This volume is Dr. Schindler's third editorial venture with CRC.

EDITORIAL BOARD

JOURNALS REPRESENTED

Cell
Development
Developmental Biology
Differentiation
EMBO Journal
Experimental Cell Research
European Journal of Immunology
Genes and Development
Genetics
Genomics
Immunology
Journal of Cell Biology
Journal of Experimental Medicine
Journal of Immunology
Journal of Neuroscience
Molecular and Cell Biology
Nature
Neuron
Nucleic Acids Research
Oncogene
Proceedings of the National Academy of Sciences of the United States
  of America
Roux's Archives of Developmental Biology
Science

TABLE OF CONTENTS

1
Developmental Genetics 1

2
Developmental Gene Expression 69

3
Developmental Cell Biology 97

4
Maternal Controls, Cytoplasmic Determinants,
and Imprinting.. 155

5
Cell Interactions 169

6
Cell Lineage and Developmental Fate 201

7
Cytodifferentiation — Cell- and Tissue-Specific Gene Expression and
Maintenance 215

8
Homeobox Genes 243

9
Morphogenesis and Pattern Formation 271

Author Index 297

Subject Index... 305

INTRODUCTION

Developmental biology is, fundamentally, the study of various biological processes, including the control of gene expression, the specification of cell fates, and the formation of pattern, all of which are characterized by necessary temporal and spatial events. Understanding these processes requires the integration of observations based on embryological, genetic, biochemical, physiological, and molecular biological investigations. Whereas molecular characterization has been the recent focus of the field, extensive biological groundwork is needed as a foundation upon which to build the molecular components.

Our current understanding of eukaryotic development suggests that a complex array of regulatory circuits must be coordinately controlled. Such control entails more than "simply" ensuring that the correct events occur in the right time and in the right place. It requires the regulation of families of genes, requires the multiple expression of individual genes, and requires the interaction of multiple gene products.

Through this complexity, some unifying themes are beginning to emerge. One area of current excitement is the evolving concept of "universality" among certain bioactive molecules. This universality suggests that a number of developmentally relevant molecules are shared by multiple organisms and perform similar functions in various developmental systems. In addition, such molecules may perform different functions in the same organism at different times.

The recent demonstration that certain biomolecules have been conserved throughout evolution suggests that these molecules may play important functional roles during eukaryotic development. Peptide growth factors are one such class of biomolecules. So called because of the assays used for their original isolation, these growth factors are known to influence cell behavior in many ways.

Several research groups, exploiting the biological properties of the frog *Xenopus laevis,* have demonstrated that certain growth factors play an important role in the induction of the basic body plan in the embryo. Thus, peptide growth factors play a significant role in specifying cell fates by induction. Coupled with other signaling molecules, like hormones or retinoic acid, they represent an essential component in the signal transducing machinery responsible for this major developmental process.

Homeobox-containing genes are a second class of such molecules. The homeobox domain has been evolutionarily conserved and has been found in the genomes of many animal species. The homeobox region encodes a DNA-binding domain of the helix-turn-helix motif, suggesting that homeobox-containing proteins can regulate developmental processes at the level of gene transcription observations in both mouse and *Drosophila* have shown that homeobox-containing proteins can bind to their own regulatory sequences and similar sequences in other homeobox-containing genes and thus have the potential to regulate each other's expression.

The helix-turn-helix motif could also serve to facilitate the interaction of several proteins, which could then function in aggregate as a transcription factor.

It has recently been shown that growth factors can induce the expression of homeobox-containing genes. Studies in *Xenopus* have shown that exposure to certain growth factors, or combinations of growth factors, leads to the induction of homeobox-containing genes. These observations thus identify putative molecular participants in the cascade of events that lead to mesoderm induction.

The extent of evolutionary conservation between *Drosophila* and murine homeobox-containing genes strongly suggests that such genes play a critical role in the developmental plans of metameric organisms. Yet we remain woefully ignorant of how these genes accomplish their biological roles.

A second area of current excitement is mammalian embryology. Although the mouse is unlikely to replace the fruit fly, *Drosophila,* as the workhorse of research in developmental biology, recent technological advances have greatly facilitated the use of the mouse as a powerful experimental model. The polymerase chain reaction (PCR) has revolutionized the field of mammalian developmental genetics. A major application of the technique is in the investigation of gene expression. The traditional method of mRNA isolation from mouse embryos, followed by Northern blot analysis with specific probes, was difficult and labor intensive. Reverse transcription of total RNA followed by PCR amplification of the resulting cDNAs can generate detectable amounts of low-abundance transcripts. With the appropriate selection of primers for the PCR reaction, it will be possible to isolate mammalian homologues of genes in various superfamilies that have been evolutionarily conserved and are presumably developmentally relevant, such as growth factor genes and homeobox-containing genes.

Insertional mutagenesis — the random integration of exogenous DNA into an endogenous gene, thus inactivating it - is one technique used for generating mouse mutant phenotypes. Subsequent investigation can determine if the affected gene is developmentally relevant. Although one cannot select for a specific mutation, it is possible to generate developmentally interesting mutations.

Another way to create mutant phenotypes is by expressing exogenous genes in transgenic animals. The appropriate design of a specific transgene construct can provide some degree of control over its expression. For example, by designing constructs with tissue-specific promoters, it is possible to express or overexpress the transgene in a nonrandom way.

A major caveat in the production of transgenic animals was the inability to know in advance if the integrated transgene would be expressed in the animal. Therefore, a cell system was needed that would allow the assessment of expression of exogenous DNA and would subsequently contribute to the embryonic cell pool and behave in a normal manner. Such a pluripotent cell type was developed and has become known as embryonic stem (ES) cells.

These ES cells can be maintained as permanent cultures and can be used to investigate the behavior of early embryonic cells. When introduced into a blastocyst, they can contribute to all the lineages of a developing embryo, forming a chimera. Appropriate breeding of such chimeras can effectively transmit the ES cell to the next generation.

With a growing effect being invested in studying the human genome, the increased availability of mouse mutants with identifiable defects should prove valuable. Close to 40 percent of the mouse genome can be matched to conserved regions of the human genome. Because the mouse genome is amenable to experimental manipulation with transgenes, it should be possible to use these transgene tags as reference points to help identify and ultimately map many mouse genes. Once the map location of the mouse gene is known, it can provide a clue as to where the homologous human gene might map.

The specific topics outlined above are by no means comprehensive and are not the only areas of developmental biology that are generating such excitement within their respective scientific communities. However, they are representative of how the complexities of animal development are giving way to some unifying themes that seem to underlie all of development.

It is difficult to speculate where developmental biology will evolve in the near future. The amount of detail that will define the integrated circuitry that directs development will continue to increase. With this increased detail will come a better sense of the fundamental rules that guide development and the classes of molecules that play by those rules. We will learn why fruit flies and humans are genetically more the same than different and gain a new and deeper appreciation for the importance of those small differences. The ultimate challenge will be to explain how those small differences can direct the development of organisms as diverse as fruit flies and humans.

Developmental Genetics 1

INTRODUCTION

Mutant phenotypes remain the driving force behind most studies in developmental genetics. Technological advances have facilitated our ability to generate such phenotypes, including generating them in a controlled, nonrandom fashion. Thus, it is easier to understand the underlying genetic basis for the observed abnormality.

Advances in our ability to handle and manipulate isolated DNA will undoubtedly lead to new and clever ways to generate interesting mutants. Of particular interest is the evolving technology on how to work with larger fragments of DNA. Manipulating such large fragments will facilitate the functional analysis of genes previously too large for such procedures.

In this chapter, selected articles focus on both the generation and characterization of mutant phenotypes. Directed mutagenesis in several experimental model systems is discussed in the context of generating mutations of known genotype. Yeast artificial chromosomes are discussed both in the context of gene transfer and gene isolation. The use of PCR in concert with other techniques is offered as novel ways to manipulate DNA fragments. Gene targeting and the use of ES cells to "construct" valuable mouse models of human disease mutations is explored.

Phenotypic characterization of a variety of mutant phenotypes is discussed. Loss of function, gain of function, suppression, and controlled mutations are all presented as interesting biological examples of how to best dissect the genetic basis of development. In aggregate, such phenotypic variation should lead to our further understanding of the underlying genetic circuitry directing eukaryotic development.

Finally, the issue of gene mapping is addressed. With the highly visible development of the Human Genome Initiative, it is important to remember the biology that will be the ultimate beneficiary of this Initiative. Uncovering genes responsible for developmental defects in humans is an auspicious goal. By using existing developmental abnormalities in animal models as a road map, such a goal may be reached sooner. Therefore, insight into the mapping strategies in model organisms is valuable.

1

Dictyostelium Erasure Mutant HI4 Abnormally Retains Development-Specific mRNAs During Dedifferentiation

B. Kraft, A. Chandrasekhar, M. Rotman, C. Klein, and D. R. Soll

Dev. Biol., 136, 363—371, 1989   1-1

The HI4 mutant of *Dictyostelium* is normal in growth and morphogenesis, but partially defective in the reverse program of dedifferentiation. Whereas wild-type cells are subject to an "erasure event", the loss of the capacity for rapid reinitiation of development during disaggregation, HI4 cells abnormally retain the capacity to rapidly reaggregate. Previous work has shown that HI4 cells retain EDTA-resistant cohesiveness and contact sites A (gp80) much further into dedifferentiation than do wild-type cells. This accounts for their ability to reaggregate, despite the fact that they lose the ability to release and respond to cAMP during dedifferentiation at the same time as wild-type cells. The present work investigated HI4 cells further.

Northern blot analysis revealed that HI4 acquired two key developmental transcripts, for gp80 and the cysteine protease gene CP2, at the same times in development and at similar levels as did wild-type cells. In contrast, after disaggregation and transfer to erasure medium, HI4 retained high levels of both CP2 and gp80 mRNAs much longer than did wild-type cells. Changes in cAMP levels in HI4 cells at the onset of aggregation and when disaggregated and suspended in erasure medium were similar to changes in wild-type cells. Just as with wild-type cells, the addition of cAMP to an erasure culture of HI4 cells resulted in an immediate and dramatic decrease in the level of gp80 transcripts.

These findings imply that the retention of the ability to rapidly reaggregate after the erasure event in HI4 cells is due, at least in part, to a defect in the programmed removal of select developmentally acquired mRNAs during dedifferentiation. It seems likely that the abnormal retention of mRNAs for gp80 and CP2 accounts for the retention of the ability to reaggregate. These abnormalities do not appear to be secondary to elevated intracellular cAMP levels.

♦ Although it is a common developmental process, dedifferentiation is not well studied. The exquisite synchrony of the developing aggregates of *D. discoideum* offers a unique opportunity to study the mechanisms of dedifferentiation. Previous work has shown that when aggregates of developing *Dictyostelium* cells are disaggregated and placed in growth medium they synchronously and irreversibly lose their ability to reaggregate at 90 min. In addition, they lose developmentally acquired traits and proteins according to a specific program of dedifferentiation. The dedifferentiation defective mutant HI4 does not lose developmentally acquired cell cohesiveness as well as the cell cohesion molecule gp80.

Now, Kraft et al. have shown that the abnormal retention of specific developmentally acquired proteins in the mutant HI4 is paralleled by stabilization of specific mRNAs. The mRNAs for cysteine protease 2 and the cell surface glycoprotein gp80 are rapidly lost after the erasure event in wild-type cells but retained in the mutant. The unusual stability of these messages is not due to a change in cellular cAMP levels because both mutant and wild-type cells rapidly lower their cellular cAMP levels when placed in growth medium. HI4 also is still capable of cAMP-induced destabilization of the gp80 mRNA. Overall, the data indicate that the aberrant mRNA stability seen during the program of dedifferentiation is due to a novel, non-cAMP dependent mechanism. The continued characterization of this mutant should provide additional important information about the process by which cells control dedifferentiation. *Stephen Alexander*

Molecular Complementation of a Genetic Marker in *Dictyostelium* Using a Genomic DNA Library
J. L. Dynes and R. A. Firtel
Proc. Natl. Acad. Sci. U.S.A., 86, 7966—7970   1-2

The analysis of mutant genes in eukaryotes often employs the technique of genomic library complementation, in which a recessive mutant phenotype is rescued by the introduction of a cloned copy of the wild-type gene. This paper shows how a partial Sau3A *Dictyostelium discoideum* genomic library, in an extrachromosomally replicating shuttle vector, was used to complement the *Dictyostelium* mutant HPS400.

A *Dictyostelium* genomic library was inserted into vectors containing a *Dictyostelium* actin promoter-bacterial neomycin gene fusion, to permit selection in *Dictyostelium* using G418, a bacterial plasmid backbone, to permit propagation in *E. coli,* and a region from the *Dictyostelium* plasmid Ddp1 that directs extrachromosomal replication. The mutant HPS400 requires thymidine or thymidylate for growth and does not yield a lysate with detectable thymidylate synthase activity.

Introduction of library DNA into HPS400 cells was accomplished by electroporation. Five G418-resistant transformants were isolated that did not require thymidine. The DNA responsible for the transformations all contained the same 3.0-kb partial Sau3A fragment, which hybridized to a single 1.2-kb mRNA from wild-type and transformed HPS400 strains. Sequence analysis of a portion of the 3.0-kb Sau3A fragment showed a single open reading frame and several features typical of *Dictyostelium* genes, but no homology to the highly conserved thymidylate synthase of other species nor to any other sequence in the database examined.

These experiments provide evidence that an easy-to-use system of shuttle vectors can be employed to rescue the thymidine-requiring HPS400 mutant. Surprisingly, the isolated rescuing gene was not homologous to thymidylate synthase from other species. This library should be useful for isolating developmentally interesting genes from *Dictyostelium*.

Complementation of a *Dictyostelium discoideum*
Thymidylate Synthase Mutation With the Mouse Gene Provides a
New Selectable Marker for Transformation
A. C. M. Chang, K. L. Williams, and A. Ceccarelli
Nucleic Acid Res., 17, 3655—3661, 1989   1-3

The neomycin resistance gene has been used as a selectable marker for transformations of *Dictyostelium discoideum;* it would be useful to have a second selectable marker for experiments involving sequential transformation. The experiments described here explored the use of the mouse thymidylate synthase gene as a selectable marker.

The highly conserved mouse thymidylate synthase gene was placed into a shuttle vector containing the actin 15 promoter fused to the *neo* gene. This vector was used to transform the thymidine-requiring mutant strain HPS400 using the calcium phosphate method. Transformants capable of growth on axenic medium without thymidine were isolated. Southern and Northern analyses suggested that the transformants contained and expressed sequences hybridizing with mouse thymidylate synthase DNA. After sporulation of the transformants, storage of spores for 3 months at 4°C, and germination, the cells did not require thymidine but in other ways were identical to the HPS400 parent strain. One transformant line still did not require thymidine after over 60 generations under nonselective conditions.

These results suggest that the mouse thymidylate synthase gene was expressed and functional in HPS400 cells. This gene therefore may provide a second selectable marker for the transformation of *D. discoideum*. The use of this gene as a selectable marker does not require the continued growth in the presence of antibiotics. These new vectors, unlike those based on the actin-*neo* fusion, seem to be maintained at times in low copy number.

Complementation of Myosin Null Mutants in *Dictyostelium*
discoideum **by Direct Functional Selection**
T. T. Egelhoff, D. J. Manstein, and J. A. Spudich
Dev. Biol., 137, 359—367, 1990   1-4

Previous reports have shown that the single conventional myosin heavy chain gene of *Dictyostelium discoideum* can be deleted, resulting in cell lines defective in both cytokinesis and sporogenesis. These myosin null mutants are viable when grown on surfaces, but become large and multinucleate, dying in suspension. In this report, experiments are described that attempted to complement myosin null cells by direct functional selection.

The transforming plasmid was constructed by fusing the myosin coding region to the actin 15 promoter and translation initiation sequences in a shuttle vector containing the *Dictyostelium* autonomously replicating sequence *Ddp*1. Electroporation of this plasmid into HS2201 and HS2211 (identical to *mhc*A⁻/B and *mhc*A⁻/A5), 2 independently constructed myosin null cell lines, was followed by transfer of cells to suspension culture. Transformants were isolated.

Western blot analysis of transformants revealed expression of myosin at similar levels to wild-type cells; doubling times of transformants in log phase growth were 9 to 11 h, similar to the parental line Ax4. Cytokinesis appeared normal in the rescued cells, as did development. Since both cytokinesis and sporulation of the mutant cells were rescued by the transformation, it is likely that the mutant phenotype was due to the absence of myosin and not to secondary mutations. Southern blot analysis showed that integration of the transforming plasmid and rearrangement of Ddp1 sequences had likely occurred.

This method may be used to introduce cloned genes into *D. discoideum* wherever a suitable selection scheme can be devised. The technique should prove useful both for studying cloned genes and for confirming the phenotypes of mutants. It can also be used in cells to which antibiotic resistance markers have already been introduced. Finally, through DNA manipulation prior to transformation, this method may be useful for the dissection of functional domains within a gene.

♦ A large number of mutant strains have been isolated which block many of the steps of development in *Dictyostelium*. However, the analysis of these mutants has been hampered by the inability to isolate the corresponding gene. Three reports now indicate that molecular complementation will now be possible in this organism as it is in yeast and mammalian cells. This should greatly extend the ability to analyze the cell and developmental biology of this organism.

Dynes and Firtel take advantage of a mutant strain HSP400 previously shown to lack thymidylate synthase activity and requires exogenous thymidine for growth. They constructed a partial Sau3A library (3-12 Kb inserts) in a modified *Dictyostelium* transforming vector and used this to obtain transformants which grew independently of added thymidine. The frequency of transformation was approximate 10^{-5}. The transformants recovered carried either a 3.9 or 3.0 Kb insert. The 3.9 Kb insert contained the smaller 3.0 Kb Sau3A fragment. The genomic clone encodes a gene which surprisingly has no homology with any of the highly conserved thymidylate synthases. The clone hybridizes to a 1.2 Kb mRNA in both parental and transformed cells. although it is expressed at a five-times-greater level in the transformants.

The data suggest that either the defect in HPS400 is not in the thymidylate synthase gene or the complementing gene is a second site suppressor of the defect. The newly cloned gene can also be used as a selective marker for DNA transformation in strains already transformed to G418 resistance.

In a related paper, Change et al. used the mouse thymidylate synthase gene in a transformation vector pTS1 to complement the defect in strain HSP400. This result indicates that the primary defect in HSP400 is in the thymidylate synthase gene. One of the useful results of this study was the observation that transformants with only a single copy insert could be obtained which is generally not true for G418 based selection.

The report by Egelhoff et al. describes the complementation of a myosin heavy chain (MHC) null mutant with the cloned heavy chain gene. Interestingly, the null mutant was generated by homologous recombination. Although not quite as difficult as complementation from a genomic library, it does show that a gene can be cloned via complementation by taking advantage of the behavior of the wild-type cells to grow in suspension in contrast to the MHC mutants which only grow attached to the substratum. *Stephen Alexander*

A New Kind of Informational Suppression in the Nematode *Caenorhabditis elegans*

J. Hodgkin, A. Papp, R. Pulak, V. Ambros, and P. Anderson
Genetics, 123, 301—313, 1989   1-5

Mutations can sometimes be restored to normal function by certain extragenic alleles. Often such suppression involves modification of the cellular machinery for transcribing or translating the mutant gene product. This is termed "informational suppression", and the best known example is nonsense suppression, in which altered tRNAs insert an amino acid at a nonsense codon that otherwise results in premature termination of translation. In this paper, independent reversions of mutations of three *Caenorhabditis elegans* genes were studied; the suppressors isolated seemed to form a new class of informational suppressors.

Suppressors of certain alleles of *unc-54*, a myosin heavy chain gene, lin-29, a heterochronic gene, and *tra-2*, a sex determination gene, were isolated. These suppressors occurred at six loci, and any of them could suppress the three mutations above as well as certain alleles of *tra-1,* another sex determination gene, and *dpy-5,* a morphogenetic gene, possibly encoding collagen. These suppressors did not act on the many other hypomorphic mutations tested. The suppressors also had a morphological phenotype: abnormal morphology of the adult male tail and of the hermaphrodite vulva. Generally, suppression was recessive and incomplete. The suppressible allele of *unc-54* is known to have a deletion in 3′

noncoding sequences. The six suppressors are named *smg 1—6,* for suppressor with morphogenetic effect on genitalia.

These findings show that *smg* mutations are allele-specific suppressors of certain alleles of various genes. The lack of common features among the suppressible genes seems to imply that the suppression is informational. The basis for the morphological effects of the mutations is unknown.

♦ Most of us are familiar with the "informational suppression" of nonsense mutations by altered tRNAs. This paper describes a novel class of informational suppressors, terms *smq* suppressors (for *suppressors with morphogenetic effects on genitalia*). As expected for informational suppressors, mutations at the six *smq* loci identified in this study act as allele-specific suppressors of mutations in a wide variety of genes: *unc-54* (a myosin heavy chain gene),. *lin-29* (a heterochronic gene), *tra-1* and *tra-2* (sex determination genes), and *dpy-5* (a morphogenetic gene). In addition, *smq* mutations cause abnormal morphogenesis of the male tail and hermaphrodite vulva. Based on what is known about the *smq*-suppressible mutations, it is likely that the *smq* suppressors act at the level of mRNA processing, transport, or stability. Further analysis of *smq* mutations and the alleles they suppress should provide valuable insight both into mRNA metabolism and into the genes affected by the suppressors. *Susan Strome*

Analysis of Gain-Of-Function Mutations of the *lin-12* Gene of *Caenorhabditis elegans*
I. Greenwald and G. Seydoux
Nature, 346, 197—199, 1990   1-6

In *Caenorhabditis elegans,* the wild-type hermaphrodite gonad contains two cells, Z1.ppp and Z4.aaa, which can become the anchor cell (AC). Intercellular communication normally results in one cell becoming the AC, while the other becomes a ventral uterine precursor cell (VU). Previous work has shown that the *lin-12* gene governs the fates of these two gonad cells, and may encode a cell-surface receptor for communication between them. Here, an investigation of gain-of-function mutations of *lin-12* is presented.

Previously, it has been shown that null mutations of *lin-12* result in both cells becoming ACs and that semidominant *lin-12 (d)* mutations result in both cells becoming VUs. When cells are heterozygous for *lin-12 (+)* and *lin-12 (d),* increased proportions of hermaphrodites have presumptive ACs that are transformed into VUs. This shows that increased *lin-12* activity results in the VU fate.

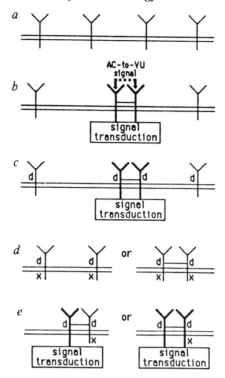

FIGURE 1-6.   A model for how dimerization leads to activation of the *lin*-12 protein. *a,* In wild type, lin 12 is a receptor that is inactive in the absence of its ligand. *b,* In wild type, the ligand ('AC-VU' signal) promotes dimerization, leading to signal transduction by means of the intracellular domain. Ligand binding to receptor monomer may cause a conformational change that results in or stabilizes dimerization; alternatively, the ligand may have a more direct role in a cross-linking mechanism. *c,* In a *lin-12(d)* homozygote, dimerization and signal transduction are ligand independent. The *lin-12(d)* mutations might result in ligand-independent activation of the *lin-12* protein by inactivating a negative regulatory domain or by causing changes in the higher-order structure. *d,* In a lin-12(n302 n653) homozygote, the n653 mutation either prevents homodimer formation or inactivates lin-12. *e,* In a *lin-12(d)/lin-12(n302 n652)* heterozygote, heterodimers can form and are active. (From Greenwald, I. and Seydoux, G., *Nature,* 346, 197—199, 1990. With permission.)

Sequence analysis of eight *lin-12 (d)* mutations showed that all were missense alterations of the putative extracellular region of the protein. This implies that all mutations modify protein activity or stability.

Laser ablation experiments with wild-type animals showed that isolated cells become AC cells, while expression of the VU fate requires both cells, with one sending and one receiving the "AC-to-VU" signal; the receiving cell becomes VU. In *lin-12 (d)* mutants, laser ablation experiments resulted in isolated cells expressing the VU fate. When heteroallelic combinations of *lin-12 (d)* mutations and other *lin-12* mutations were made and analyzed, sometimes greater activity than in *lin-12 (+)* resulted.

These findings resulted in the postulation of a model (Figure 1-6), based on the idea that the *lin-12 (+)* protein is a self-associating receptor for the AC-to-VU signal. The receptor is activated by binding its ligand, which induces self-association. This model explains the *lin-12 (d)* mutation as resulting in receptor self-assocation in the absence of ligand. Further biochemical experiments will be necessary to confirm or disprove this model.

♦ During the development of the *C. elegans* gonad, one of two equipotential somatic gonad cells becomes the anchor cell (AC); the other becomes a ventral uterine precursor cell (VU). *lin-12* plays a pivotal role in the AC/VU decision. The gene appears to function as the receptor of an AC-to-VU signal, which is normally required to specify the VU fate. Gain-of-function *lin-12(d)* mutations lead to the production of two VUs at the expense of the AC. In this paper, Greenwald and Seydoux show that in *lin-12(d)* mutants, the VU cell fate can occur independently of an AC-to-VU signal. Their sequencing of eight *lin-12(d)* mutations, all of which map to a small non-EGF extracellular region, suggests an explanation for this finding. *lin-12* receptor is hypothesized to be activated by dimerization. In wild-type worms this requires ligand (perhaps AC-to-VU signal) binding, whereas in *lin-12(d)* worms it can occur in a ligand-independent manner. Analysis of the *Drosophila Notch* protein, which is similar to *lin-12,* supports this model. Testing of the model now requires biochemical approaches. *Susan Strome*

A Second *Trans*-Spliced RNA Leader Sequence in the Nematode *Caenorhabditis elegans*
S.-Y. Huang and D. Hirsh
Proc. Natl. Acad. Sci. U.S.A., 86, 8640—8644, 1989   1-7

Many genes of *Caenorhabditis elegans* have at their 5′ end a stretch of 22 nucleotides that is *trans*-spliced from a 100-nucleotide precursor RNA. This spliced leader (SL) is found in the genomes of all nematodes thus far examined. Here, a different trans-spliced leader, termed SL2, is described.

Primer extension sequencing of the 5′ ends of *C. elegans* glyceraldehyde-3-phosphate dehydrogenase (GAPHD) mRNAs revealed that transcripts from 3 of the 4 GAPDH genes contained SL, while the 22 nucleotides at the 5′ end of the *gpd-3* message differed from SL. These 22 nucleotides were not found adjacent in the genome to *gpd-2* or to *gpd-3*. SL2 differs from SL1 in only eight nucleotides. SL2 seems to be present in the genome in four copies and in two variants, α and β. SL2 RNA originates from a 110- or 110-nucleotide transcript. When SL2 DNA was hybridized to total RNA from *C. elegans,* a smear of RNAs of a wide range of sizes was detected. SL2 was found in RNAs from *C. elegans* var. Bergerac and *C. briggsae,* but not from the nematodes *P. redivvivus* or *H. contortus,* or from *Dictyostelium* or humans.

These experiments suggest that a second *trans*-spliced leader exists in *C. elegans*, with some similarities and some differences from the original SL. Transcripts of different genes may be *trans*-spliced with only one specific spliced leader. The functional significance of *trans*-splicing remains unknown.

◆ Several years ago, Krause and Hirsh discovered that mRNAs from three of the four actin genes received a 22-nucleotide trans-spliced leader sequence, termed the spliced leader or SL. Since that report the list of known mRNAs that are trans-spliced has grown steadily. The paper of Huang and Hirsh describes the finding that a different 22-nucleotide SL sequence, called SL2, is present on the transcripts from one of the four glyceraldehyde-3-phosphate dehydrogenase (GAPDH) genes; transcripts from the other three GAPDH genes receive the original SL, SL1. The functional significance of trans-splicing is not yet understood, and now we must also wonder how the specificity of SL selection is regulated and whether the different leader sequences confer different properties on their respective mRNAs. *Susan Strome*

The Identification and Suppression of Inherited Neurodegeneration in *Caenorhabditis elegans*
M. Chalfie and E. Wolinsky
Nature, 345, 410—416, 1990   1-8

Programmed cell death is a normal part of development of *C. elegans*, with affected cells becoming refractile and condensed as they die. In this report, a mutation involving an abnormal type of late-onset cell death is described.

The dominant mutation *deg-1 (u38)* was found in mutagenized animals screened for touch insensitivity. In these mutants, the two PVC interneurons degenerate during the second and third larval stages. The degeneration is first seen as a vacuole surrounding the nucleus; over time, and in a temperature-dependent fashion, the vacuole enlarges and the nucleus degenerates.

The *deg-1* mutation was rare and frequently reverting. Two partial intragenic suppressor mutations for *u38* were isolated, which delayed neuronal death. Gene dosage experiments suggested that the *u38* mutation resulted in an abnormal, toxic gene product. Animals with loss of *deg-1* function appeared normal.

The *deg-1(u38)* mutation differed genetically from programmed cell death mutations in that mutations that prevent all programmed cell deaths did not prevent the *deg-1* degenerations. In contrast, mutations in the *mec-6* gene suppress the degenerations caused by both the *mec-4(d)* and *deg-1(u38)* mutations. Dominant *mec-4* mutations have previously been shown to result in degeneration of touch receptor neurons, probably due to production of a toxic product within the affected cells.

Sequence analysis of *deg-1* cDNAs resulted in an inferred polypeptide sequence that was not strongly similar to other sequences in a protein database. The polypeptide has a hydrophobic regions, two possible sites of *N*-linked glycosylation, and a cysteine-rich region. These features are consistent with those of a membrane protein.

The neuronal degeneration associated with *deg-1* resembles that caused by *mec-4* This type of neurodegeneration is distinct from programmed cell death, but perhaps can serve as a model for human genetic diseases, such as Huntington's disease, in which neuronal death occurs late in life.

♦ Insights into human disorders may be obtained by studying animal models with similar disorders. This paper describes a potential animal model for the selective neural degeneration that accompanies a variety of human diseases. A dominant mutation in the *C. elegans deg-1* gene results in late-onset degeneration of a small number of neurons, resulting in insensitivity to touch. The *deg-1* gene product is likely to be a membrane protein, although it is not homologous to any of the ion channels or receptors known to function in neurons. Genetic analysis suggests that the *deg-1* gene is nonessential, perhaps because it belongs to a gene family, and that the dominant *deg-1* mutation encodes an abnormal, toxic gene product. Interestingly, mutations in another mechanosensory gene, *mec-6,* suppress the neural degeneration caused by the dominant *deg-1* mutation. This result suggests that wild-type *mec-6* activity is required for the neurodegenerative process. Hopefully, analysis of *deg-1* and *mec-6* will provide some clues about the cause and process of neurodegeneration; eventually, this could aid in developing methods for preventing or curing neurodegenerative diseases in humans. *Susan Strome*

The FLP Recombinase of Yeast Catalyzes Site-Specific Recombination in the Drosophila Genome

K. G. Golic and S. Lindquist
Cell, 59, 499—509, 1989   1-9

One reason why yeast is such an attractive experimental organism is that techniques have been developed for site-specific recombination in its genome. Here, a system for site-specific recombination in *Drosophila melanogaster* is presented, borrowed in part from the yeast *Saccharomyces cerevisiae.*

The heart of this system is the FLP recombinase of yeast. This recombinase acts on copies of the FLP recombination target (FRT) found on the *S. cerevisiae* 2-µ plasmid in inverted repeats. FLP catalyzes recom-

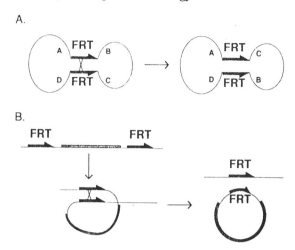

FIGURE 1-9.   FLP recombination products. The results of recombination between (A) the inverted repeats of FRT in the 2-μm circle and (B) direct repeats of FRT. The FRTs are indicated by solid half-arrows. (From Golic, K. G. and Lindquist, S., *Cell*, 59, 499—509, 1989. With permission.)

bination between these FRTs, resulting in the inversion of part of the plasmid with respect to the rest (Figure 1-9, A). In the presence of direct repeats of FRTs, FLP catalyzes excision of a circle containing the intervening DNA and one FRT (Figure 1-9, B).

D. melanogaster were transformed, using P-element mediated transformation, with FLP coding sequences under the control of the heat-inducible *Drosophila* hsp70 promoter. Other flies were transformed with an allele of the *white* gene flanked by direct FRT repeats. When the flies were crossed, eye color provided an assay of FLP activity.

FLP activity resulted in either excision or amplification of *white*. Recombination occurred in somatic tissue, resulting in mosaic eyes. The patterns of mosaicism in the eye varied with the developmental stage at which FLP was induced. Recombination also occurred in the germline, resulting in progeny with all white or dark red eyes. The frequency of recombination varied with the severity of the heat shock used to induce FLP. Recombination due to a single copy of hsFLP was easily detectable. Southern blot analysis was used to confirm recombination events at a molecular level.

These experiments showed that the yeast FLP recombinase can catalyze site-specific recombination in both somatic and germline cells of *D. melanogaster*. This system should be useful for the study of development and of developmentally interesting genes in fruitflies, and might be used to turn genes on or off.

♦ Homologous recombination in yeast provides a powerful genetic tool that has greatly enhanced the utility of this organism as an experimental system. Homologous recombination in higher eukaryotes has either not been demonstrated (e.g. *Drosophila)* or is much rarer and difficult to control (e.g., mammalian cells in culture). Although the results described in this paper do not directly address homologous recombination, they do demonstrate the feasibility of directing at will the integration and excision of DNA into and from specific sites on the chromosome.

The approach that was used in this paper is simple and is derived from yeast. The yeast 2-µm plasmid encodes a recombinase (termed FLP) that acts on a specific target DNA sequence (termed FRT) present in the 2-µm plasmid as 599 bp inverted repeats (*in vitro,* the minimal functional FRT is 28 bp). Importantly, recombinase purified from transgenic *E. coli* and the target FRT DNA sequence, are sufficient to promote recombination *in vitro,* suggesting that no other host factors are required from FRT-directed recombination to occur. To demonstrate that the FLP/FRT system works in *Drosophila,* the gene coding for FLP recombinase was placed under the control of the strong and inducible *hsp70* heat-shock promoter and this construct (termed *hspFLP)* was integrated into the *Drosophila* genome by P-element mediated transformation. Thus, synthesis of FLP recombinase can be induced at will in all tissues of the fly simply by subjecting the *hspFLP* flies to a heat pulse. The authors chose the *white* eye color gene to assay FLP activity. This gene has a number of important advantages. It is not required for viability or reproduction, it is cell autonomous (it acts only in the cell where the gene is expressed), and its activity can be easily scored by simple visual inspection of the fly. No *white* activity yields white eyes, one dose of *white* yields orange eyes and two doses of *white* yield red eyes. The compound eye of *Drosophila* has about 800 ommatidia. Thus, gain or loss of the white gene in the eye of only a fraction of the precursor cells results in clonal patches of altered eye color. The white gene was placed between two FRT sequences (termed *>white>* where ">" represents the flanking FRT) and this construct was transformed into flies that otherwise lacked an active *white* gene. By crossing *hspFLP* and *>white>* flies, the two transgenes were brought into the same fly. The FLP-dependent gain and loss of the white gene (generation of white and red eye patches over an orange background) could be easily demonstrated. Moreover, genetic evidence indicated that recombination also occurs in the germ line, since heat-induced flies generated progeny with completely white or red eye color. Loss and gain events vary with the severity of heat induction (loss was more frequent than gain), higher temperatures favoring loss of *>white>*, and lower temperatures favoring gain. Other evidence suggests that FLP is active only in dividing cells. Molecular analysis (Southern blots) indicated that gain of *white* was associated with the generation of two tandem copies of the gene, probably by unequal sister chromatid exchange or by integration of a previously excised *>white>* into the sister chromatid.

Because the recombinase acts in the eye and in the germ line, it is likely to be active in most tissues of the fly. These findings suggest a number of useful potential applications. The system could be used to produce somatic and germ line mosaics of any cloned gene by placing it between two FRTs and transforming it into the organism. Constructs could be devised to turn specific genes on rather than off. FLP activity could be restricted to specific tissues by fusing the gene to appropriate promoters. Finally, the system is also likely to work in other organisms. For instance, P-element vectors have provided such a powerful tool to study *Drosophila* but appear to be specific for this species. The results reported in this paper open the possibility that the FLP-based system can be devised for introducing cloned genes into other animals, such as insect vectors for human disease. The ability of genetically modifying insects of medical and economic importance is of obvious importance. *Marcelo Jacobs-Lorena*

Targeted Gene Mutations in *Drosophila*
D. G. Ballinger and S. Benzer
Proc. Natl. Acad. Sci. U.S.A., 86, 9402—9406, 1989   1-10

In *Drosophila,* the ability to target mutations to previously cloned, interesting genes would be useful. In this paper, a method to do this, based on the polymerase chain reaction (PCR) and on mutagenesis via insertion of P-elements, is presented. Use of this technique requires sequence information from a small portion of the gene of interest.

The heart of this method is the detection of the insertion of a P-element into the gene of interest via the PCR. PCR employs two oligonucleotide primers; a DNA sequence becomes geometrically amplified only if it is flanked by the two primers. As used here, one primer contained a sequence from the gene of interest, while the other contained the terminal sequence of the P element (Figure 1-10).

One fly containing a defective P element in the vermillion gene was mixed with 100 wild-type flies lacking the P element, the DNA was isolated from the 101 flies, and 40 cycles of PCR were performed. The predicted DNA amplification product was detectable by size on an ethidium bromide-stained gel and by hybridization to a vermillion gene probe on a Southern blot. The same method was then applied to two genes of unknown function represented by cDNA clones and expressed in adult compound eyes. Flies containing random P-element insertions were produced by introducing high levels of transposase activity into a strain of flies containing multiple defective P elements; a PCR screen for mutagenesis of these two genes yielded amplification products for both among 6316 mutagenized flies. When *in situ* hybridizations of the amplification products were carried out, the product from only one gene was found to be authentic, and the other was considered spurious.

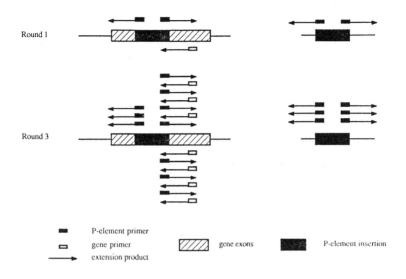

Round 1

Round 3

■	P-element primer
□	gene primer
▨	gene exons
■	P-element insertion
→	extension product

FIGURE 1-10. Use of PCR to detect insertion of a *P*-element transposon into a specific gene. Two segments of genomic DNA are shown. In one case (left), a *P*-element transposon is shown inserted near a gene of interest; in the other (right), it is at a random unlinked site. Two oligonucleotide primers are shown, one containing a sequence from the gene to be targeted □, and the other containing the terminal sequence of the *P* element (■). (upper) DNA extension products after one cycle of replication. (lower) DNA extension products after three cycles. Exponential amplification occurs when the newly synthesized strand initiated by each of the primers incorporates sequences complementary to the other — in this example, only when a *P* element has inserted near the gene oligonucleotide. Other DNA extension products increase only linearly and, after multiple rounds of amplification, represent only a small proportion of the total DNA synthesized. (From Ballinger, D. G. and Benzer, S., *Proc. Natl. Acad. Sci. U.S.A.*, 86, 9402—9406, 1989. With permission.)

This method for site-directed, P-element-mediated mutagenesis of cloned *Drosophila* genes was sensitive and practical. One advantage of this technique is its ability to detect mutations in the heterozygous state. Elimination of spurious amplification products from further study requires confirmation by *in situ* hybridization. This method should be applicable to other organisms in which appropriate integrating vectors are available.

"Site-Selected" Transposon Mutagenesis of *Drosophila*
K. Kaiser and S. F. Goodwin
Proc. Natl. Acad. Sci. U.S.A., 87, 1686—1690, 1990   1-11

This paper presents a method of mutagenesis useful for the analysis of *Drosophila* genes that have been cloned but for which no phenotypic information is available. It is based upon the detection of P-element insertion into particular genes using the polymerase chain reaction (PCR).

The PCR, as employed here, uses one primer specific for the P-element and another specific for the gene of interest.

The singed *(sn)* locus was used here as a model system. Two strategies for inducing random P-element mutagenesis were employed. In the first, males bearing P elements were mated with females lacking them. In the second, flies with many defective but mobilizable P elements were crossed with flies that contain a nonmobilizable source of transposase from an engineered P element. Batches of the eggs of mutagenized flies were then subjected to PCR. When amplification products were detected, the population was subdivided in successive stages, permitting the isolation of the rare fly containing a mutation in the *sn* gene.

When serial dilutions of CFL5 (containing *sn*) DNA into Oregon R (containing *sn+*) DNA were made and subject to PCR, detection of the equivalent of 1 mutant fly in 1000 was achieved by hybridization to Southern blots. This detection method was insensitive to spurious amplification products. When mutant flies were produced by both strategies described above, they were detected using PCR and hybridization methods. The mutations were confirmed by phenotypic analysis.

This method of "site-specific" mutagenesis, as tested with the *sn* locus of *Drosophila,* seems successful, simple, fast, and general. It permits the ready screening of 10^5 flies. It allows the isolation of mutations in the heterozygote and of those resulting in weakly detectable or obscrue phenotypes. It can be used with flies that are virtually wild type, and should be adaptable for use with other organisms.

♦ The method described in these papers combine two recent and very powerful research tools: P elements and the polymerase chain reaction (PCR). The basic approach is to mutagenize the fly's genome by mobilization of P elements and to detect insertions in the gene of interest by PCR. Given the size of the *Drosophila* genome and the rate of P element transposition, it can be calculated that an insertion event will occur within a 2-kb DNA fragment (this is approximately the size limit for detection by PCR) in 1 out of every 10,000 flies. Moreover, the authors have demonstrated that an insertion event can be easily detected even if only one fly in a few hundred carries a P element in the gene of interest. It follows that after appropriate pooling, there is a good likelihood that an insertion in the gene of interest will be found after analysis of no more than 50 pools. Once a positive pool is identified, the fly carrying the mutation is isolated by analysis of progressively smaller pools.

Even in an organism like *Drosophila,* for which relatively sophisticated genetic tools are available, there are situations where determining the phenotype of certain mutations is either not possible or very laborious when using conventional methods. Possible examples are genes cloned by homology with those of other organisms, genes cloned by screening cDNA libraries with an antibody that recognizes a particular protein, or genes cloned by the "enhancer trap" technique (see report in last year's volume). The only requisite for the present approach is the knowledge of partial DNA sequence (a fragment as small as 50 nucleotides may be sufficient). By use of appropriate primers, more than one gene can be screened per experiment. Since the screen can easily be done with heterozygous flies, the analysis of the phenotype is done only after the appropriate strain of flies is isolated. Thus, the method is applicable for essential genes (i.e., those that yield lethal or sterile mutants) and prediction of phenotype (as is essential for conventional genetic screens) is not required. The only phenotype that needs to be scored is the band of a PCR product on a gel. In cases where a particular insertion does not lead to a detectable phenotype or where the phenotype is only a partial loss-of-function, more severe mutations can be easily generated by promoting the excision of the inserted P element (P-element excision is frequently accompanied by deletion of neighboring fly DNA). Finally, it should be pointed out that this method is applicable to cells in culture or to other organisms, even those for which methods of genetic analysis are less developed. The only requirement is the availability of appropriate transposons or the ability to generate small DNA deletions between sites complementary to two PCR primers. *Marcelo Jacobs-Lorena*

Drosophila P Element Transposase Recognizes Internal P Element DNA Sequences
P. D. Kaufman, R. F. Doll, and D. C. Rio
Cell, 59, 359—371, 1989   1-12

While much is known about the genetics and the structure of P elements, the biochemistry of their transposition and regulation is not understood. This report describes the purification and characterization of P element transposase.

The P-element transposase from an overproducing *Drosophila* cell line was fractionated chromatographically, including one step of high-affinity, transposase binding-site DNA chromatography. A polypeptide of molecular weight approximately 92,000 was isolated that was immunoreactive with transposase-specific antibodies and that co-migrated with an immunoaffinity-purified transposase protein.

DNAse I footprint analysis suggested that the purified transposase bound specifically to P-element nucleotides 48 to 68, near but outside the terminal 31 basepair inverted repeat, partially overlapping the promoter region (Figure 1-12). A second binding site also seemed to lie near but outside the 3' 31 terminal repeat, at nucleotides 2855 to 2871 (Figure 1-12). Both these sites are previously known to be important for transposition *in vivo*. The two binding sites have a ten-nucleotide consensus sequence. Mutations known to reduce transposition activity *in vivo* were

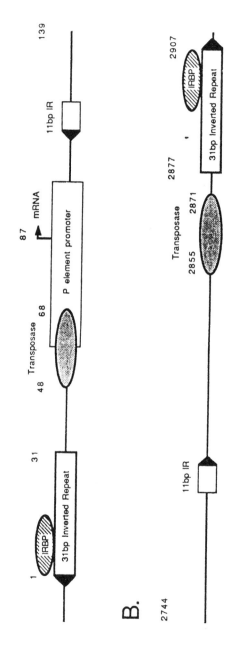

FIGURE 1-12. A diagram of molecular interactions at the P element ends. (A) The 5' end of P element DNA is diagrammed. P element nucleotide numbers are shown above the line representing the DNA. The sequences represented, P element nucleotides 1 to 139, are those required for transposition (O'Hare and Rubin, 1983; Rubin and Spradling, 1983). IrBP stands for inverted repeat binding protein, which binds to the outer 16 bp of the terminal 31-bp inverted repeats. Also, the internal 11-bp inverted repeat is shown. Transposase (stippled oval) is shown interacting with its binding site, nucleotides 48 to 68. This site overlaps sequences required for *in vitro* P element motor activity, nucleotides 58 to 103 (stippled box). The P element transcription start site at nucleotide 87 is indicated. (B) The 3' end of P element DNA is diagrammed as above. The sequences represented, P element nucleotides 2744 to 2907, are those required for transposition (O'Hare and Rubin, 1983; Rubin and Spradling, 1983; Mullins et al., 1989). Transposase (stippled oval) is shown interacting with its binding site, nucleotides 2855 to 2871. (From Kaufman, P. D., Doll, R. F., and Rio, D. C., *Cell,* 59, 359—371, 1989. With permission.)

found to abolish site-specific transposase binding *in vitro*. The transposase also has a high nonspecific affinity for DNA.

The purified P-element transposase seems to specifically bind a 10 nucleotide consensus sequence at the 3′ and 5′ ends of the P-element DNA and does not interact directly with the terminal 31 basepair inverted repeats. Proteins encoded by the P-element DNA may regulate P-element transcription. In addition to the transposase, the binding of additional *Drosophila* proteins to the terminal repeats of P elements may also be required for P-element transposition.

Regulated Splicing of the *Drosophila* P Transposable Element Third Intron *In Vitro*: Somatic Repression
C. W. Siebel and D. C. Rio
Science, 248, 1200—1208, 1990   1-13

In eukaryotic cells, alternative splicing of messenger RNAs is used to regulate the control of gene expression; in *Drosophila,* alternative splicing seems to regulate the pathway of somatic sexual differentiation and of the expression of *suppressor of white apricot*. So far, the biochemical mechanisms of these phenomena have generally been unclear. In the experiments reported here, the biochemical basis of the alternative splicing of the *Drosophila* P transposable element was explored.

Previous molecular genetic approaches have shown that expression of P-element transposition is different in different tissues, depending on the splicing of the P-element third intron (Figure 1-13). The P-element pre-mRNA contains three introns. In germline tissue, all three introns are excised, and an 87-kDa transposase is produced. In somatic tissue, only the first two introns are excised, resulting in premature termination of translation and production of a 66-kDa polypeptide.

When human (HeLa) cell nuclear extracts were added to P-element pre-mRNAs, accurate excision of the third intron (IVS3) occurred. Nuclear extracts from somatic *Drosophila* tissue culture cells did not support splicing of IVS3. When extracts from *Drosophila* somatic cells were mixed with Hela cell extracts, excision of IVS3 did not occur, but splicing of the *Drosophila fushi tarazu* intron and the rabbit β-globin second intron were not inhibited. A 97-kDa protein was isolated from *Drosophila* cell extracts that specifically interacted with IVS3. This protein interacts with IVS3 pre-mRNA within the 5′ exon sequences. The degree of binding of this protein correlated with the *in vitro* inhibition of splicing of various mutant IVS3 pre-mRNAs.

These findings suggest that the alternative splicing of IVS3 may involve specific somatic inhibition. This inhibition may be based, at least in part, on the interaction of a 97-kDa protein and 5¢ exon sequences previously implicated in the control of IVS3 splicing *in vivo*. It is possible that

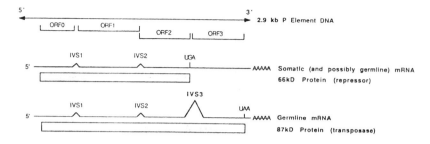

FIGURE 1-13. Tissue-specific splicing of the third intron of P element pre-mRNA. Transcription of full-length 2.9-kb P element DNA yields a single pre-mRNA containing 4 long open reading frames (ORFs). The first three ORFs are joined in the mature mRNA after splicing of the first two introns (IVS1 and IVS2) in both the soma and germline. In contrast, IVS3 is removed only in the germline, thereby restricting production of the 87-kDa transposase protein (indicated by the larger shaded box) to this tissue. In somatic cells, IVS3 is retained, and translation halts at a UGA stop codon within IVS3 to yield a 66-kDa protein (smaller shaded box). (From Siebel, C. W. and Rio, D. C., *Science,* 248, 1200—1208, 1990. With permission.)

this negative control system may also regulate expression of genes involved in germline differentiation.

♦ Transposable elements occur in most organisms and have received much attention because of their ability to cause extensive alterations of gene expression and inheritance. The P family of transposable elements of *Drosophila* is one of the best-characterized eukaryotic transposons. Not only have P elements served as an enormously important genetic tool, but they have also provided the basis for the construction of vectors for germ-line transformation. Having been extensively characterized from a molecular point of view, P elements are a valuable tool for the study of eukaryotic transposition and gene regulation. The intact 2.9-kb P element encodes a single primary transcript with 4 exons and 3 introns. The first two introns (IVS1 and 2) are spliced in all tissues to generate a mRNA encoding a 66-kDa protein believed to act as a repressor of transposition. IVS3 is spliced in the germ line, and only in this cell lineage, to generate a mRNA encoding an 87-kDa protein that is the active transposase. The two articles from Rio's laboratory concern two questions relating to P-element transposition: which P element DNA sequences does the 87-kDa transposase recognize and what is the basis for the regulated and tissue-specific removal of IVS3?

All P elements have 31-bp terminal inverted repeats that are essential for transposition. However, an additional 120 bp are required at each P element end for efficient transposition to occur. These include an internal 11-bp inverted repeat and the P-element promoter situated around the transcription initiation site 87 bp away from the 5' end. To investigate the site of interaction of the transposase with the P element, this protein was overexpressed in *Drosophila* cultured cells from a gene driven by the metallothionein promoter and lacking IVS3. The protein was purified by conventional and DNA-affinity chromatography. DNAase I protection ("footprinting") analysis revealed that the purified transposase did not bind the 31-bp inverted repeat elements, but rather to a sequence internal to these elements. The transposase probably recognizes a 10-bp sequence present at both ends of the P element and protects an additional 5 bp on each side from DNAase digestion. A set of linker-scanning mutations centered around the transposase binding site was assayed both for transposase binding and for the ability to promote P-element transcription in cell-free extracts. These assays revealed that the sites required for transposase binding and transcription initiation overlapped, although they did not coincide. However, the effect of the transposase on the efficiency of transcription was not measured. It is possible that in addition to promoting transposition, the transposase (and/or the related 66-kDa repressor protein) have a regulatory role on P-element transcription.

In prokaryotes, host factors, in addition to transposon-encoded factors, are required for transposition. The highly conserved nature of some of the noncoding P-element sequences and previous mutational analyses suggest that host factors are required also for P-element transposition. One candidate is a non-P-element-encoded protein that recognizes the outer 16 bp of the highly conserved 31-bp repeat (see report in the previous volume of this series). Future work may clarify some of these issues.

It has been known that IVS3 is spliced only in the germ line. Mutational analysis (cf. previous volume in this series) has narrowed the required cis-acting sequences to a region within the exon sequences adjacent to the splice site and extending to within 26 nucleotides of IVS3. However, little was known about the mechanisms involved in this regulating splicing. Formally, regulation could involve repression in somatic tissues, activation in germ line cells, or a combination of both. Now, Siebel and Rio report that a precursor P-element transcript can be accurately spliced in a mammalian cell-free system but not in a *Drosophila*-derived system. This provided an assay for identifying *Drosophila* proteins that prevented splicing of the precursor in the mammalian extract. First, the authors demonstrated that this inhibition is specific for P-element sequences, since the *Drosophila* extract did not prevent other precursors to be spliced. Second, using UV-crosslinking experiments, the authors demonstrated that the *Drosophila* extracts contained a 97-kDa protein that could bind preferentially to P-element sequences. The protein recognizes a stretch of about 30 nucleotides in the exon immediately adjacent to IVS3. Importantly, addition of an excess of this oligonucleotide to a splicing extract can relieve the inhibition caused by the *Drosophila* extract. Based on this and other evidence the authors propose that the failure of the P-element precursor transcript to be spliced in somatic tissues is due to the negative regulation of splicing by the 97-kDa protein. It is possible that other proteins are also required for this regulation. Moreover, these findings do not rule out that in germ cells, P-element RNA splicing also requires positively-acting regulatory factors. The 97-kDa protein was purified from *Drosophila* cells that do not carry P elements. It is likely that the 97-kDa protein has defined cellular functions, such as to prevent the expression of germline-specific genes in somatic cells. Finally, it should be noted that negative regulation of splicing has been identified in a number of other systems, notably in the control of sex differentiation in *Drosophila* (see separate report in this volume). *Marcelo Jacobs-Lorena*

Constructing Deletions with Defined Endpoints in *Drosophila*
L. Cooley, D. Thompson, and A. C. Spradling
Proc. Natl. Acad. Sci. U.S.A., 87, 3170—3173, 1990   1-14

Small chromosomal deletions are useful tools in *Drosophila* genetics. Using present methods, neither the size nor the endpoints of created deletions can be controlled. This paper presents a method for the construction of specifically designed chromosomal deficiencies using P transposable elements. The basis of this method is the high frequency of rearrangements with endpoints corresponding to the sites of preexisting P elements that occurs in P/M hybrid dysgenesis.

These experiments tested whether P-element-induced rearrangements could be controlled by using a single source of transposase and a small number of transposon targets. The first step in generating a defined rearrangement was to construct strains with defective P elements at the desired endpoints. These strains involved the ebony and rosy genes; mutants bearing P-element insertions into these genes were constructed that were either in *cis* (on the same homolog) or in *trans* (Figure 1-14).

The next step was the introduction of a transposase-producing element into the *cis* and *trans* progenitor strains, by crosses with transposase-producing stock (Figure 1-14). Progeny flies were scored for ebony and rosy. Excision of the right boundary element measured transposase activity, detected by lack of a functional rosy gene.

Ebony flies were recovered that were both rosy and rosy[+]. No ebony flies were recovered from the *trans* experiment among 4017 chromosomes scored, but 45 ebony flies were found from a *cis* experiment in which 3884 chromosomes were scored. Cytological examination of 12 ebony flies generated in the cis experiments showed that 10 contained the predicted deletion and 2 had complex rearrangements. Southern blot analysis showed that the deletion endpoints rarely coincided precisely with the P-element termini.

These experiments show that "designer deletions" can be created in *Drosophila* using P elements. Many single element insertions lines useful

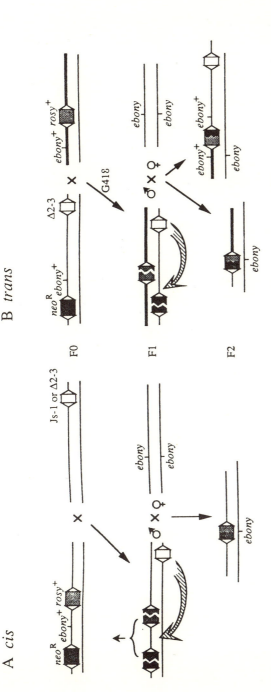

FIGURE 1-14. Strategies for deleting DNA between P elements located in *cis* (on the same homolog) (A) or in *trans* (B). (A) A stock containing in *cis* both a pUChsneo [pUChsneo (10) element (*neo*[R])] at locus 93B and a *rosy*[+] P element at 93F was crossed to a transposase-producing (Js-1 or D2-3(99B)] stock (F[0] generation). The F[1] males were mated to *rosy*, *ebony* virgin females. F[2] progeny were scored for *rosy* (probable excisions of only the right element) and for *ebony* (putative deletions spanning from the right to the left elements). (B) A stock was constructed just prior to its use that contains a recombinant chromosome harboring both the *neo*[R](93B) element and D2-3(99B). The *neo*[R](93B), D2-3(99B) stock was crossed in flanking elements, followed by reciprocal religation of the chromosome fragments, would give to rise to both deletions including *ebony* and duplications of the wild-type ebony gene (F[2]). Only the deletions would be detected in this experiment. Horizontal lines represent chromosomes. Hexagons represent P elements. Broken hexagons indicate transposase-induced breaks. The large, hatched arrows indicate transposase. (From Cooley, L., Thompson, D., and Spradling, A. C., *Proc. Natl. Acad. Sci. U.S.A.*, 87, 3170—3173, 1990. With permission.)

for this purpose are currently available for the production of *cis* progenitor stains.

♦ Deletion mutations are powerful molecular tools for understanding the structure of any genome. A major problem in deletion analysis is the difficulty of establishing the precise endpoint of each deletion. In *Drosophila,* P elements transpose at high frequency leading to an equally high frequency of chromosomal rearrangements whose endpoints correspond to the sites of preexisting P elements. By controlling the number of transposon targets and restricting transposase to a single source, the authors have developed a method to generate chromosomal deletions with precisely determined endpoints within the preexisting P elements. With the wealth of available P element *Drosophila* stocks, it could now be possible to generate an extensive array of deletion mutations with precisely determined endpoints. *Joel M. Schindler*

Isolation and Characterization of an Olfactory Mutant in Drosophila with a Chemically Specific Defect

S. L. Helfand and J. R. Carlson
Proc. Natl. Acad. Sci. U.S.A., 86, 2908—2912, 1989   1-15

Little is known about the molecular mechanisms of olfaction. The work reported here was designed to identify *Drosophila* mutants exhibiting a defective response only to a subset of chemical odorants. These mutants with specific anosmias were isolated using behavioral assays.

An olfactory T maze (Figure 1-15), was used to screen ethyl males of 1150 lines, each bearing a unique X chromosome mutagenized with ethyl methanesulfonate. The line 3D18 showed a significantly lower response to benzaldehyde than did wild-type control flies; 3D18 flies showed a normal response to 3-octanol. When tested in an olfactory jump assay, 3D18 flies showed a significant defect in their response to benzaldehyde, but not to ethyl acetate nor to propionic acid. These flies also had a visible cuticular pigmentation phenotype, more severe in females, with extremely high penetrance. In addition, homozygous 3D18 flies had low fertility due to a defect in fertilization transmitted through the maternal genotype.

Both meiotic recombination mapping and duplication and deficiency mapping of the mutation localized the olfactory and pigmentation phenotypes to the 7F1 to 8A4-5 interval of the X chromosome. The pentagon gene is located in this interval, and its map position agreed well with that determined for the 3D18 pigmentation phenotype. Independently derived pentagon alleles also had olfactory, pigmentation, and fertility defects. Complementation analysis showed that two recessive pentagon alleles tested failed to complement 3D18 in the olfactory jump assay and a third pentagon allele failed to complement the fertility defect of 3D18.

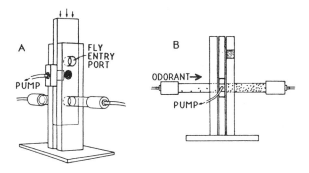

FIGURE 1-15. Olfactory T maze. (A) fifty to 100 flies are entered through the entry port into a chamber in the central, vertical sliding plate. (B) The plate is slid down into the bottom position, from which flies can escape either to the left or the right. The collecting tube on the left contains chemical vapors; the collecting tube on the right contains control air. Air is continuously drawn through the system by a pump (Cole-Parmer KNF 7056-25) at a rate of 1 l/min. Air entering the system from the left is drawn over the surface of a chemical odorant solution in a sidearm flask; control air entering from the right passes over diluent. Air leaves the system through a set of holes too small for a fly to pass through. After a period of 1 min, the number of flies in the two tubes are counted. A response index is computed by subtracting the number of animals on the odorant side from the number on the control side and dividing by the total. Flies are tested only once and are then discarded. (From Helfand, S. L. and Carlson, J. R., *Proc. Natl. Acad. Sci. U.S.A.*, 86, 2908—2912, 1989. With permission.)

These findings suggest that the olfactory mutation described here has a common origin with the pigmentation and infertility phenotypes co-isolated. These three defects may be due to a lesion at the pentagon locus. The specificity of the olfactory defect implies that the mutants may be defective in receptor, transduction, or processing molecules specific to a particular subset of chemicals.

A Simple Chemosensory Response in *Drosophila* and the Isolation of *acj* Mutants in Which It Is Affected
M. McKenna, P. Monte, S. L. Helfand, C. Woodard, and J. Carlson
Proc. Natl. Acad. Sci. U.S.A., 86, 8118—8122, 1989   1-16

The olfactory system of *Drosophila* has been largely unexplored, at least in part due to a lack of techniques for measuring olfactory function. Here, a chemosensory behavior that is easily measurable, using a single fly, is described. The behavior was used to isolate a set of mutants in which it was defective.

Flies were found to jump when exposed to vapors of certain chemicals. This behavior was the basis of the assay used here: the jump response. Flies were inserted into a vertical tube and permitted to crawl up the wall (Figure 1-16). When certain chemical vapors were passed through the

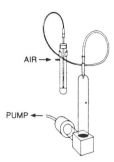

FIGURE 1-16. Apparatus for testing jump response. A single fly is introduced into the tube with a mouth aspirator and is allowed to walk approximately one third to one half of the way up the side. Air is drawn through the system continuously by means of a pump. When the Teflon tubing from the sidearm vessel is inserted into the hole at the top of the tube, air is drawn over the surface of the chemical placed in the sidearm vessel and through the tube. The tubing is inserted into the tube for 3 s, giving the animal a 3-s exposure to chemical vapors. A jump event is scored if the fly lands on the screen at the base of the tube during this period. (From McKenna, M., Monte, P., Helfand, S. L., Woodard, C., and Carlson, J., *Proc. Natl. Acad. Sci. U.S.A.*, 86, 8118—8122, 1989. With permission.)

tube, the flies jumped from the wall to a nylon screen below. A jump response was scored if the jump occurred within 3 s after the introduction of the vapors.

The jump response had a very high signal-to-noise ratio of 19. It was elicited by specific aldehydes, organic acids, and acetate esters, and was not proportional to the vapor pressure of the chemical. The jump response exhibited dose dependence. Surgical removal of the third antennal segments markedly attenuated the jump response to three chemicals tested.

When 200 lines of flies with mutagenized X chromosomes were tested for defective jump responses, 9 mutants were isolated. These mutants were normal in giant fiber physiology. Two mutants had abnormal electroreginograms that indicated defective transmission of impulses from retinal photoreceptor neurons. Four of the mutants had recessive defects in the specific response to ethyl acetate; these mutants defined three complementation groups. One mutant was found to be defective in the antennal change in electrical potential following odorant stimulation.

The jump assay for measuring olfactory function appears to be quick and easy and should provide the ability to test a single fly, eliminating the need to establish cell lines to screen for mutants. Further physiological and genetic investigations of the mutants described here are in progress.

♦ Insects possess an extremely sensitive olfactory system that allows them to detect and discriminate airborne chemicals. Such chemicals play important roles in determining location of food, oviposition sites, and sexual behavior. The insect olfactory system is presently poorly understood. This is in contrast to the visual system, for which much information has recently accumulated. The use of genetics to dissect the development and function of the eye was a major determining factor for the progress achieved with the visual system. The fact that many mutations that interfere with eye formation or visual pathways do not affect viability provided an added advantage. In principle, it should be possible to use the sophisticated genetics of *Drosophila* to study the olfactory system in a similar way.

These two papers attempt to develop genetic approaches for the study of the olfactory system. An important requisite for this type of study is the availability of a simple and efficient assay for the screening of mutants defective in the olfactory system. Two such assays are described. One is Helfand and Carlson's "olfactory T maze" assay. Fifty to 100 flies are placed in a horizontal cylindrical chamber. Air is drawn from the middle, thus pulling the test chemical from one end and solvent-saturated air from the other. If the chemical acts as a repellent, flies will avoid the chemical side; conversely, if it is an attractant, the flies will migrate toward that side of the chamber; finally, if the test substance is neutral there will be no preference in fly distribution. Thus, by measuring the distribution of the flies between the two sides of the chamber the behavioral response to different substances can be determined. Olfactory mutants will demonstrate a behavior that deviates from that of wild-type flies.

A second assay is the "jump response" assay of McKenna et al. In this assay single flies are placed in a vertical cylinder with air flowing from top to bottom. The fly is allowed to crawl half way up the wall of the tube at which point the flow of air is replaced by a 3-s-flow of test chemical. If the fly is sensitive to the chemical it reacts by jumping to the bottom of the cylinder; it not, there will be no reaction. A fly mutant for the detection of a jump-inducing substance will not jump when exposed to this substance. Since this behavioral assay is not lethal to the fly, it allows mutant screening in the F1 generation, thus obviating the need of first establishing F2 lines for each chromosome to be tested. Moreover, the authors have shown that the test can be performed multiple times in rapid succession, allowing the rapid retesting of any putative mutant fly.

Helfand and Carlson characterize a mutant — 3D18 — that is defective in its response to the repellent benzaldehyde as measured by both assays. Interestingly, the defect is specific, since the responses to three other chemicals was normal. 3D18 flies were also normal in a number of other behavioral, neurophysiological, and morphological tests. 3D18 turned out to be allelic to *pentagon,* a mutant first described by Bridges in 1922. In addition to the olfactory phenotype, mutant flies have reduced fertility and defects in pigmentation of the cuticle (a hyperpigmented pentagonal spot appears in the dorsal part of the thorax; hence the mutant name). In a way, this pleiotropy might render more complicated the detailed characterization of the 3D18/*pentagon* locus.

Using the jump response assay, McKenna et al. isolated nine independent jump response mutants. Some mutants exhibited selectivity between ethyl acetate and benzaldehyde, while others did not. In some jump-response mutants the visual system was also affected. Electrophysiological recordings from the antennae were defective for one of the mutants. The third antennal segment is believed to be the principal olfactory organ in flies. To determine whether the jump response was mediated by the antennae, McKenna et al. assayed flies that had parts of their antennae either surgically or genetically removed. The latter experiments used *Nasobemia* and *Antennapedia B* homeotic mutant flies. These experiments demonstrated that the jump response depends to a large extent on the third antennal segment. However, since these flies still demonstrated a residual response, it is hypothesized that other parts of the fly also participate in chemosensory reception.

Overall, these studies demonstrate the feasibility of obtaining mutants that are defective in their responses to airborne chemicals. The cloning of some of these genes may provide important new insights. P-element-tagged mutants could facilitate enormously the task of molecularly characterizing such genes but, unfortunately, such mutants have not yet been obtained (cited in McKenna et al.). Although much more work remains to be done, it is likely that a genetic approach will provide new insights on insect chemosensory mechanisms. *Marcelo Jacobs-Lorena*

Pregnancies from Biopsied Human Implantation Embryos Sexed by Y-Specific DNA Amplification
A. H. Handyside, E. H. Kontogianni, K. Hardy, and R. M. L. Winston
Nature, 344, 768—770, 1990   1-17

For some X-linked genetic diseases, prenatal diagnosis is not available. In this case, the sex of the fetus may be determined prenatally, with the implication that male fetuses, affected or nonaffected, will be aborted. If, instead of fetuses, preimplantation embryos were sexed, then only female embryos could be transferred to the uterus and no abortions would be necessary. Here, the establishment of two pregnancies after *in vitro* fertilization and sexing of the embryo is reported.

All women in five couples at risk of transmitting recessive X-linked diseases had had previous terminations of affected fetuses; most had had difficulties conceiving, and all preferred this approach to a specific diagnosis performed later in pregnancy. Superovulation after pituitary suppression resulted in an expected number of oocytes collected (11.2 per treatment cycle) and normally fertilized (6.3 per treatment cycle). After 3 d of *in vitro* development, most fertilized embryos had reached the 8-cell stage and were biopsied. A hole was drilled through the zona pellucida, and acid Tyrodes (pH 2.4) applied locally; one or two cells were biopsied from each embryo (the development of human embryos is known to proceed normally after removal of one or two cells at this stage).

Cell biopsies were analyzed by polymerase chain reaction-based amplification of a short fragment of a Y-chromosome-specific repeat sequence. Male DNA from a single cell was easily detected and was distinguishable from maximum background contamination. In a few cases, analysis was impossible, although lysis of cell biopsies after removal often resulted in intact nuclei, with DNA amplification unaffected.

The results after uterine transfer suggest that pregnancy rates after biopsy of embryos at this stage might be equal to those following transfer of intact embryos to infertile women. Twin pregnancies have been established in two women, and chorionic villus sampling followed by karyotyping showed that both women have fetuses with normal female karyotypes.

Enzymatic Amplification of a Y Chromosome Repeat in a Single Blastomere Allows Identification of the Sex of Preimplantation Mouse Embryos
M. W. Bradbury, L. M. Isola, and J. W. Gordon
Proc. Natl. Acad. Sci. U.S.A., 87, 4053—4057, 1990   1-18

Although several approaches have been developed for the identification of the sex of preimplantation mammalian embryos, none is optimally rapid, easy, and accurate. The work described here resulted in a technique requiring less than 12 h to identify the sex of a preimplantation mouse embryo, and was based on polymerase chain reaction (PCR) analysis of the DNA of a single blastomere.

The 102-base oligonucleotide primers used for the PCR were derived from a 1.5-kilobase DNA fragment of Y-chromosome-specific repeated sequences. These primers were used to score the DNA of single blastomeres derived from embryo biopsy, using PCR. The primer sequence was probably present in XX DNA, but in far fewer copies than in XY DNA. After DNA amplification, blastomeres with high intensities of the 102-bp band were considered male, with low-intensity bands considered female, and those with intermediate-intensity bands considered of undetermined sex.

72% of embryos were classed as males or females. Among eight pregnancies which arose from transfer of biopsied embryos, the sex had been correctly predicted in all eight.

These results suggest that use of PCR permits the rapid, accurate determination of sex from a single mouse embryo blastomere. As used here, the technique fails to make a determination of sex in 28% of cases. It seems likely that improved selection of oligonucleotide primers for the PCR will lead to a reduction in ambiguous results.

♦ As discussed in the 1990 Year Book, the polymerase chain reaction (PCR) has had a dramatic effect on experimental approaches to mamma-

lian development. An obvious extension of this work is the potential for using PCR for detection of birth defects. The results reported in the two papers cited above show that it is possible to identify the sex of mouse and human embryos by amplifying Y-specific sequences. The advantage of PCR over more conventional biopsy techniques is that it can be applied to a single blastomere of the preimplantation embryo, and it does not depend on expression of the gene in question. Furthermore, the whole procedure requires less than 12 h, thereby eliminating the need to freeze embryos before results are available. A potential problem would be the inability to make a diagnosis because of lack of amplification or presence of background amplification. However, it is important to note that not all embryos need to be correctly diagnosed but rather that the diagnosis be certain on those embryos selected for transfer. Clearly, the main goal of PCR genotyping in the future would be the ability to evaluate allelic differences for both autosomal and sex-linked loci. The data reported in these papers demonstrate that such an evaluation may be possible. Furthermore, the authors cite results from tissue culture studies indicating the feasibility of detecting allelic differences. Attainment of this goal would constitute a significant advancement in prenatal diagnosis and, thus, could be a viable option for many families carrying genetic defects. The possibility of genotyping for genetic disorders after *in vitro* fertilization would eliminate the need for pregnancy termination after implantation.
Terry Magnuson

Dwarf Locus Mutants Lacking Three Pituitary Cell Types Result from Mutations in the POU-Domain Gene *pit-1*
S. Li, E. B. Crenshaw, III, E. J. Rawson, D. M. Simmons,
L. W. Swanson, and M. G. Rosenfeld
Nature, 347, 528—533, 1990   1-19

The pituitary-specific transcription factor, Pit-1, binds to *cis*-active elements, and may transactivate the cell-specific expression of the rat prolactin and growth hormone genes in the lactotrophs and somatotrophs, respectively. In the rat, the expression of the *pit-1* gene precedes that of growth hormone and prolactin; in the mature pituitary, *pit-1* gene expression occurs in somatotrophs, lactotrophs, and thyrotrophs.

pit-1 is a member of a large gene family of transcription factors, called POU-domain proteins, each containing a region similar to a homeobox and a POU-specific domain. If *pit-1* encodes a growth hormone transcription factor, a mutation in the gene might result in a dwarf phenotype. The Snell dwarf mouse, *dw,* is a recessive mutation, characterized by the absence of synthesis of growth hormone, prolactin and thyroid-stimulating hormone, and the absence of thyrotrophs, lactotrophs, and somatotrophs in the mature mouse. A second dwarf mutation is the dwarf Jackson mouse mutation, *dw^J*, allelic to dw; a third mutation called Ames dwarf, df, is nonallelic to the others. Mice homozygous for either *dw*ᴶ and df also lack the same three anterior pituitary cell types. The experiments described here tested the hypothesis that defects in *pit-1* result in a dwarf phenotype by analysis of the three dwarf mutations.

By somatic cell genetics, the pit-1 gene was mapped to the same mouse chromosome (16) as that to which the dw was previously mapped. Restriction mapping showed that either an inversion or an insertion of a DNA segment of greater than 4 kb occurred in the *pit-1* gene of the *dw*ᴶ mouse. Sequence analysis of pit-1 cDNA from Snell dwarf pituitary glands revealed a point mutation that would alter a tryptophan residue conserved in all known homeobox domains in organisms from mammals to yeast. In all three classes of mutants, the expression of *pit-1* mRNA and protein was abnormally low or undetectable. In the Ames dwarf, the lack of detectable *pit-1* gene expression showed that the Ames mutation is epistatic to the expression of *pit-1*.

These experiments show that both alleles of the *dw* locus consist of mutations of *pit-1*. The findings imply that *pit-1* activates a part of the pituitary developmental program and provide further evidence that *pit-1* is a pituitary-specific, growth hormone transcription factor.

♦ The merging of the mouse molecular map with the genetic map is occurring at ever-increasing speed and this is once again demonstrated by the results reported in this paper. Three autosomal recessive dwarf mutations have been described in mice. Two of these mutations, the Snell dwarf (*dw*) and the dwarf Jackson (*dwj*). are allelic and map to chromosome 16. The third mutation, the Ames dwarf (*df*), is nonallelic to the other two and maps to chromosome 11. All three mutations result in a similar phenotype which includes hypoplasia of the anterior portion of the pituitary gland with an absence of mature somatotrophs, lactotrophs, and thyrotrophs. Consequently, homozygous animals lack growth hormone, prolactin, and thyroid-stimulating hormone. More recently, a pituitary-specific transcription factor (pit-1), known to transactivate the genes which encode prolactin and growth hormone, has been described. *pit-1* is a member of a large gene family of POU-domain transcription factors. The stage-specific pattern of *pit-1* expression is consistent with a role in regulating expression of prolactin and growth hormone during development. Based on this possibility, Li et al. predicted that a mutation in the *pit-1* gene would result in a dwarf phenotype due to the absence of growth hormone expression. An analysis of the three known dwarf mutations verified this prediction. Somatic cell mapping indicated that *pit-1* and the *dw* locus map to the same chromosome. Molecular analysis then revealed an inversion or an insertion of a DNA segment of >4 kb in *pit-1* of *dw*ʲ mice, and a G-to-T alteration in *pit-1* of *dw* mice that would convert a tryptophan in the POU homeodomain to a cysteine residue. This particular tryptophan residue is invariant among the POU and homeodomain transcription factors so far described and is necessary for high-affinity DNA binding. Pit-1 transcripts or protein were not detectable in homozygotes carrying either the *dw* or the *dwʲ* allele. These data directly link Pit-1 to cell-type-specific differentiation in the anterior pituitary. Although *pit-1* and *dw* are probably the same locus, mice homozygous for the *df* mutation also lack detectable *pit-1*. This suggest that *df* may act upstream of *pit-1*. Future work is likely to concentrate on the role of the *df* locus in regulating *pit-1* and also what genes act downstream of *pit-1*. Terry Magnuson

---

Multiplex Gene Regulation: A Two-Tiered Approach to Transgene Regulation in Transgenic Mice
G. W. Byrne and F. W. Ruddle
Proc. Natl. Acad. Sci. U.S.A., 86, 5473—5477, 1989   1-20

Transgenic mice have been useful for studying gene function, based on the use of inducible or tissue-specific promoters. These mice exhibit partially unregulated expression of the gene of interest, or they can be induced to express the gene of interest only in limited tissues or at limited times. The system presented here was devised in an attempt to provide an inducible method of gene expression for transgenic mice that is inducible in a highly versatile way. It also provides the ability to study genes that may be lethal in early development.

The multiplex gene regulatory (MGR) system is based on the transactivation of the herpes simplex virus immediate-early (IE) genes by the transactivator virion polypeptide VP16. For example, one transgenic mouse line, the transresponder, was created that contained the bacterial reporter gene CAT regulated by an IE promoter (Figure 1-20). A second transactivator mouse line contained the HSV-1 transactivator VP16, under the control of the neurospecific NF-L murine promoter. When these lines are crossed, mice that contain both transgenes should express CAT in tissues and at times in which NF-L gene is expressed.

Eight transgenic mouse lines were produced that contained an IE promoter linked to the CAT gene. Southern blot analysis showed that 6 of these lines contained from 2 to 20 head-to-tail concatameric repeats of the IE-CAT transgene. In 4 lines tested, no CAT activity was found in any tissue examined. Infection of mice from these 4 lines with HSV-1 resulted in induction of CAT at various levels. Transgenic mice also were produced containing VP16 under the control of a fragment of the NF-L promoter; one mouse induced CAT when mated to an animal containing IE-CAT transgene. Strong CAT gene expression was found in brain and spinal cord, with weak expession in heart, and no detectable expression in liver, spleen, or intestine.

These experiments provide evidence that mouse lines expressing VP16 can be established. Expression of VP16 was not lethal to the developing embryo. Tissue-specific expression of VP16 may be inferred from

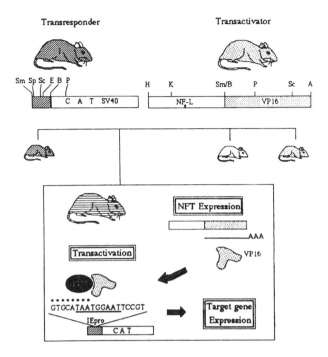

FIGURE 1-20. A diagrammatic representation of the two-tiered multiplex regulatory system. Two transgenic mouse lines are represented. One transgenic line, the transresponder (left), contains a target gene (CAT) regulated by the HSV-1 IE promoter element (hatched area). In the transresponder line, there is no expression of the target gene. The second transgenic line, the transactivator (right), contains the HSV-1 transactivator VP16. In this example, VP16 is regulated by a neurospecific promoter element from the murine NF-L gene. The two lines are crossed. The offspring inherit the transresponder (left), transactivator (far right), neither (small white animal), or both transgenes (boxed region). In the double transgenic offspring, VP16 is expressed in a neurospecific pattern. In the cells expressing VP16, VP16 forms a protein-protein complex with the ubiquitous octamer binding protein (oval). This protein complex specifically activates the IE promoter, resulting in neurospecific expression of the target gene. One of the TAATGARAT domains (underlined) in the 360-bp ICP4 promoter is shown. Immediately 5′ of the TAATGARAT sequence (*) is a potential octamer binding site. B, *Ban*HI; E, *Eco*RI; K, *Kpn* I; P, *Pvu* II; Sp, *Sph* I; Sc, *Sac* II, Sm, *Sma* I. (From Byrne, G. W. and Ruddle, F. W., *Proc. Natl. Acad. Sci. U.S.A.*, 86, 5473—5477, 1989. With permission.)

these findings. The use of the MGR system permits the induction of more than 1 IE-regulated transgene in a single embryo, permitting the study of interactions between more than one developmental regulatory gene.

♦ One approach being used to understand the developmental significance of stage-specific genes is to misexpress them in the developing embryo by making transgenic mice. Abnormal expression of the transgene has the potential of resulting in a dominant-like phenotype. Such phenotypes can provide important information regarding the mechanisms of action of the gene product in question. Examples include *Hox 1.1* which results in craniofacial abnormalities when driven by an actin promoter (Balling et al., *Cell*, 58, 337—347, 1989) and *Hox 1.4* which results in megacolon when overexpressed from its own promoter (Wolgemuth et al., *Nature*, 337, 464—467, 1989). The problem with this approach is that abnormal expression resulting in dominant phenotypes often can be lethal, making it impossible to establish transgenic lines carrying these constructs. The results reported in the paper cited above present evidence for an inducible method of gene regulation. First, transgenic lines are established carrying genes of interest regulated by the promoter fragment of an immediate early (IE) gene of herpes simplex virus. The IE-regulated transgenes are then activated in specific tissues during development by mating with a second transgenic mouse that contains an IE-transactivator regulated by a promoter with appropriate tissue specificity. This allows one to activate the transgene in question at will, thus making it possible to maintain the transgenic line in the inactivated state. The exciting aspect of this approach is the possibility of inducing expression of two or more transgenes within a single embryo simultaneously by using different transresponder and activator pairs. This would make it possible to study the interactions between different gene products. Thus, the multiplex gene regulation system as described by Byrne and Ruddle adds genetic versatility to the transgenic approach of producing dominant-like phenotypes. The usefulness of such a system will certainly be demonstrated with time, and publication of additional regulating schemes similar to that described within this article will surely appear in the coming year. *Terry Magnuson*

Efficient Rescue of Integrated Shuttle Vectors From Transgenic Mice: A Model for Studying Mutations *In Vivo*
J. A. Gossen, W. J. F. de Leeuw, C. H. T. Tan, E. C. Zwarthoff, F. Berends, P. H. M. Lohman, D. L. Knook, and J. Vijg
Proc. Natl. Acad. Sci. U.S.A., 86, 7971—7975, 1989   1-21

One promising method for the study of mutagenesis *in vivo* in mammals is based on the use of shuttle vectors. These vectors contain a marker gene that can be introduced into mammalian cells and then retrieved for selection of mutated genes in bacteria. Here, a method of studying mutations *in vivo* is presented that employs both shuttle vectors and transgenic mice.

Bacteriophage lambda shuttle vectors containing the selectable bacterial *lacZ* gene were constructed. These vectors were integrated into the genome of transgenic mice in a head-to-tail arrangement. To obtain high-effficiency rescue of the vector from total genomic mouse DNA, the genomic DNA was digested with a restriction enzyme whose recognition site is not present in the vector. The fragments containing contcatamers of the lamba vector were purified from an electrophoresis gel. *In vitro* packaging of this DNA was followed by propagation of the phage in a LacZ⁻ strain of *E. coli* C.

Transgenic mice were produced containing between 3 and 80 copies of the integrated vector per haploid genome. The mouse lines that resulted showed no evidence of large rearrangements of the vector DNA. A maximum of 1.4×10^6 vectors were rescued in a single experiment with 100 µg of liver DNA from transgenic mice. When vectors were scored by the presence of colorless plaques when the bacteria were grown in the presence of the chromogenic indicator X-Gal, no mutations in the *lacZ* gene were detected among 138,816 vectors isolated from brain DNA nor from 168,160 vectors isolated from liver DNA. When transgenic mice were treated with *N*-ethyl-*N*-nitrosourea, a dose-dependent increase in the frequency of mutated vectors isolated from brain DNA resulted. DNA sequence analysis of four mutant phages rescued from brains and livers of mutagenized animals showed that GC→AT transitions had occurred in all cases.

This transgenic mouse model seems to be a practical method for study of gene mutations *in vivo* in mammals. It should permit the determination of spontaneous mutation frequencies in different organs and tissues at different genomic sites.

♦ The potential to generate useful animal models for studying various biological processes *in vivo* has been a strong rationale for the use of transgenic technology. As production techniques continue to be modified, efficiencies improved, and interesting genes isolated, the potential for generating valuable animal models increases. In this article, the authors generate a unique transgenic mouse not to study a specific gene or genetic disease, but to investigate an important biological process — mutagenesis. The authors couple the use of shuttle vectors with transgenesis to generate a mouse model that can serve as an *in vivo* model for studying spontaneous mutation frequencies. Using the integrated shuttle vector as an assay, they use a known mutagen, *N*-ethyl-*N*-nitrosourea (EtNU), to demonstrate a dose-dependent increase in the frequency of mutation in selected organs. This specific mouse model, or other designed in a similar fashion, will have enormous impact on our ability to perform useful initial screens to assess the toxicity and/or mutagenicity of a spectrum of compounds. *Joel M. Schindler*

Correction of Murine Mucopolysaccharidosis VII by a Human β-Glucuronidase Transgene

J. W. Kyle, E. H. Birkenmeier, B. Gwynn, C. Vogler, P. C. Hoppe, J. W. Hoffmann, and W. S. Sly

Proc. Natl. Acad. Sci. U.S.A., 87, 3914—3918, 1990   1-22

Mucopolysaccharidosis type VII (MPSVII) is a disease of humans that results from deficiency of β-glucuronidase activity. Recently, *gus*ᵐᵖˢ/*gus*ᵐᵖˢ mice have been described that are also nearly completely deficient in β-glucuronidase. Mice with this genotype are dwarfed, with skeletal deformities, and they succumb to premature death. The lysosomes of these mice have excessive levels of undegraded glycosaminoglycans, and, as with humans who suffer from MPSVII, their leukocytes have inclusion bodies. The human gene for β-glucuronidase has recently been isolated. The experiments presented here investigated the expression of this gene in transgenic *gus*mps/*gus*mps mice.

Transgenic male mice were established that contained the human β-glucuronidase gene, *GUSB*. These heterozygotes were mated to female *gus*mps/*gus*mps mice. Offspring carrying the human transgene were mated to siblings without the human transgene to generate mice homozygous for *gus*mps and heterozygous for the human transgene *GUSB*.

The transgenic mice containing the human β-glucuronidase gene expressed very high levels of β-glucuronidase in all tissues examined in a relative tissue distribution similar to that of the mouse enzyme in normal mice. These transgenic mice no longer showed the secondary elevations of other lysosomal enzymes found in the *gus*mps/*gus*mps mutants. The transgenic animals also had normal levels of hexuronic acid, unlike the *gus*mps/*gus*mps mutants. Further, the pathological findings in many tissues of the mutants were also corrected in transgenic mice. Grossly, the dwarfism, skeletal deformities, shortened lifespan, and other abnormalities of *gus*mps/*gus*mps mice were absent in mice that expressed the *GUSB* transgene.

These results provide evidence that the human b-glucuronidase gene is expressed in these transgenic *gus*mps/*gus*mps mice. The expression of this gene seems to correct all aspects of the murine mucopolysaccharidosis storage disease of the mutant animals.

♦ One of the most exciting practical outcomes from our increased understanding of developmental genetics is the potential of gene therapy as a viable clinical procedure. While this possibility is still years away, limited clinical trials have already begun. Many questions about the viability of such therapy still remain. The most obvious biological question is whether the introduction of a "good" gene can correct the deficit resulting from a defective endogenous gene. This article demonstrates that such deficits can be reversed through the introduction of a normal gene. In this particular case, a mouse model for a human disease was used as the recipient of the human gene for that disease. The resulting transgenic animal was phenotypically normal. This observation dramatically shows that an endogenous genetic defect can be reversed by the integration and expression of an appropriate normal gene. In addition, these results show that a mouse can successfully express a human gene to correct the deficit. Together, these observations provide strong support for the potential success of gene therapy as a viable therapeutic protocol. *Joel M. Schindler*

A Growth-Deficiency Phenotype in Heterozygous Mice Carrying an Insulin-Like Growth Factor II Gene Disrupted by Targeting

T. M. DeChiara, A. Efstratiadis, and E. J. Robertson
Nature, 345, 78—80, 1990   1-23

The expression of insulin-like growth factor II (IGF-II) may be important in the embryonic development of mice. To examine this hypothesis, this work was designed to introduce mutations at the IGF-II gene locus in the mouse germ line. To do this, a single allele of cultured mouse embryonic stem (ES) cells was disrupted by gene targeting; affected cells were used in the construction of chimeric animals.

To disrupt the IGF-II gene of ES cells, a replacement vector was introduced to cells by electroporation. The vector contained fragments of the mouse IGF-II gene and transcriptionally competent cassettes for the genes *neo,* encoding bacterial neomycin resistance, and *tk,* encoding the thymidine kinase of herpes simplex virus. Successful targeting, or homologous recombination, of this DNA would abolish the function of the IGF-II gene and add G418 resistance to the cells. Since the *tk* cassette would not be integrated during homologous recombination events, the ES cells would retain their insensitivity to gancyclovir.

About 7% of G418-resistant colonies were also resistant to gancyclovir. Of these, polymerase chain reaction assays showed that about 3% were homologous recombinants. These cells were injected into host blastocysts, which could be scored as chimeras based on coat pigmentation. The frequency of chimera formation was high, but the extent of chimerism varied with the genetic background of the blastocyst.

Some chimeric animals were of only 60% the body size of wild-type pups. All these had a disrupted IGD-II allele, while none of the normal-sized pups had this disrupted gene. The reduction in body size seemed to be expressed earlier than at embryonic day 16, for growth rates of these mice were normal after day 16. These heterozygous embryos had only one-tenth the mRNA from IGF-II as did wild-type embryos. These embryos often matured into animals otherwise normal and fertile.

These results suggest that the elimination of one copy of the IGF-II gene results in a growth-deficiency phenotype in the embryos. This defect persists after birth. It is not known whether the relationship between the gene disruption and the phenotype is direct or indirect.

Targeting of Nonexpressed Genes in Embryonic Stem Cells Via Homologous Recombination

R. S. Johnson, M. Sheng, and M. E. Greenberg
Science, 245, 1234—1236, 1989   1-24

Gene targeting to murine embryonic stem (ES) cells has been reported for several genes. The targeting technique relied on the abilities to maintain the pluripotency of blastocyst-derived ES cells *in vitro* and to select for homologous recombination events. The use of this method may be possible only for genes expressed in ES cells. The experiments presented here tested this hypothesis.

♦ Three genes were tested for their ability to be targeted to ES cells: the adipsin and aP2 genes, transcriptionally active predominantly in murine adipocytes, and c-*fos*. Northern blot analysis showed that c-*fos* was expressed by ES cells, while messenger RNAs for adipsin and aP2 were undetectable even after prolonged exposure of autoradiographs.

Gene targeting vectors were constructed for each of the three genes of interest. The vectors contained the herpes simplex virus thymidine kinase gene, *tk*, and the neomycin resistance gene, *neo*. The *neo* gene was inserted into the genomic clones in a way that disrupted their transcription or translation or both. While cells containing randomly integrated vectors would be resistant to G418, expression of *tk* would render them sensitive to gancyclovir. In contrast, cells containing vectors that integrated via homologous recombination would be resistant to both G418 and to gancyclovir, since tk sequences would not have been integrated. In this case, the gene of interest would be replaced by a *neo*-disrupted homolog.

About 0.1% of all cells electroporated with vectors were resistant to G418. Southern blot analysis showed that 5 to 10% of cells resistant to both G418 and gancyclovir were homologous recombinants. The number of recombinants over the number of G418-resistant cells was comparable for the 3 genes of interest (from 3×10^{-4} to 4×10^{-5}), and was similar to those previously reported for the expressed gene hypoxanthine-guanine phosphoribosyl-transferase (2.5×10^{-5}). There was no evidence that any recombinant cell lines expressed adipsin or aP2. Injection of recombinant cells into mouse blastocysts resulted in the production of chimeric mice.

These findings provide evidence that genes that are not expressed or are expressed at exceedingly low levels in ES cells may still be disrupted by homologous recombination. This process seems as efficient as targeting of genes expressed in these cells, and seems not to result in the loss of the cells' pluripotency.

Inactivating the b$_2$-Microglobulin Locus in Mouse Embryonic Stem Cells by Homologous Recombination
B. H. Koller and O. Smithies
Proc. Natl. Acad. Sci. U.S.A., 86, 8932—8935, 1989   1-25

In mice, β$_2$-microglobulin is a polypeptide required for the expression of many histocompatibility class I proteins and of the cell surface antigens Qa/tla. The roles of these β_2m-associated antigens during development is unclear. To investigate their function, the work here was designed to disrupt the β_2m gene in mice.

The first step of the strategy used was the creation of murine embryonic stem (ES) cells with disrupted β_2m genes. These were produced by gene targeting using a vector containing about 5 kilobases of the β_2m gene disrupted by *neo*. Electroporation of the vectors was followed by selection for G418-resistant colonies. Homologous integrants were selected from these by a polymerase chain reaction assay that used primers whose binding sites would be juxtaposed only if homologous recombination had occurred. Two PCR-positive clones were identified, expanded, and then confirmed by Southern blot analysis to contain one copy of the β_2m gene disrupted as predicted by the insertion of *neo*.

Once ES cell lines containing mutated β_2m genes were produced, some of these cells were injected into blastocysts, and embryos were reimplanted into pseudopregnant females. Coat color allowed ready identification of the production of chimeric pups. More than 70% of mice produced in this way were chimeras, and Southern blot analysis confirmed that these animals contained mutated β_2m genes.

These results provide evidence that gene targeting resulted in the production of ES cells containing disrupted β_2m genes, and that these cells could be introduced into blastocysts resulting in the production of chimeric animals. Cells heterozygous for mutations in β_2m genes seem viable in the intact mouse. To better understand the function of β_2m and its associated antigens, experiments creating homozygous mutations in the β_2m gene are planned.

Germ-Line Transmission of a Planned Alteration Made in a Hypoxanthine Phosphoribosyltransferase Gene by Homologous Recombination in Embryonic Stem Cells

B. H. Koller, L. J., Hagemann, T. Doetschman, J. R. Hagaman, S. Huang, P. J. Williams, N. L. First, N. Maeda, and O. Smithies
Proc. Natl. Acad. Sci. U.S.A., 86, 8927—8931, 1989   1-26

Recent work has established that homologous recombination can create planned mutations in murine embryonic stem (ES) cells *in vitro,* and that these mutant cells, injected into blastocysts, result in chimeric animals. Some of these animals are chimeric in their gonads and can transmit the mutation to their progeny. A previously reported attempt to combine gene targeting with germ-line transfer resulted in a male mouse chimera that transmitted a corrected hypoxanthine phosphoribosyltransferase (HPRT) gene to its progeny, but a deletion occurred during or after

the recombination event. The work presented here was designed to accomplish germ-line transmission of targeted gene mutations without complications.

Homologous recombination was induced by previously published methods in ES cells derived from an HPRT⁻ male whose HPRT gene contained a single deletion of the promoter and the first two exons. The introduction of a correcting plasmid to these cells was followed by selection in hypoxanthine/aminopterin/thymidine-containing medium. Of 19 resulting colonies, 18 were shown by Southern blotting to contain the predicted correction of HPRT; 5 of these seemed to contain a single copy of the correcting plasmid inserted without evidence of gene conversion. These latter five lines were used for further experiments.

Corrected cells were injected into blastocyts and chimeric pups were produced, as seen by alterations from the parental coat color. When male chimeras were mated to females of the same genotype as that from which the original blastocysts were derived, pups resulted whose coat color was derived from genes of the original ES cell line. Two of these four fertile male chimeras transmitted a correctly altered HPRT gene to some of their female progeny, as documented by Southern blot analysis of the pups.

This work provides evidence that gene targeting can correct a deletion mutation of HPRT in ES cells and that such corrected cells can be used to produce chimeras that transmit the corrected gene through the germ line. Various technical manipulations are helpful in the success of this process. If embryonic stem cells can be cultured, this procedure might be applicable to other species.

Normal Development of Mice Deficient in β₂-M, MHC Class I Proteins, and CD8⁺ T Cells

B. H. Koller, P. Marrack, J. W. Kappler, and O. Smithies
Science, 248, 1227—1230, 1990   1-27

Various functions have been ascribed to the major histocompatability class I proteins; little is known about the function of the nonpolymorphic class I related proteins, Qa/Tla. Both may be involved in cell-cell interactions during development. Since both types of protein require β_2-microglobulin (β_2M) for their expression, the work described here was designed to investigate these issues by studying embryos and mature mice that lacked β_2M.

Mouse chimeras were produced that contained mutated β_2M; these chimeras were engineered via gene targeting into embryonic stem cells, which were then injected into blastocysts. Some of these chimeras exhibited germline transmission of disrupted β_2M genes; after the appropriate matings, animals with homozygous disruptions of β_2M were produced.

These animals were healthy and appeared normal. Their lymph nodes, spleens, and thymuses were of normal size and contained normal numbers of cells. Lymph node cells from heterozygous mice had low levels of class I antigens, while those from homozygous mutants showed no detectable cell surface expression of native heavy chain complex.

Thymocytes from 4- or 5-week-old homozygous mutant mice had about one tenth the normal proportion of $CD4^-CD8^+$ cells as normal or heterozygous mice. Homozygous mutants had thymocytes, spleen cells, and lymphocytes with normal profiles of $\alpha\beta$ T cell receptor (TCR) positive T cells. In contrast to normal animals or heterozygous mutants, homozygotes completely lacked $\alpha\beta$ T cells that expressed CD8. The $\gamma\delta$ T cell receptor positive cells of homozygous mutants included about 25% that were $CD8^+$.

These findings suggest that mice homozygous for a disrupted β_2M gene develop normally. The only defect found in these animals was a complete deficiency in $CD8^+$ $TCR\alpha\beta$ T cells, cells that normally mediate cytotoxic T cell function. Further work should elucidate the role of deficiency of these effector cells on immune competence and the development of neoplasms.

Germ-Line Transmission of a c-*abl* Mutation Produced by Targeted Gene Disruption in ES Cells
P. L. Schwartzberg, S. P. Goff, and E. J. Robertson
Science, 799—803, 1989   1-28

Germ-line transmission of mutations of interesting genes is a powerful tool for studying the functions of the gene products. Previously, the only gene reported as targeted for mutation and subsequently transmitted in the germ line was that encoding hypoxanthine-guinine phosphoribosyl transferase, for which selection methods exist. Since this laboratory is interested in the function of c-*abl,* the cellular homolog of the Abelson murine leukemia virus oncogene, they conducted experiments designed to result in germ-line transmission of mutations of c-*abl,* a nonselectable autosomal gene.

The first stage of this work was the introducion of a substitution mutation into the c-*abl* locus of mouse embryonic stem (ES) cells. Because c-*abl* is ubiquitously expressed, it was feared that a null mutation might result in severe deleterious effects or lethality. To minimize this possibility, experiments were designed to engineer a c-*abl* defective in the coding sequence's carboxy terminus, a domain that is not needed for tyrosine kinase activity but is important for transformation of lymphocytes. A DNA construct was produced that contained a promoterless *neo* gene fused to c-*abl* sequences. Only homologous recombination of c-*abl* or random integration adjacent to a cellular promoter would result in G418 resistance of cells that integrated this construct.

After introduction of the construct into the CCE ES cell line by electroporation, drug-resistant colonies were isolated. Southern blotting showed that 7 independent homologous recombination events had occurred out of 239 colonies screened. Cells that contained one disrupted copy of c-*abl* were injected into blastocysts; chimeras were scored on the basis of alterations from parental coat pigmentation and were recovered in high frequencies (32 to 52% of live-born animals examined).

When phenotypically male chimeras were mated with tester females, some progeny resulted whose coat color indicated that they contained genes from the CCE ES cell line. Some of these progeny of the chimeric males were shown by Southern analysis to carry the c-*abl* mutation. These heterozygotes were phenotypically normal at 10 weeks of age, with no detectable tumors or lymphomas.

These experiments provide evidence that transmissible mutations can be introduced into nonselectable genes in ES cells by homologous recombination. Such a mutation in the carboxy terminus of the c-*abl* gene was created. Analysis of more severe mutations that abolish the kinase activity of this gene, or study of homozygous mutants in the carboxy terminus mutation, should help explain the function of this gene.

The *Wnt*-1 (*int*-1) Proto-Oncogene Is Required for Development of a Large Region of the Mouse Brain
A. P. McMahon and A. Bradley
Cell, 62, 1073—1085, 1990   1-29

The *Wnt*-1 (*int*-1) proto-oncogenegene was originally identified as a gene induced in mouse mammary tumors by mouse mammary tumor virus. Since then, this gene has been found to be expressed normally in mouse development in spermatids and in the fetal central nervous system (CNS). The gene is homologous to the *Drosophila* segment polarity gene *wingless* (*wg*). The products of both *wg* and *Wnt*-1 seem to be secreted proteins and may be involved in cell-to-cell signalling. The work described here tested the hypothesis that *Wnt*-1 regulates pattern in the developing mouse CNS. To do this, null mutations in *Wnt*-1 were generated and their phenotypes analyzed.

Six independent embryonic stem cell lines were generated in which insertion of a *neo*R gene by homologous recombination inactivated a *Wnt*-1 allele. To enrich for recombinants, negative selection against the herpes simplex virus *tk* gene was also employed, with homologous recombinants but not random integrants resistant to the pyrimidine derivative FIAU. 1 in 3.6×10^7 treated cells were correctly targeted. Germ line chimeras were produced by injecting mutated cells into host blastocysts; matings between heterozygous parents yielded progeny that were further analyzed.

Animals homozygous for the *Wnt*-1 null mutation had a phenotype detected only within the developing CNS. The phenotype was apparent at 9.5 d of gestation. Most of the midbrain and the contiguous rostral metencephalon failed to develop in homozygous mutants. The area deleted in mutant mice at 9.5 d extended further than the region in which *Wnt*-1 is normally expressed. The caudal hindbrain and spinal cord appeared normal in the homozygous mutants, perhaps due to redundancy among other members of the *Wnt*-1 gene family. Homozygous mutants were born, but none lived more than 24 h.

These results provide evidence that *Wnt*-1 normally functions in determination or subsequent development of a large and specific portion of the early CNS. While the midbrain and cerebellum do not seem to be required for *in utero* development, *Wnt*-1 expression seems necessary for the survival of mouse pups.

Targeted Disruption of the Murine *int*-1 Proto-Oncogene Resulting in Severe Abnormalities in Midbrain and Cerebellar Development
K. R. Thomas anfd M. R. Capecchi
Nature, 346, 847—850, 1990   1-30

The experiments in this paper explored the function of the *int-1* protooncogene by disrupting the *int-1* gene in mouse embryonic stem cells. The disruption was mediated by a vector containing *neo* and *tk* cassettes, and was detected by positive-negative selection with G 418 and gancyclovir, and confirmed by Southern blot analysis. The mutant cell line was used to establish a chimeric mouse that transmitted the mutation through its germ line. By breeding mice from this line, both heterozygotes and homozygotes for null *int-1* alleles were available for phenotypic analysis.

Six homozygous mutants were detected, with only one surviving to adulthood. This mouse exhibited loss of balance and coordinated movement, suggesting a defective cerebellum. When embryos from heterozygous matings were examined, malformations in the mesencephalon and metencephalon were aparent in homozygous, but not heterozygous mutants. By 17.5 d, homozygous mutants had mild midbrain hydrocephaly. The single surviving homozygote had hydrocephaly in the caudal region of the cerebral hemispheres and in the midbrain.

These findings suggest that the disruption of a single *int-1* allele results in normal mice, but that the disruption of both copies causes severe defects in midbrain and cerebellar development. Apparently, the *int-1* disruptions studied here have variable penetrance.

Gene Targeting in Normal and Amplified Cell Lines
H. Zheng and J. H. Wilson
Nature, 344, 170—173 , 1990   1-31

Gene targeting in mammalian cells has the promise to become an important tool for studying development. In mammalian cells, however, this technique results in a much lower frequency of homologous recombination, compared to random integration, than it does in yeast. If the rate-limiting step in this process could be identified, the frequency of targeting events might be improved.   If the rate-limiting step is the search for homology, the difference in genome size between yeast and mammals might explain the difference in the frequency of homologous recombination. The experiments in this paper were designed to test this hypothesis.

Frequency of homologous recombination events were compared between normal CHO cells, containing 2 copies of the dihydrofolate reductase gene, and amplified CHOC 400 cells containing 800 copies. Since the two cell lines were otherwise identical genetically, this experiment should show increased frequency of homologous recombination in the CHOC cells if the concentration of the target gene is critical.

The targeted gene was a 4.6-kb segment of the DHFR gene, interrupted with the neo gene and with the tk gene at one end. After introduction of this vector, cells were subject to positive-negative selection by growth in G418- and FIAU-containing medium. Targeted recombinants, *neo*[+] and *tk*[-], should be resistant to both drugs, while random recombinants should be resistant to G418 only.

In three experiments, the frequency of targeted recombinants, as assessed by Southern blot analysis, ranged from 30 to 100% of doubly resistant clones. The efficiency of homologous recombination in CHOC 400 and CHO was comparable, with ratios of 2.1, 1.5, and 1.3 in the 3 experiments.

These experiments suggest that cells with a 400-fold difference in the number of the target gene undergo homologous recombination at similar frequencies. This implies that the search for homology is not rate limiting in this process. The difference in frequencies of homologous recombination between mammalian cells and yeast does not appear to be due to the difference in genome size.

Germ-Line Transmission of a Disrupted β_2-Microglobulin Gene Produced by Homologous Recombination in Embryonic Stem Cells
M. Zijlstra, E. Li, F. Sajjadi, S. Subramani, and R. Jaenisch
Nature, 342, 435—438 , 1989   1-32

The β_2-microglobulin (β_2-m) protein is a part of the murine major histocompatibility complex class I molecules, which have many roles in the immune system. It has also been suggested that these MHC class I molecules have other roles outside the immune system, perhaps involving endocrine or pheromonal responses. The β_2-m protein itself has been proposed to induce collagenase in fibroblasts. The work described here was designed to investigate the functions of β_2-m by creating a disruption mutation in the β_2-*m* gene of mice.

A replacement-type vector was constructed containing 10 kilobases of β_2-m sequences interrupted in the second exon by a *neo* cassette and with a tk gene at the 5' end. Initial selection was in medium containing G418. Half of all selected clones had lost *tk* sequences. Further selection relied on pooling of clones, growth of pools into eight-cell colonies, halving of pools, and polymerase chain reaction (PCR) assays on pools. Single cells of positive pools were subcultured and later identified by PCR.

Correctly targeted cells were recovered with a frequency of about 1 in 2.5×10^6 input cells or about 1 in 25 of G418-resistant clones. These cells were injected into mouse blastocysts, and chimeric mice were recovered. Three males transmitted the disrupted β_2-m gene as predicted by Mendelian genetics. No phenotype was detected in animals heterozygous for the mutation.

These findings provide evidence that mouse lines containing disruption mutations in β_2-*m* have been created. These heterozygotes are therefore capable of embryonic development. Further experiments are planned to examine the immune function of the heterozygotes and to create and analyze homozygous mutant mice.

β_2-Microglobulin Deficient Mice Lack CD4⁻8⁺ Cytolytic T Cells

M. Zijlstra, M. Bix, N. E. Simister, J. M. Loring, D. H. Raulet,
R. Jaenisch
Nature, 344, 742—746, 1990   1-33

Continuing the investigation described in the preceding paper, the experiments presented here were designed to study the function of the β_2-microglobulin (β_2-m) protein in mice. Mice heterozygous for a disruption mutation in β_2-*m* were intercrossed, and normal, heterozygous, and homozygous F2 animals, as judged by Southern blots, were analyzed.

9 of 33 embryos and 23 of 101 adults were investigated. All progeny, including homozygous mutants, appeared normal on visual inspection and upon autopsy. All animals were fertile and reared litters successfully.

Based on Northern blot analysis, homozygous mutants seemed to have extremely low levels of β_2-*m* mRNA of a size smaller than normal and consistent with the disruption mutation. Immunoprecipitation experiments suggested that no β_2-m protein was present in embryonic fibroblasts of homozygotes, while heterozygous mutants had normal levels of β_2-m.

Homozygotes appeared to completely lack detectable surface expression of most major histocompatibility class I molecules and of a functional Fc receptor.

Analysis of lymphoid organs of homozygotes showed 100- to 150-fold reductions from normal levels of T-cell antigen receptor (TCR) $\alpha\beta^+CD4^-8^+$ T cells, although heterozygotes seemed unaffected. In contrast, the populations of TCR $\alpha\beta^+$ $CD4^+8^-$ cells, MHC class II-restricted $CD4^+8^-$ cells, surface Ig^+B cells, and $\gamma\delta^+$ T cells were normal in homozygotes. Precursors to cytotoxic T lymphocytes (CTL-p) were absent in homozygous mutants, and spleen cells from homozygotes were unable to simulate CTL-p cells or to elicit significant CTL responses by responder cells from two different lines.

These experiments suggest that mice lacking detectable β_2-m protein can develop normally. This implies that class I molecules may not have an essential role in development. Encounters with class I MHC molecules seems required for the differentiation of TCR ab^+ $CD4^+8^-$ cells, but not necessary for the differentiation of other defined thymocyte subsets.

♦ In the 1990 Year Book, several reports were cited demonstrating gene targeting in mouse embryonic stem (ES) cells. ES cells are pluripotent and have been demonstrated to retain the ability to contribute to the germ line of blastocyst-injection chimeras. These reports generated considerable excitement because of the possible genetic experiments that now could be undertaken in mouse. Reports appearing during this last year demonstrate that preplanned alterations in specific genes can be made in the germ line of mice by homolgous recombination in embryonic stem cells. Although the frequency of site-directed mutagenesis varies considerably between genes, expression of the target gene in ES cells does not appear to be a factor. Thus, the potential of creating mutations in mice at will is a reality. It is now possible to begin to mutate a vast array of genes that are known to be expressed during development but whose functions are poorly understood. In addition, the potential of creating animal models for human diseases now exists, and this is particularly important with the increasing number of human disease genes being cloned each year.

Some of the targeted genes that have been passed through the germ line in mice include *C-abl, IGF-II, β_2, int-1,* as well as the gene which encodes hypoxanthine phosphoribosyltransferase. So far, only β_2 and *int-1* mutants have been bred to homozygosity. β_2-deficient animals appear normal except for deficiencies in class I histocompatibility antigens and in $CD4^-CD8^+$ T cells. These results show that neither β_2 nor cell-surface expression of the major histocompatibility class I proteins play a vital role in embryogenesis. These mice now provide a valuable model to assess the role of cytotoxic T cells in the immune response to a number of pathogens. The *int-1* homozygotes show a variable phenotype ranging from prenatal death to survival with severe ataxia. Phenotypic analyses suggest a rule for *int-1* in development of the mesencephalon and metacephalon. However, many of the sites of expression in the developing embryo seem to be unaffected.

Although the *IGF-II* mutants had not been bred to homozygosity at the time of publication, mice heterozygous for the mutation show an interesting growth-deficiency phenotype. All heterozygous progency from male chimeras were smaller than ES cell-derived littermates, suggesting that *IGF-II* plays a role in embryonic growth. The heterozygous phenotype described is associated with male transmission of the mutant gene which raises the possibility of imprinting. It will be interesting to see if a similar phenotype appears with female transmission.

One final note of interest is that gene-targeting experiments have concentrated on disrupting the gene to produce "knock-out" mutations. In addition to these types of mutations, the ability to introduce point mutations would be of considerable importance. Overall, the ability to mutate genes in mice has generated much interest and a number of labs are now attempting to use this approach. Several components still need to be optimized; these include vector construction, selection scheme, and strains used for ES cells and host embryos. It would now be helpful to have a catalogue of variables that have been tried with recommendations for conditions most likely to succeed. *Terry Magnsuon*

The Murine Mutation Osteopetrosis Is in the Coding Region of the Macrophage Colony Stimulating Factor Gene

H. Yoshida, S.-I. Hayashi, T. Kunisada, M. Ogawa, S. Nishikawa, H. Okamura, T. Sudo, L. D. Shultz, and S.-I. Nishikawa
Nature, 345, 442—444, 1990   1-34

The recessive murine mutation osteopetrosis (*op*) results, in homozygous mutants, in severe deficiencies in mature macrophages and osteoclasts and restricted capacity for bone remodeling. Although macrophages and osteoclasts originate from a common hematopoietic progenitor, the inability of transplants of normal bone marrow cells to cure *op/op* mice suggests that the defects may be due to an abnormal hematopoietic environment. The experiments presented here investigated the molecular and biochemical bases of the defects in these mice.

Primary lung fibroblast cell lines were established from *op/op* mice and tested for their ability to produce factors that support the proliferation of normal macrophage precursors. Fibroblasts from *op/op* mice were deficient in stimulating proliferation of co-cultured bone marrow cells, resulting primarily in granulocytes. In contrast, fibroblasts from normal mice stimulated massive proliferation of hematopoietic cells, primarily macrophages. Culture supernatant from op/op fibroblasts had only one fifth the macrophage precursor colony-stimulating activity of normal fibroblasts, and that activity was abolished by antiserum to GM-CSF. In contrast, 80% of the colony stimulating activity of supernatant from normal fibroblasts was present after treatment with anti-GM-CSF.

Northern blot analysis showed that mRNA from the macrophage colony-stimulating factor gene *Csfm* was present in normal levels in fibroblasts from op/op mice. Polymerase chain reaction sequencing of 4 independent cDNA clones of Csfm from o*p/op* mice revealed a thymidine insertion resulting in a TGA stop codon.

Homozygous *op/op* mice seem to have defective production of M-CSF, apparently due to a nonsense mutation in the *Csfm* gene. It is possible that normal bone remodelling of *op/op* animals during fetal life may depend on maternal M-CSF.

♦ There are numerous mutations in the mouse that have arisen spontaneously or have been induced in experiments involving chemicals or radiation. Although many of these mutations show very interesting recessive phenotypes, it has been difficult to progress beyond a phenotypic analysis given the problems associated with cloning a gene not surrounded by closely linked markers. The goal of the mouse gene mappers is to increase the resolution of the molecular genetic map to the 1 cM level and to do so within the next 5 years. If this were achieved, it should be possible with the developing YAC technology to clone almost any gene in the mouse. In the meantime, a few of the mutant loci are being identified by luck through the colocalization of a cloned gene with already known mutations. For example, *Pax-1* was found to be the same locus as *undulated,* and the c-*kit* protoonocogene was found to be the same locus as the dominant *white spotting* gene (both of these have been discussed in the 1990 Year Book). The latest mutation to be added to this category is *op (osteopetrosis).* This mutation has been mapped to mouse chromosome 3, and recessive mice are found to be deficient in mature macrophages and osteoclasts, and have a restricted capacity for bone remodeling. It turns out that the *op* mutation results in a single base pair insertion in the coding region of the macrophage colony stimulating factor gene. This produces a stop codon 21 base pairs downstream from the insertion. Thus, production of functional protein product (M-CSF) is defective in homozygous mice. The results reported in this paper indicate that production of the factor is not necessary for maintenance of various types of M-CSF-responsive cells other than osteoclasts. These results also raise the question of whether bone remodeling during fetal development is independent of M-CSF. Another possibility is that maternally derived M-CSF is enough to result in normal development of the fetus. These results demonstrate that the biological importance of a gene can only be determined by correlating gene product with a mutant phenotype, and the ability to do so is likely to lead to some surprising results. For example, the *limb deformity* gene is expressed throughout the developing fetus but causes specific limb defects. In addition, although c-*kit* is expressed in a number of places in the developing fetus and adult organism, when mutated, the effects seem to be localized to specific stem cell populations. These types of observations indicate that much is to be learned regarding the nature of interacting gene products and their effects on specific developmental processes. *Terry Magnuson*

Establishment of the Mouse Chromosome 7 Region With Homology to the Myotonic Dystrophy Region of Human Chromosome 19q

J. S. Cavanna, A. J. Greenfield, K. J. Johnson, A. R. Marks,
B. Nadal-Ginard, and S. D. M. Brown
Genomics, 7, 12—18, 1990   1-35

The most common form of adult muscular dystrophy is myotonic dystrophy (DM). The gene for this disease is inherited in an autosomal dominant fashion, and is located on chromosome 19q.13.2. A number of genetic markers have been mapped to this region of chromosome 19, but the gene itself has not been isolated. The work reported here was designed to help find the DM gene by genetic mapping of the homologous mouse chromosomal region.

In addition, human chromosome 19q13.2/13.1 is the locus for the human genetic condition, susceptibility to malignant hyperthermia (MHS). A candidate gene for the MHS locus in humans is the ryanodine receptor (RYR) gene, which encodes the Ca^{2+} release channel in the sarcoplasmic reticulum of skeletal muscle cells. Genetic mapping of the homologous mouse chromosomal region should also provide evidence supporting or refuting the possible identification of RYR as the MHS gene.

Mapping of DNA probes, including ATP1A3, TGFB, CKMM, and PRKCG, was based on a mouse Mus domesticus/Mus spretus interspecific backcross segregating for the genetic markers pink-eye dilution and chinchilla. 86 backcross progeny were analyzed for pigmentation mutations and by hybridization of mouse chromosome 7 and human chromosome 19 DNA probes.

Previous work had suggested a close homologous relationship between human chromosome 19q and proximal mouse chromosome 7. These analyses showed that the order of human probes in the 19q region is inverted with respect to the centromere, but otherwise identical to the order of mouse probes in proximal chromosome 7 (Figure 1-35). This conservation of order suggests that the human DM gene maps between CKMM and D19S51. The data also predict that RYR is close to or identical with MHS.

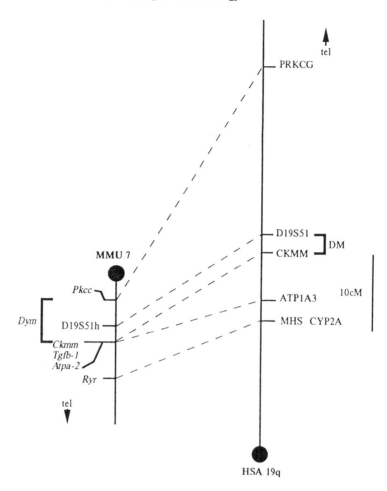

FIGURE 1-35. Comparative genetic maps of mouse chromosome 7 and human chromosome 19q. Comparative genetic maps of proximal mouse chromosome 7 (MMU7) and human chromosome 19q (HSA19q). Human genetic distances, apart from PRKCG-CKMM, were derived from summary genetic maps (male recombination distances) of the Report of the Committee on Linkage and Gene Order. For the PRKCG-CKmm distance, the sex is not determined. The figure shows the conservation of order of loci in the syntenic groups on these two chromosomes. Also shown are the large differences in genetic distances between the loci on the two chromosomes. The two human disease loci myotonic dystrophy (DM) and susceptibility to malignant hyperthermia (MHS) are shown on the human chromosome 19q map along with their putative homologs (*Dym* and *Ryr*, respectively) on mouse chromosome 7. (From Cavanna, J. S., Greenfield, A. J., Johnson, K. J., Marks, A. R., Nadal-Ginard, B., and Brown, S. D. M., *Genomics*, 7, 12—18, 1990. With permission.)

A Molecular Genetic Linkage Map of Mouse Chromosome 4 Including the Localization of Several Proto-Oncogenes

J. D. Ceci, L. D. Siracusa, N. A. Jenkins, and N. G. Copeland
Genomics, 5, 699—709, 1989   1-36

Molecular genetic linkage maps can be useful for identifying candidate genes for interesting phenotypic mutations or diseases. The characterization of chromosomal segments conserved in different species can help in the construction of linkage maps and may also assist in the development of animal models for human diseases. To these ends, this paper reports the construction of a molecular genetic linkage map of mouse chromosome 4. This chromosome is known to exhibit synteny to the human chromosome 1p, which contains several proto-oncogenes.

The mapping method used was the rapid and efficient system of interspecific backcross analysis. This method uses the detection of restriction fragment length polymorphisms that are readily detected between the distantly related species *Mus domesticus* and *Mus spretus.* Various DNA probes were used for the analysis by Southern blotting of 122 F2 animals.

A 64-centimorgan span of mouse chromosome 4 was mapped. A comparison of these mapping data to those obtained from inbred stain mapping did not show evidence of inversions or deletions in this region. A region of synteny between mouse chromosome 4 and human chromosome 8 was defined that includes *Lyn,* a member of the *Src* gene family of proto-oncogenes, and Mos. *Tsha,* a glycoprotein hormone that maps to 6q12-q21 in humans, was localized to this region of mouse chromosome 4, providing the first example of homology between these two chromosomes. Certain loci of human chromosomes 9 and 1 were mapped to mouse chromosome 4. Among the human genes mapped to this region of chromosome 4 are several oncogenes and the sites of several translocations commonly found in human diseases.

An Interspecific Backcross Linkage Map of Mouse Chromosome 8

J. D. Ceci, M. J. Justice, L. F. Lock, N. A. Jenkins, and N. G. Copeland
Genomics, 6, 72—79, 1990   1-37

This paper reports on the continued effort by members of this laboratory to construct a molecular genetic map of mouse chromosomes that should be useful in predicting the location of many human genes. Based on interspecific backcross analysis, the work presented here was designed to map mouse chromosome 8. Some genes of interest that were previously assigned to mouse chromosome 8 by somatic cell analysis include insulin receptor, plasminogen activator, adenosine phosphoribosyltransferase (APRT) pseudogene-1, haptoglobin (HP), metallothionein-1 (MT-1), mitochondrial uncoupling protein, and zinc finger protein-4.

The 67-centimorgan molecular genetic map developed covered 84% of the predicted length of mouse chromosome 8. Homologies betwen regions of mouse chromosome 8 and human chromosomes 8 and 19 were found. A new region of synteny between mouse chromosome 8 and human chromosome 16 was described. The gene order MT-1, HP, APRT is the same in humans and in mice.

Based on these findings, mouse chromosome 8 is predicted to contain the homologous region of human chromosome 16q that is associated with myelomonocytic leukemia plus eosinophilia. If an association in the mouse can be found between myelomonocytic leukemia and chromosome 8 abnormalities, a murine model for this disease might be possible.

An Interspecific Backcross Linkage Map of the Proximal Half of Mouse Chromosome 14
J. D. Ceci, D. M. Kingsley, C. M. Silan, N. G. Copeland, and
N. A. Jenkins
Genomics, 6, 673—678, 1990   1-38

Comparative gene mapping between species can be used to help predict the location of a gene or a mutation, and might be useful in the establishment of animal models of human diseases. Comparative maps also assist in the identification of candidate genes for genetic diseases. The work presented here was designed to construct a molecular genetic linkage map of the proximal half of mouse chromosome 14. The technique used was interspecific backcross analysis.

The segregation pattern of various DNA probes was determined by Southern blot analyses of 198 progeny of mouse backcrosses. Using *hr* (hairless) as the anchor locus, the gene order determined was centromere-*Hap* (a retinoic acid receptor-related gene)-*Plau* (urokinase plasminogen activator)-[*Psp-2* (a parotid secretory protein-related sequence), *Rib-1* (ribonuclease-1)-*Tcra* (T-cell receptor α)-*Ctla* (cytotoxic T-lymphocyte-associated protein-1)-*Bmp-1* (bone morphogenetic protein-1)-*hr*. *Bmp-1, Hap,* and *Plau* were localized to chromosome 14 for the first time.

The human *Hap* has been designated retinoic acid receptor β, but it is possible that the murine *Hap* is not the mouse homolog to this member of the gene family. The localization of *Plau* establishes homology between mouse chromosome 14 and human chromosome 10. While *Plau* might be important for cellular involution, migration, or metastasis, the murine gene did not map near any mutation known to affect any of these processes. It is possible that *Bmp-1* is a candidate gene for osteosarcoma susceptibility.

Radiation Hybrid Mapping: A Somatic Cell Genetic Method for Constructing High-Resolution Maps of Mammalian Chromosomes

D. R. Cox, M. Burmeister, E. R. Price, S. Kim, and R. M. Myers
Science, 250, 245—250, 1990   1-39

Mapping of the human genome has been aided by the use of restriction fragment length polymorphisms (RFLP) in conjunction with genetic linkage analysis, *in situ* hybridization, the ability to separate human chromosomes from one another, and pulsed-field gel electrophoresis. Still, using present methods, it is difficult to order DNA sequences more than a few hundred kilobases apart. The work presented here was performed in an attempt to construct high-resolution, contiguous maps of human chromosomes. Using the techniques described, a high-resolution map of the proximal 20-megabase pairs of the long arm of human chromosome 21 was created.

The method used was radiation hybrid (RH) mapping, a somatic cell genetic approach to ordering DNA markers spanning megabase pairs, at a resolution of 500 kilobases. In this method, high doses of X-rays are used to generate large fragments of a chromosome of interest. The fragments are recovered in rodent cells, and rodent-human hybrid clones are analyzed for the presence of specific human DNA markers. The frequency of separation of markers indicates the frequency of breakage and the distance between them. Analogous to meiotic mapping, marker order can then be determined.

To map a region of human chromosome 21, a Chinese hamster-human somatic cell hybrid was used that contained a single human chromosome 21 and little other human DNA. Irradiation of this cell line with 8000 rad of X-rays resulted in an average of 5 human chromosome fragments per cell. To ensure continued growth of these irradiated cells, they were fused with nonirradiated hamster recipient cells and donor-recipient hybrids were selected by growth in HAT medium. 103 independent hybrid clones were isolated and each was assayed by Southern analysis for the presence of 14 DNA markers from human chromosome 21.

Construction of a RH map was performed, and computer analysis provided an estimate of the degree of likelihood of the marker order. Pulsed-field gel electrophoresis (PFGE) mapping was used to confirm the RH map; that method determines physical linkage of markers after electrophoresis, transfer to a membrane, and hybridization or large DNA fragments. The PFGE and RH maps had markers in identical order separated by similar distances. Both maps were in good agreement with previously described physical and meiotic maps of this region.

The use of RH can often complement PFGE in mapping. In RH mapping, nonpolymorphic DNA markers, which are useless in meiotic mapping, can be ordered. Additional advantages of this technique include the fact that all probes in every cell line give information, and the ability to vary the range of resolution by varying the dosage of X-rays used to generate chromosome fragments.

Comparison of the Physical and Recombination Maps of the Mouse X Chromosome

C. M. Distech, G. K. McConnell, S. G. Grant, D. A. Stephenson, V. M. Chapman, S. Gandy, and D. A. Adler

Genomics, 5, 177—184, 1989   1-40

While the mapping of mouse chromosomes promises great usefulness, the mapping of the murine X chromosome is of special interest because the X-linked genetic loci associated with the inactivation of a single X chromosome of females are known to be conserved. In this paper, *in situ* hybridization was used to map five random mouse genomic DNA markers and five cloned genes to the mouse X chromosome. The genes mapped were those for clotting factors VIII and IX (*Cf-8* and *-9*), Duchenne muscular dystrophy (*Dmd*), phosphoglycerate kinase-1 (*Pgk-1*), and α-galactosidase (*Ags*). The physical map obtained was then compared to the recombination map obtained from the analysis of progeny of interspecific mouse backcrosses.

The physical and recombination maps had good agreement in the ordering of genes. The order, from the centromere was *Ags, Pgk-1, Dmd, Cf-8,* and *Cf-9*. Distances were also in good agreement, except near the telomere where increased frequencies of meiotic crossovers often occur. In this region, loci were spaced farther on the recombination map than on the physical map.

This effort adds ten more markers to the physical map of the mouse X chromosome. For three regions of the X chromosome, there seems to be conservation between humans and mice in physical distances. In contrast, the recombinational map of the human chromosome suggests that the human X chromosome is of larger total genetic length.

Molecular Mapping Within the Mouse Albino-Deletion Complex

D. K. Johnson, R. E. Hand, Jr., and E. M. Rinchik

Proc. Natl. Acad. Sci. U.S.A., 86, 8862—8866, 1989   1-41

Radiation-induced germ line mutations in the mouse permit the identification of genes essential for mammalian development by associating mutant developmental phenotypes with specific genomic subregions. The work presented here was designed to create a strategy of molecular analysis of developmentally interesting genes. The strategy involves the use of complexes of overlapping deletion mutations to map DNA clones to chromosomal subregions associated with specific phenotypes. The albino (*c*) locus was used for analysis here because deletions in this region have already been associated with various developmental phenotypes such as early embryonic, perinatal and juvenile lethality, male sterility, and inner-ear abnormalities.

For molecular probes of the c-region of chromosome 7, random anonymous DNA clones were obtained from a library made from flow-sorted chromosomes. These clones were screened on DNAs from *Mus spretus-Mus musculus* interspecific hybrids carrying c-region deletion chromosomes. When clones corresponded to the deletion in a particular F1 hybrid animal, hybridization bands for the particular deleted RFLP were absent.

There were 72 informative clones identified. One, 11B2, mapped within the Fp1 deletion and defined the locus *D70R1*. This clone was not deleted in the Ai-, B-, or D-groups of prenatally lethal deletions, suggesting that it lay between *c* and *Hbb* on the standard genetic map; this was confirmed by standard three-point linkage analysis.

Since *D70R1* is near *sh-1,* it should be useful for investigating this mutation of inner-ear anomalies. Since D70R1 is also near the distal breakpoints of the Bp, Dp, and Dq deletions, it should facilitate molecular "walks" across the breakpoint regions and investigations into the subregion required for development of the preimplantation embryo. The general strategy for mapping anonymous DNA clones to regions of interesting deletion mutations should be useful in the mapping of regions associated with other mutant developmental phenotypes.

Genetic Mapping in the Region of the Mouse X-Inactivation Center

J. T. Keer, R. M. J. Hamvas, N. Brockdorff, D. Page, S. Rastan, and S. D. M. Brown
Genomics, 7, 566—572, 1990   1-42

In mammals, one of the two X chromosomes is randomly inactivated. Various kinds of evidence suggest that a locus on the mouse X chromosome, the X-inactivation center, affects this process. The locus has been mapped to a point just distal to the T16H breakpoint. The work described here was designed to better define genetically the region of the mouse X-inactivation center.

The progeny of two interspecific *Mus domesticus/Mus spretus* crosses were subjected to pedigree analysis using Southern blotting to a number of probes distal to the T16H breakpoint. In addition, several genetic loci, microclones, and *EagI* linking clones distal to the breakpoint were mapped.

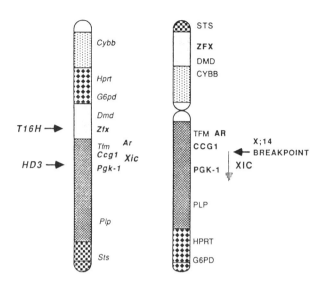

FIGURE 1-42. Comparative map of the mouse and human X chromosomes. The position of the mouse X-inactivation center (Xic) and the human X-inactivation center (XIC) with respect to the various linkage groups conserved between the two X chromosomes is indicated. Boldface type highlights those loci most important in determining the relevant positions of the mouse and human X-inactivation centers. (From Keer, J. T., Hamvas, R. M. J., Brockdorff, N., Page, D., Rastan, S., and Brown, S. D. M., *Genomics, 7,* 566—572, 1990. With permission.)

In all, 12 loci were mapped to the mouse X-chromosome by this analysis. In humans, the X-inactivation center has been localized to a small region in band Xq13. By combining mouse and human mapping data, 2 locations for the mouse inactivation center, both flanking the Ta^{25H} deletion, seem possible (Figure 1-42). An analysis of the syntenic region suggests that the more likely site of the mouse inactivation center is in the region of *Pgk-1* distal to *Ccg-1*. The linking clone EM13 and the microclone DXSmh44 have been mapped to this region by the work presented here. Both probes should be useful for a finer analysis of this region and a better definition of the position of the mouse X-inactivation center.

A Genetic Map of Mouse Chromosome 1 Near the Lsh-Ity-Bcg Disease Resistance Locus
B. Mock, M. Krall, J. Blackwell, A. O'Brien, E. Schurr, P. Gros, E. Skamene, and M. Potter
Genomics, 7, 57—64, 1990   1-43

The *Lsh-Ity-Bcg* locus of mouse chromosome 1 is involved in susceptibility to the three pathogens *Leishmania donovani, Salmonella typhimurium,* and *Mycobacterium bovis.* Although the phenotype of this locus has been well described, neither the coding sequences nor the gene product has been identified. The work presented here was undertaken to aid the reverse genetic approach to isolation of the locus.

Eight genetic markers were mapped using isozyme and restriction fragment length polymorphism analyses of backcross progeny, recombinant inbred strains, and congenic strains of mice. In progeny from the crosses between BALB/cAnPt and DBA/2NPt mice, or in progeny from backcrosses between C57BL/10ScSn and B10.L-*Lsh*$^{r/s}$ mice, the co-segregation of allelic isoforms or RFLPS with the *Lsh-Ity-Bcg* resistance phenotype was determined. In addition, RFLP analysis was performed on BXD recombinant inbred strains.

The gene order of eight markers closely linked to *Lsh-Ity-Bcg* was determined. Similar recombination frequencies were determined from all three test systems.

The most closely linked marker was found to be villin (*Vil*), which may be within 1 to 4 centimorgans of the *Lsh-Ity-Bcg* gene. The markers nebulin (*Neb*) and D2S3, which map to a homologous region of human chromosome 2q, were not found to be closely linked to *Lsh-Ity-Bcg*. These results provide a groundwork for a finer analysis of genes surrounding the *Lsh-Ity-Bcg* locus.

Efficient Linkage of 10 Loci in the Proximal Region of the Mouse X Chromosome

L. J. Mullins, D. A. Stephenson, S. G. Grant, and V. M. Chapman
Genomics, 7, 19—30, 1990   1-44

Previous work from this laboratory established a series of anchor loci spanning a distance of 50 centimorgans (cM) on the X chromosome of the mouse. The work described here extended this analysis by mapping six additional loci in the proximal portion of the X chromosome. Besides random genomic DNA probes, the loci mapped here include the gene responsible for chronic granulomatous disease, cytochrome $b_{245}\beta$-chain (*Cybb*), and the gene encoding the neuron-specific phosphoprotein synapsin (*Syn-1*). A cDNA probe from B lymphocytes encoding the X-linked lymphocyte-regulated gene family (*Xlr*) was also mapped.

The mapping technique used here involved interspecific Mus crosses (Figure 1-44). The first 3 steps, through analysis of backcross progeny, resulted in the establishment of anchor loci over a 50-cM distance, as previously reported. For the new loci studied in the present work (P in Figure 1-44), the most useful restriction length fragment polymorphisms (RFLPs) were identified. Once found, the segregation of these new RFLPs in the relevant recombinant backcross progeny were analyzed. In this

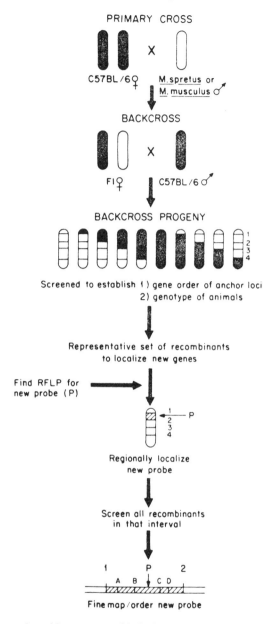

FIGURE 1-44. An outline of the strategy used for both the initial characterization of the backcross offspring and their subsequent utilization to locate and map a new X-linked locus. The primary cross between C57BL/6JRos and either *M. spretus* or *M. musculus,* the backcross of F_1 females to C57BL/6JRos males, and the genetic analysis of the backcross offspring were as previously described. For the purposes of this study, the "anchor loci" are defined as *Otc, Hprt, G6pd,* and *Ags* (1 through 4, respectively). Localization of an unascribed locus (P) to the *Otc/Hprt* region is defined by demonstrating coincident segregation with the two anchor loci. Detailed mapping of the new locus is then defined in relation to the anchor loci by screening all known recombinants within the prescribed interval. (From Mullins, L. J., Stephenson, D. A., Grant, S. G., and Chapman, V. M., *Genomics, 7,* 19—30, 1990. With permission.)

way, only those backcross progeny recombinant between successive anchor loci required further analysis.

Using this method, several new loci were added to the map of the proximal X chromosome. By minimizing multiple recombination events across a 20-cM region, the order of loci was defined as *Cybb, Otc, Syn-1/ Timp, DXWmh141/Xlr-1, DXSmh172, Hprt, Xlr-2,* and *Cf-9.* The regions of the human chromosome homologous to the sites of *Xlr-1* and *Xlr-2* may contain candidate human genes for a lymphoproliferative disorder and a condition of elevated IgM levels.

Perspective: A Common Language for Physical Mapping of the Human Genome

M. Olson, L. Hood, C. Cantor, and D. Botstein

Science, 245, 1434—1435, 1989   1-45

In this paper, the authors espouse the use of short tracts of single-copy DNA sequence as landmarks for use in the construction of a map of the human genome. By using polymerase chain reaction, such sequences can be easily recovered. Mapping by the determination of the order and spacing of unique DNA sequences will provide a common language for data from multiple sources and eliminate the need for large clone archives.

The 1988 plan issued by the National Research Council Committee on the Mapping and Sequencing of the Human Genome involved a hybrid physical map based on both restriction mapping (specifying the sites of specific restriction endonuclease cleavages) and contig mapping (specifying the overlap relationships among a set of clones). Here it is proposed that all types of mapping landmarks be "translated" into the common language of sequence-tagged sites (STSs). This would require sequencing a short tract of DNA from clones that define landmarks on maps. The overwhelming advantage of using STSs is that no access to biological materials would be required to assay a DNA sample for the presence of landmark information. Only raw sequence data and instructions for a PCR assay for that sequence (oligonucleotide primers to be used) would be required.

Adoption of an STS standard would facilitate comparisons of maps constructed in different laboratories and would enable the mapping of the human genome to begin with a quickly constructed crude map that could evolve smoothly into a more refined product. It is recommended that existing sets of mapped DNA probes be converted to STSs. Once accomplished, these data would provide a direct precursor to a low-resolution physical map. The authors urge adoption of an STS standard for the human genome project.

Molecular Cloning of the *t complex responder* Genetic Locus

L. L. Rosen, C. D. Bullard, L. M. Silver, and J. C. Schimenti
Genomics, 8, 134—140, 1990   1-46

The *t complex responder* (*Tcr*) locus of the mouse is a small region in the center of t haplotypes that is required for male transmission ratio distortion (TRD) of *t* haplotypes. These variant forms of the *t* complex carry recessive mutations causing male sterility and embryonic lethality, but when *Tcr* is present, heterozygous male mice transmit the *t* chromosome to nearly all of their offspring. *Tcr* has been mapped to the T66B region of *t* haplotypes by analyses of partial *t* haplotypes, products of rare recombination events between *t* haplotypes and wild-type forms of the *t* complex. The T66B locus is a member of a family of large, dispersed duplicated blocks of DNA, some of which encode transcripts found specifically in male germ cells. One of these transcripts is encoded by the T66B-a gene. This points to the T66B-a gene as a candidate for *Tcr,* but this conclusion is strongly dependent on the exact location of the boundaries of the T66B locus. The work described here was designed to accurately delineate the *Tcr* interval.

The DNA between breakpoints of recombinant chromosomes defining the responder locus was cloned and mapped. This resulted in the localization of *Tcr* to a 150- to 220-kilobase (kb) region of *t* haplotypes that contains the T66B-a gene.

The T66 region of *t* haplotypes arose by duplication of large genomic segments. The first, involving perhaps 100 kb around the Tu66 probe, resulted in the generation of the α, β, and γ subfamilies of T66 elements. This work provides evidence suggesting that an ancestral α-β unit of about 220 kb was triplicated to form the present arrangement of the first 6 T66 elements of t haplotypes. It is possible that this triplication permitted the evolution of the "mutant" *Tcr* gene, T66B-a, while the transcriptionally active T66A-a and T66C-a genes maintained the wild-type function.

Comparison of Linkage Maps of Mouse Chromosome 12 Derived from Laboratory Stain Intraspecific and *Mus spretus* Interspecific Backcrosses

M. F. Seldin, T. A. Howard, and P. D'Eustachio
Genomics, 5, 24—28, 1989   1-47

The use of interspecific crosses between laboratory mice and those from various *Mus* species has greatly assisted genetic analysis; this is due to the many polymorphisms between the species at various loci. However, these local polymorphisms might be accompanied by differences in chromosomal organization that have not yet been discovered in studies of small linkage groups. The present investigation mapped eight markers spanning most of chromosome 12 in an attempt to compare marker order and spacing of a large genetic interval.

This analysis tested progeny from 198 intraspecific laboratory back-crosses (C57B11/6J x SWR/J) and 115 interspecific backcrosses (C3H/HeJ-*gld/gld* x *M. spretus*). Animals were typed for polymorphisms at four or five loci based in anonymous DNA clones and three genes (α1-antitrypsin, the μ constant region gene of the *IghC* complex, and *Lamb-1*).

Progeny of the interspecific *M. spretus* backcross showed small but significant deviations from 1:1 segregation of alleles. Maps constructed from both types of backcrosses showed similar marker order, but the interspecific *M. spretus* map was only 82% as long as the interspecific map. Strong positive interference was seen in the interspecific backcross data.

This study suggests that gene order on chromosome 12 is well conserved between laboratory strains and *M. spretus*. The data also provide evidence that small but significant differences in marker spacing occur between the species. These differences were not seen in maps when they were compared interval-by-interval. It seems likely that multiple small rearrangements and deletions have occurred between the two kinds of chromosomes.

A Molecular Genetic Linkage Map of Mouse Chromosome 2

L. D. Siracusa, C. M. Silan, M. J. Justice, J. A. Mercer, A. R. Bauskin, Y. Ben-Neriah, D. Duboule, N. D. Hastie, N. D. Copeland, and N. A. Jenkins

Genomics, 6, 491—504, 1990   1-48

Mouse chromosome 2 contains many loci involved in developmental, morphological, and neurological abnormalities. To help understand the biology of these loci, previous work from this laboratory employed an interspecific backcross of C57BL/6J and *Mus spretus* to construct a molecular genetic linkage map of the distal portion of chromosome 2. Here, continued molecular analysis of the backcross is presented, in the proximal region of mouse chromosome 2.

Genomic DNAs from backcross progeny were analyzed by Southern blot hybridization to probes from *AB1, Acra, Ass, C5, Cas-1, Fshb, Gcg, Hox-5.1, Jgf-1, Kras-3, Ltk, Pax-1, Prn-p,* and *Spna-2*. Three previously unmapped loci, *Jgf-1, Kras-3,* and *Ltk,* were mapped to chromosome 2 in this analysis; *Acra,* which was previously mapped to mouse chromosome 17, was here firmly mapped to chromosome 2.

Combined with previous work, a molecular genetic map including 25 loci on chromosome 2 was constructed. This map seemed consistent and colinear with a previously published chromosome atlas cytological map constructed by M. F. Lyon.

The *Kras-3* locus may provide molecular access to the developmental mutations Wasted and Ragged. *Acra* may be a candidate gene for mdm (muscular dystrophy with myositis); *I1-a* and *I1-1b* may be candidate genes for the *tsk* (tight skin) mutation. Alternatively, *Acra* and *I1-1a* and *I1-1b* might provide molecular access to *mdm* or *tsk*.

Homologies to human chromosomes 9q, 2q, 11p, 15q, and 20 were seen. The data permit the prediction that the *Sey* (small eye) mutation of the mouse lies between *Cas-1* and *Fshb* and is the mouse homolog of human aniridia.

This molecular genetic linkage map of mouse chromosome 2 should provide candidate genes for, or molecular access to, several interesting mutations. In addition, it is or will be useful in providing mouse models for several human diseases.

Linkage Map of Mouse Chromosome 17: Localization of 27 New DNA Markers

V. Voncek, H. Kawaguchi, K. Mizuno, Z. Zaleska-Rutczynska,
M. Kasahara, J. Forejt, F. Figueroa, and J. Klein
Genomics, 5, 773—786, 1989   1-49

Mouse chromosome 17 contains the two extensively studied gene complexes, *H-2* and the *t* complex, involved in the control of the immune response and embryonic and male germ cell differentiation, respectively. Because of the importance of these complexes, workers in this laboratory began systematic genetic mapping of chromosome 17. Here, the results of the first phase of this project are reported.

Using a LINE 1 repetitive sequence as a probe, 52 anonymous DNA clones from chromosome 17 were isolated and 27 were found to display restriction fragment length variation among common inbred laboratory strains. Accordingly, these 27 probes, free of repetitive sequences, were mapped using recombinant inbred strains, congenic strains, F2 segregants, or intra-*t*-recombinants.

Together with previously mapped markers, a map was constructed from these data that comprise 125 DNA loci of chromosome 17. The loci span 71 centimorgans, probably the entire length of the chromosome. Most markers lie in the proximal part of the chromosome in which the *t* and *H-2* complexes lie. The distal portion of chromosome 17 contains a few markers whose exact locations are poorly mapped.

♦ The ability to achieve a 1-cM resolution of the mouse molecular map is critical for cloning any developmental gene not otherwise marked by a molecular tag. Genetic linkage maps are based on co-inheritance of allele combinations across multiple polymorphic loci. Interspecific crosses exploit the differences between genetically diverse *Mus* species and standard inbred strains. Because such hybrids are multiply heterozygous across all chromosomes, gametic allele combination is easier to discern through polymorphisms. Thus, using interspecific backcross mapping, high resolution, multilocus linkage maps for the mouse genome are being established. Potential problems with this approach include chromosomal rearrangements such as translocations and inversions which would provide spurious results. Therefore, mapping by *intra*specific as well as recombinant inbred crosses is important for assessing the generalities of the *inter*specific map. Direct molecular analysis by PCR of single gametes for allelic differences has the potential of increasing the number of gametes that can be scored, thereby establishing more precise recombinational distances. In conjunction with the genetic map, a physical map is also being developed. For example, for any cloned genes that are mapped, DNA sequence information can be used to convert these probes into sequence-tagged sites (STSs). The goal is to establish an ordered set of STSs space at intervals of 100 kb. This would enable the cloning of any gene mapping between two anchor sites without the need for maintaining standard reference libraries. Another approach for saturating the map with STSs is through regional mapping. Chromosome-specific or chromosome-enriched libraries are being made for particular regions of interes' with respect to genetics and biological function. Examples include the *t*-complex (chromosome 17), the albino-deletion complex (chromosome 7), the dilute-short ear deletion complex (chromosome 9), and the agouti complex (chromosome 2). In addition to regional mapping, the generation of an overlapping contig map of the entire mouse genome is also a goal. The strategy is to characterize large tracts of DNA in the form of YACs, cosmids, and phage vectors. The contig map can be merged with the genetic map by using STSs or by taking probes containing genetically mapped polymorphisms and identifying the contigs that contain them. The development of the genetic and physical map, together with targeted mutagenesis, will revolutionize mammalian developmental genetics. Virtually any gene will be clonable. The focus will likely be on identifying the key regulator genes and determining how they function. *Terry Magnuson*

A Method for Difference Cloning: Gene Amplification following Subtractive Hybridization

I. Wieland, G. Bolger, G. Asouline, and M. Wigler
Proc. Natl. Acad. Sci. U.S.A., 87, 2720—2724, 1990   1-50

It is often desirable to understand the difference between two genomes, as, for example, when one contains a mutation. Here, a method for genomic difference cloning is presented that is at least as powerful as the only other published method, first described by Lamar and Palmer.

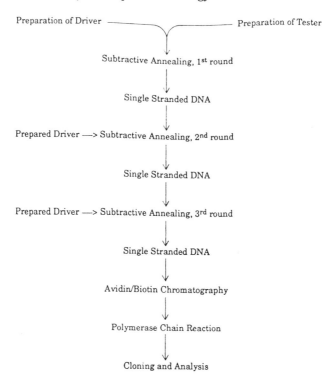

FIGURE 1-50. Flow diagram for genomic difference cloning. (From Wieland, I., Bolger, G., Asouline, G., and Wigler, G., *Proc. Natl. Acad. Sci. U.S.A.*, 87, 2720—2724, 1990. With permission.)

The "tester" DNA population contains "target" DNA sequences that are absent in "driver" DNA. In this method of subtractive hybridization, tester DNA is cleaved by restriction enzymes, biotinylated, and ligated to "template" oligonucleotide (Figure 1-50). It is then mixed with an excess of sonicated driver DNA, melted, and annealed. Single-stranded DNA is isolated by hydroxylapatite; this DNA is enriched in target sequences but also contains large amounts of unreannealed driver. Therefore, this procedure is repeated twice. Finally, tester DNA highly enriched with "target" sequences are separated from driver DNA by avidin/biotin affinity chromatography. Amplification of target sequences using polymerase chain reaction and cloning and analysis are the last steps of this method.

To model the gain of information in which a cell is infected by a single pathogenic organism, the first test involved isolating single copy levels of λ DNA from placental DNA. Three tests resulted in approximately 300- to 750-fold enrichments of λ sequences. To model loss of information, tester was placental DNA and driver was from immortalized lymphocytes of a patient with Duchenne's muscular dystrophy (DMD). Analysis of isolated, PCR-amplified products resulted in two clones that hybridized to

DNA from human-hamster-hybrid cells retaining only the human chromosome X. These DNAs also failed to hybridize to DNA from the patient with DMD. The calculated enrichment for the isolated sequence was 100-fold.

This method, employing PCR to amplify low yields of DNA and render it clonable, and biotinylation of the tester DNA to separate it from driver DNA after hybridization, seems to result in enrichment of about 100- to 700-fold of target sequences. This should be useful for analysis of target sequences of a few megabases in size, making it suitable for isolation of sequences lost in large deletions. It may also be useful for analysis of small deletions in organisms with small genome size.

♦ Small differences within two entire genomes may have significant biological consequence. Therefore, it would be valuable to develop a method that could easily identify and isolate such differences. While several methods exist to distinguish such differences in cDNA populations, only one has addressed this problem for genomic DNA. This article proposes a new approach that couples several clever molecular tricks. The authors biotinylate their "tester" DNA and then perform several cycles of subtractive hybridization with the second "driver" DNA population. The unannealed "tester" DNA, enriched for the target sequences desired, is preferentially recovered using avidin/biotin affinity chromatography. This recovered DNA fraction is subjected to PCR amplification and subsequently cloned for further analysis. This method should enrich several hundredfold for specific differences between two otherwise identical genomes. *Joel M. Schindler*

Systematic Screening of Yeast Artificial-Chromosome Libraries by Use of the Polymerase Chain Reaction
E. D. Green and M. V. Olson
Proc. Natl. Acad. Sci. U.S.A., 87, 1213—1217, 1990   1-51

Standard methods for screening large clone libraries involve hybridization of probes to filters, which contain immobilized DNA molecules released from lysed cells or virus particles. These colony hybridization techniques are effective, but become cumbersome if, for example, a 60,000 clone yeast artificial-chromosome (YAC) human genomic library is screened with a few thousand probes (this situation might occur in the systematic analysis of the human genome). This paper describes an alternative approach for screening large, ordered libraries of YAC clones. The technique is based on assaying DNA samples derived from pools of clones using the polymerase chain reaction (PCR).

The human genomic library screened was stored as about 23,000

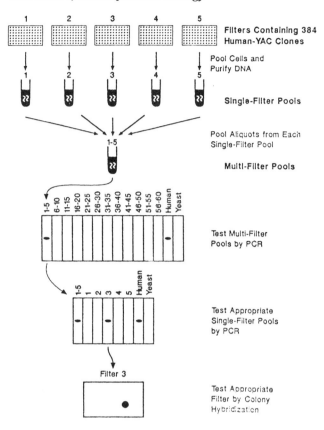

FIGURE 1-51. Schematic representation of the PCR-based strategy for screening human YAC libraries. (From Green, E. D. and Olson, M. V., *Proc. Natl. Acad. Sci. U.S.A.*, 87, 1213—1217, 1990. With permission.)

individually picked clones in 96-well microtiter plates. This library should contain, on average, two copies of all single-copy sequences. The first step in screening this human YAC library involved growing individual clones in arrays of 384 colonies per filter, requiring 60 filters for the entire library (Figure 1-51). Cells from each filter were pooled and a single DNA preparation was made from each filter. DNA from five filters was mixed to yield multi-filter pools. Each multi-filter pool was screened by PCR, electrophoresis, and ethidium bromide staining, with a positive (total human genomic DNA) and negative (a YAC clone containing a yeast DNA insert) control included. When a positive multi-filter pool was detected, each constituent single-filter pool was analyzed using the same method. Finally, the location of a positive clone on 384-clone-containing filter was determined by colony hybridization using the radiolabeled PCR product as a probe.

This strategy was applied to the isolation of the human tumor necrosis factor β gene and the human membrane cofactor protein gene. These isolations occurred successfully despite the generation of inappropriate PCR products, which occurs when pools of YAC DNA are used as templates.

This PCR-based approach for screening large human YAC libraries is sensitive and permits the determination within a day of whether a particular DNA sequence is present in a library. Only one tenth the number of filters needs to be prepared and handled as in colony hybridization screening methods.

♦ The Human Genome Initiative (HGI) has recently received a great deal of attention. While both pros and cons for the Initiative have been well expressed, the long-term value for understanding aspects of the genetic basis of human development is substantial. The current focus of the Initiative involves technology development. One area of particular interest is how to successfully isolate specific DNA sequences from among a large population of clones containing large fragments of DNA. The authors propose a systematic screening method that uses pooled aliquots of DNA and PCR amplification to determine which individual clone contains the specific DNA sequence of interest. This approach should allow for the isolation of a single copy gene from an entire library of cloned DNA. *Joel M. Schindler*

Transfer of a Yeast Artificial Chromosome Carrying Human DNA from *Saccharomyces cerevisiae* into Mammalian Cells

V. Pachnis, L. Pevny, R. Rothstein, and F. Costantini
Proc. Natl. Acad. Sci. U.S.A., 87, 5109—5113, 1990   1-52

It would be helpful for the study of the regulation of gene expression if DNA segments greater in size than those that can be contained in bacterial cloning vectors could be transferred into mammalian cells. This paper presents a method to transfer yeast artificial chromosomes (YACs) carrying 450-kilobase DNA insertions into mouse fibroblasts in culture.

The YAC HY 19, which carries inserts of about 450 kilobases of human DNA, was modified by the addition of the *neo* gene. Yeast cells carrying HY 19-neo chromosomes were stripped of their cell walls with Zymolyase, and the resultant spheroblasts were fused to L cells in suspension in the presence of polyethylene glycol 1500.

About 50 L-cell colonies resistant to G418 were obtained from every 10^6 cells input. Southern blot analysis confirmed the presence of the *neo* gene in these colonies. Based on *Alu* fingerprint analysis of 19 of the resultant L-cell clones, 47% contained a largely intact copy of the HY 19-neo YAC. Subsequent Southern blot analyses, using restriction fragments of the YAC probed with neo sequences, suggested that no gross rearrangements of the YAC had occurred. Furthermore, DNA from some clones tested positive with probes from both YAC arms. *In situ* hybridization using a biotinylated *neo* probe showed a single site of integration into mouse chromosomes. Some clones showed stability of G418 resistance after 30 to 40 generations, although in others the *neo* gene seemed to have been inactivated.

These results suggest that high-frequency transfer of YACs containing large inserts of human DNA into mouse L cells can occur. These transfers seem to involve stable integration of large parts of the YACs into mouse chromosomes. This system might be applicable for transfer of DNA into pluripotent embryonic stem cells that can be used for the generation of transgenic animals.

♦ A primary biological assay for the function of a particular sequence of DNA is the introduction of that DNA into a cell. This assay is used to investigate whether the DNA of interest can "rescue" a mutant phenotype or confer a new dominant phenotype on the treated cell. One limitation of this assay has been the size restriction of transferred DNA. This restriction is the result of limitations imposed by existing cloning vectors. As a result, large or complex genes could not be analyzed in this biological assay. It is now possible to clone much larger segments of chromosomal DNA using yeast artificial chromosomes (YACS). In this article, the authors demonstrate that such large fragments of cloned DNA can be transferred to mammalian cells. This successful technique now opens the way for the introduction of these large cloned DNA fragments into various cell phenotypes for further analysis and characterization. *Joel M. Schindler*

Developmental Gene Expression 2

INTRODUCTION

Normal development demands the accurate temporal and spatial expression of genetic information. The fundamental rules which underlie the developmental control of gene expression seem to be fairly consistent among most eukaryotic species. However, the specifics of that regulation do vary in concert with genetic diversity.

Throughout eukaryotic development, one is likely to find at least a single example of every possible type of regulation imaginable for developmental gene expression. However, certain general mechanisms seem to prevail. While transcriptional regulation predominates, posttranscriptional, translational, and posttranslational modifications of gene products do exist. In addition, the components of "molecular cascades " that regulate developmental expression are abundant and diverse.

During the past year, much effort has been placed on characterizing the many factors responsible for controlling developmental gene expression. The "factorology" has lead to interesting models explaining how different components can interact with each other at the protein level and the protein-DNA level to explain both the temporal and the spatial nature of developmental gene expression.

Articles included in this chapter include examples of multiple promoters for single genes, genetic rearrangements, and alternative splicing strategies as well as transcription factors that function in a tissue specific manner. All are viable means of regulating developmental gene expression.

The Cyclic Nucleotide Phosphodiesterase Gene of *Dictyostelium discoideum* **Contains Three Promoters Specific for Growth, Aggregation, and Late Development**
M. Faure, J. Franke, A. L. Hall, G. J. Podgorski, and R. H. Kessin
Mol. Cell Biol., 10, 1921—1930, 1990   2-1

In *Dictyostelium discoideum*, the cyclic nucleotide phosphodiesterase is required for cellular aggregation and proper formation of fruiting bodies during late development. Previous work has shown that two distinct phosphodiesterase mRNAs are found in *D. discoideum:* a 1.9-kilobase (kb) mRNA specific for growth, and a 2.4-kb mRNA specific for aggregation. These mRNAs, which differ only in their 5'untranslated regions, are transcribed from the single phosphodiesterase gene via two distinct promoters. In this report, a further characterization of the phosphodiesterase mRNAs and their regulation and promoters is presented.

Using Northern blot analysis, a third phosphodiester mRNA of 2.2 kb was found during the preslug stage of late development. This mRNA was transcribed from a promoter adjacent to the phosphodiesterase coding sequence, unlike the distal promoters used to direct transcription of the 1.9- and 2.4-kb mRNAs. Using constructs containing the chloramphenicol acetyltransferase (CAT) reporter gene fused to each of the three putative phosphodiesterase promoters, the transcriptional activity of each promoter was analyzed individually. By measuring the levels of CAT produced, it was found that each promoter had the same temporal regulation as the mRNA it transcribed. The 1.9-kb mRNA accumulated maximally in vegetative cells, the 2.4-kb mRNA accumulated during aggregation, and the 2.2-kb mRNA accumulated during late development.

These results provide evidence that the phosphodiesterase gene of *D. discoideum* has 3 independent promoters, all contained in a region extending 4.1 kb upstream of the ATG codon. Each directs the transcription of an mRNA of a different length with a different temporal pattern of expression. The aggregation-specific promoter is inducible by cAMP; the late-development-specific promoter induces transcription detected only in prestalk cells. The three promoters may regulate the temporal and spatial expression of the phosphodiesterase gene by responding to different signals.

♦ At the onset of development, amoebae of *Dictyostelium discoideum* chemotax towards cAMP and thereby construct a multicellular assembly. The level of cAMP is modulated by a specific cAMP phosphodiesterase (PDE) so that the cAMP receptors do not become saturated. Mutants that do not express PDE are unable to aggregate, but can be rescued by the addition of PDE or transformation of the PDE gene. Previous experiments have shown that the PDE gene was transcribed from two different promoters — one for growth and a second cAMP-stimulated promotor specific for aggregation. The present report by Faure et al. shows that a third specific promotor, proximal to the structural gene, is used for expression of PDE late in development with cAMP probably acts to guide morphogenetic movements of the cells in the multicellular assemblies. Interestingly, the culmination specific PDE is expressed exclusively in prestalk cells. This suggests that PDE functions in reducing cAMP levels in these cells which is consistent with current models of cell-type differentiation

and pattern formation. Further experiments should elucidate the signals and second messenger pathways which result in differential promotor utilization of this important gene as well as the role it plays in patterning.
Stephen Alexander

The *unc-5, unc 6,* **and** *unc-40* **Genes Guide Circumferential Migrations of Pioneer Axons and Mesodermal Cells on the Epidermis in** *C. elegans*
E. M. Hedgecock, J. G. Culotti, and D. H. Hall
Neuron, 2, 61—85, 1990   2-2

Mutants with abnormal cell or axon trajectories during development can be used to help identify the molecules that guide cell migrations and those that are required for cell motility. In this report, mutations in three genes of *Caenorhabritis elegans* are described that are required to guide migrating cells and pioneer axons along the dorsoventral axis of the epidermis (Figure 2-2).

The genes *unc-5, unc-6,* and *unc-40* comprise a global, circumferential guidance mechanism. In mutant animals, circumferential movements of cells still occur, but in a misguided manner; longitudinal movements are unaffected. The gene *unc-5* affects dorsal migrations, *unc-40* primarily affects ventral migrations, and *unc-6* affects migrations in both directions. While all three genes affect circumferential migrations throughtout the animal, the relative importance of these cues varies for different classes of cells. These genes do not affect cell motility per se, but only cell guidance, specifically for pioneer axons growing on the epidermis.

Analysis of single and double mutants showed that all known *unc-6* functions also require *unc-5 , unc-6* or both, and that all *unc-5* and most *unc-40* functions require *unc-6.* The dorsal and ventral guidance functions of the *unc-6* gene were found to be separately mutable, raising the possibility that a single *unc-6* protein has two separate structural domains, or, alternatively, that two distinct *unc-6* proteins assist dorsal and ventral guidance.

These three genes may encode guidance molecules or receptors for such molecules. The genes may identify opposite, adhesive gradients that guide dorsal and ventral migrations on the epidermis. Recent findings suggest that *unc-5* encodes a transmembrane protein of basal lamina, homologous to two families of adhesion proteins. It is possible that *unc-5* is a laminin B2 receptor present on cells during dorsal migrations, and that *unc-40* is a laminin B2 receptor found on cells during ventral migrations.

♦ Of the approximately 30 genes that are known to affect cell and axon migrations in *C. elegans, unc-5, unc-6,* and *unc-40* are specifically re-

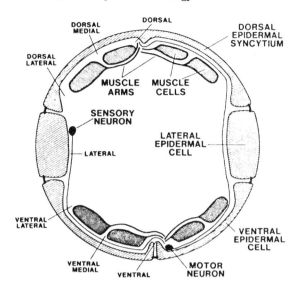

FIGURE 2-2. Nematode body wall. Schematic cross section of a newly hatched larva showing the arrangement of epidermis, muscles, and neurons (after Sulston et al., 1983; White et al., 1986). The epidermal basal lumina (not shown) separates ectoderm, e.g., epidermis and neurons, from mesoderm, e.g., muscles. Longitudinal nerves run at the dorsal, dorsal medial, dorsal lateral, lateral, ventral lateral, ventral medial, and ventral positions. A typical motor axon with cell body on the ventral epidermal ridge must grow dorsally across ventral, lateral, and dorsal epidermal cells to reach the dorsal nerve cord. A typical sensory axon with cell body on the lateral epidermal ridge must grow ventrally across lateral and ventral epidermal cells to reach the ventral nerve cord. Arms from the dorsal and the ventral muscle cells project to motor axons in the dorsal and the ventral nerve cords, respectively. (From Hedgecock, E. M., Culotti, J. G., and Hall, D. H., *Neuron*, 2, 61—85, 1990. With permission.)

quired to guide migrating cells along the dorsoventral axis of the epidermis. Mutations in *unc-5* affect dorsal migrations, whereas those in *unc-40* mainly affect ventral migrations; *unc-6* affects migrations in both directions. The migration of both mesodermal and neural cells is affected by all three genes. Since both mesodermal and neural cells contact and may be guided by the basal lamina of the epidermis (neural cells also contact the basolateral surface of the epidermis), an attractive hypothesis is that the three *unc* genes encode basal lamina components and/or receptors. Indeed, *unc-6* is now known to encode laminin B2, a major component of all basal laminae, and *unc-5* appears to encode a transmembrane protein that may serve as a laminin receptor. This paper provides a great deal of useful background information on nematode neurobiology and gonad development. More importantly, it significantly advances the analysis of extracellular matrix molecules, their receptors, and their roles in cell migration. *Susan Strome*

Proper Expression of Myosin Genes in Transgenic Nematodes
A. Fire and R. H. Waterston

EMBO J., 8, 3419—3428, 1989   2-3

Caenorhabditis elegans has four genes encoding the heavy chain of myosin. The work presented here was directed at understanding myosin gene expression. Using recently developed gene transfer techniques, clones of two myosin genes, *myo-3* and unc-54, were introduced at low copy number into the nematode germ line. The effects of these gene transfers were determined.

Plasmids containing the *myo-3* gene (pSAM, Figure 2-3, left) were injected into oocytes of animals carrying an amber mutation in the *tra-3* gene. The pSAM plasmids also contain the amber suppressor tRNA gene, *sup-7*. Selection for expression of *sup-7* resulted in three independent transformed lines, each containing different stably transformed loci. Southern blot analysis confirmed the presence of low copy number integration of the injected DNA. Each of the three transformed chromosomes rescued the lethal mutation *ste378,* a strong candidate for a mutation in the myosin heavy chain A structural gene.

Transformation with *unc-54,* the myosin heavy chain B gene, was performed using a different protocol involving direct selection for function of the myosin gene. Cosmids containing the *unc-54* gene were injected into animals homozygous for a deletion in *unc-54* (these animals are viable with a slow phenotype as larvae, and paralysis as adults) (Figure 2-3, right). Some of the resulting animals were germ line transformants with improved movement as both larvae and adults. Transformation with *unc-54* occurred at both low number in an integrated fashion and in very high copy number tandem arrays. Animals with very high copy numbers of this gene had disrupted muscle structure and function.

These findings provide evidence that two myosin genes can be introduced into the germ line of nematodes, where they can function normally when integrated at various chromosomal loci. These transformations were effected by both direct and indirect selection protocols. The reintroduced genes were expressed in the proper cell types and their protein products assembled to produce the appropriate muscle filament structures.

♦ Transformation of *C. elegans* by microinjected DNA has in recent years become an extremely powerful approach to studying many aspects of worm development. This paper illustrates the usefulness of this technology for two types of studies: analysis of gene promoters, and identification of cloned gene sequences by transformation rescue of worms mutant for the gene. Fire and Waterston and also Spieth et al. (*Dev. Biol.,* 130, 285—293, 1988) demonstrated that microinjected *C. elegans* genes become stably integrated into the genome and show proper stage- and tissue-

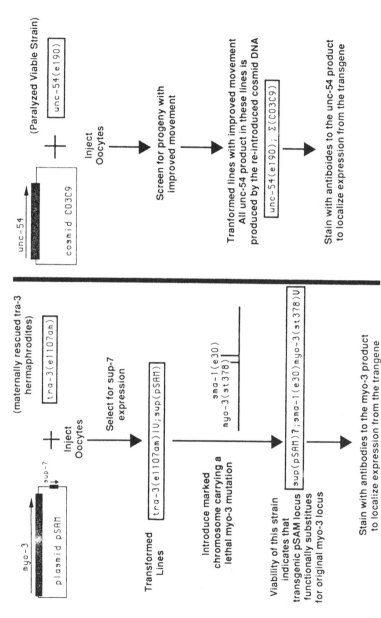

FIGURE 2-3. Strategies for introducing the myo-3 and unc-54 genes into the C. elegans germline. The myo-3 gene was introduced passively by co-selection for a linked marker (sup-7) while the unc-54 gene was introduced using a direct selection for its expression. (From Fire, A. and Waterston, R. H., *EMBO J*, 8, 3419—3428, 1989. With permission.)

specific expression. There appears to be little effect of position of integration on expression. Thus the nematode transgenic system can be used to identify DNA sequences required for regulation of gene expression; such analyses are in progress in many labs. Fire and Waterston also demonstrated rescue of myosin mutants by microinjected myosin genes. (Also see Way and Chalfie, *Cell,* 54, 5—16, 1988; abstracted in the 1989 *Year Book of Developmental Biology.*) Such transformation rescue is now being used by many investigators to clone their genes of interest. In regions where the physical map of the genome (Coulson et al., *Proc. Natl. Acad. Sci. U.S.A.,* 83, 7821—7825, 1986) has been correlated with the genetic map, genes can be identified by microinjecting cloned sequences in the region and screening for rescuing activity. This approach, along with transposon tagging of genes (see Collins et al., *Nature,* 328, 726— 728, 1987; abstracted in the 1989 *Year Book of Developmental Biology*), makes cloning of many *C. elegans* genes quite straightforward. *Susan Strome*

Developmentally Regulated Alternative Splicing of Drosophila Integrin PS2 α Transcripts
N. H. Brown, D. L. King, M. Wilcox, and F. C. Kafatos
Cell, 59, 185—195, 1989   2-4

The position specific (PS) antigens of *Drosophila melanogaster* are members of the integrin family. These cell surface receptor proteins are heterodimers of unrelated α and β subunits. While different types of receptors frequently share the same β subunit, each type of receptor seems to have its own α subunit, which presumably specifies ligand binding. Two different PS α subunits, PS1 α and PS1 β, and a single β subunit, have been identified in *D. melanogaster.* This paper describes the PS2 α gene and its RNA products.

Using a PS2 α cDNA probe and a combination of restriction mapping, Southern analysis and selective sequencing, the PS2 α gene was found to comprise 12 exons extending over 31 kilobases (kb). After splicing out of 11 introns, mature mRNAs of 5.7 kb result.

An analysis of 25 PS2 α cDNA clones from libraries of 4 to 8 h, 8 to 12 h, and 12 to 24 h embryonic tissues and of imaginal discs suggested that all mRNAs shared common 5' and 3' termini. Two major forms of PS2 α mRNA seem to exist: the canonical form and a mRNA lacking exon 8. When cDNA libraries from several different stages during the life style and from selected isolated tissues were probed with PS2 α exon 5 and exon 8, hybridization to both probes was seen at all stages and tissues examined. However, the ratio of the hybridization to the two probes varied dramatically.

A sequence comparison of PS2 α with six human integrins showed that the alternately spliced region of PS2 α is diverse among different integrins. In contrast, this region was well conserved between PS2 α of *D. melanogaster* and the distantly related Mediterranean fruitfly *Ceratitis capitata*.

This work provides evidence that PS2 α of *D. melanogaster* is a large gene comprising 12 exons. The mRNAs transcribed from this gene appear to be alternatively spliced in different tissues and at different times in development. The variably spliced region may be important in the determination of the specificity or affinity of this receptor protein for its ligand.

♦ Developmental regulation of gene expression can occur at many points from the gene to the gene product. While evidence suggests that most developmental regulation occurs at the transcriptional level, examples of other regulatory points can be found. One such example is described in this article. The authors demonstrate that the integrin PS2 α gene is transcribed into two major mRNA forms which differ from each other as a result of the alternative splicing of a single exon. The authors speculate that the alternatively spliced region of the encoded protein is directly involved in ligand binding. This possibility offers the interesting prospect that the alternative splicing of transcripts from a single gene could lead to related gene products with differing developmental roles expressed at different times and in different locations. *Joel M. Schindler*

The Drosophila Gene *tailless* Is Expresed at the Embryonic Termini and Is a Member of the Steroid Receptor Superfamily
F. Pignoni, R. M. Baldarelli, E. Steingrimsson, R. J. Diaz,
A. Patapoutian, J. R. Merriam, and J. A. Lengyel
Cell, 62, 151—163, 1990   2-5

Zygotic expression of the *Drosophila* gene *tailless* (*tll*) is required for the establishment of nonmetameric structures from both anterior and posterior portions of the blastoderm fate map (Figure 2-5). It is likely that *tll* is activated by the maternal terminal gene pathway, and that, once activated, *tll* represses segmentation and activates terminal-specific genes in the terminal domains. To further the understanding of the *tll* gene, the work described here was directed to a molecular analysis of the *tll* gene.

By chromosome walking around a cytologically visible breakpoint, a candidate transcription unit was identified; this DNA was used to probe RNA blots, permitting the isolation of a cDNA clone which was used to identify genomic DNA. The transcription unit identified was used in a P-element transformation vector that was found to rescue *tll* mutants.

Expression of *tll* peaked in the 2- to 4-h embryo. *In situ* hybridization showed that *tll* was first expresed at nuclear cycle (NC) 11 and was expressed during NC11 and NC 12 in the nuclei of the termini (Figure 2-

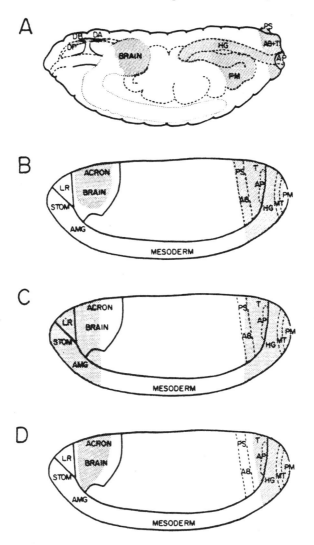

FIGURE 2-5. Comparison of domains requiring and domains expressing *tll*. (A) Outline of stage 17 embryo showing internal organs, redrawn from Campos-Ortega and Hartenstein (1985). Structures and organs missing from *tll* embryos are shaded. (B, C, and D) Fate map of blastoderm embryo, redrawn from Jürgens et al. (1986), Jürgens (1987), and Hartenstein et al (1985). (B) The anlagen giving rise to the structures deleted in *tll* embryos (from [A]) are shaded. (C) Domains expressing *tll* RNA at NC12 are shaded (average of measurements on five embryos). (The NC14 fate map is shown for orientation, but is not meant to imply that nuclei are determined at this stage.) (D) Domains expressing *tll* at NC14 (average of measurements on 10 embryos) are shaded. AMG = anterior midgut, AP = anal pads, DA = dorsal arms, DB = dorsal bridge of cephalopharyngeal skeleton DP = dorsal pouch, HG = hindgut, LR = labrum, MT = Malpighian tubules, PM = posterior midgut, PS = posterior spiracles, STOM = stomodeum, T = telson. (From Pignoni, F., Baldarelli, R. M., Steingrimsson, E., Diaz, R. J., Patapoutian, A., Merriam, and Lengyel, F., *Cell,* 62, 151—163, 1990. With permission.)

5). This symmetrical expression pattern was transient. After the cellular blastoderm stage, *tll* expression was found only in the anterior, especially the developing brain.

A *tll* cDNA clone was sequenced. The predicted protein has a molecular weight of about 50 kDa. This protein is similar to members of the steroid receptor superfamily, with domains similar to both the DNA binding and ligand binding domains of other proteins in this superfamily.

This work has identified 10 kilobases of DNA sufficient to rescue *tll* mutants. This gene, a member of the steroid receptor superfamily, was found to have the spatial and temporal expression generally expected from the mutant phenotype.

♦ The similarity of structural motifs among various biomolecules has suggested that those biomolecules share functional similarities as well. Such shared motifs have provided support for the suggestion that certain classes of related biomolecules have been evolutionarily conserved and perform similar functions in many different organisms. In this article, the authors show the developmental expression of a specific *Drosophila* gene and demonstrate structural similarity of the protein encoded by that gene with an entire class of molecules, the steroid receptor superfamily. This similarity leads to the suggestion that the *Drosophila* gene product functions as a transcription factor with DNA and ligand binding domains that function similarly to the steroid receptors. If true, this protein could play an interesting role in the genetic circuitry related to the developmental regulation of gene expression in *Drosophila*. *Joel M. Schindler*

Two Distinct *Xenopus* Genes with Homology to MyoD1 Are Expressed Before Somite Formation in Early Embryogenesis
J. B. Scales, E. N. Olson, and M. Perry
Mol. Cell Biol., 10, 1516—1524, 1990   2-6

Members of the MyoD gene family participate in the determination and diferentiation of skeletal muscle cells *in vitro*. These genes are usually expressed only in skeletal muscle cells preceeding terminal differentiation. The proteins encoded by these genes are homologous with proteins in the Myc family. In this work, regulatory factors participating in myogenic determination of vertebrate embryos were sought.

A cDNA library derived from *Xenopus laevis* embryos at the tailbud stage was screened simultaneously with two probes spanning the highly conserved, *myc*-like regions of MyoD1 and myogenin, two members of the MyoD gene family. Several cDNAs were isolated that had extensive sequence similarities to each other and to MyoD1. Two of these clones, Xlmf1 (*Xenopus laevis* myogenic factor 1) and Xlmf25, were used as probes in Northern blot hybridization to RNA from various adult frog tissues and from oocytes, fertilized eggs, and embryos at various stages in early development. These experiments indicated that Xlmf1 and Xlmf25 were expressed in the adult only in skeletal muscle, were absent from oocytes and fertilized eggs, and were first expressed during gastrulation. Maximal induction of these transcripts occurred in the early neurula and was maintained at least until stage 32, the last stage examined. The expression of these transcripts seemed to precede that of cardiac α-actin. Xlmf1 cDNA activated myogenesis when used to transform pluripotent mouse stem cells of line C3H10T$_1$/$_2$, but with a frequency 1 to 2 orders of magnitude less than achieved using MyoD1 or myogenin genes.

These experiments provide evidence that transcription from genes related to the MyoD family of myogenic factors occurs in midgastrula embryos of *X. laevis* well before the segregation of somites. This gene expression occurs at the time of commitment of mesodermal cells to the myogenic lineage and before their overt differentiation. It seems likely that Xlmf1 and Xlmf25 function in determination of muscle cells in *X. laevis* development.

MyoD Expression in the Forming Somites Is an Early Response to Mesoderm Induction in *Xenopus* Embryos

N. D. Hopwood, A. Pluck, and J. B. Gurdon

EMBO J., 8, 3409—3417, 1989   2-7

MyoD is a regulator of muscle development, extensively studied *in vitro,* whose exact role is unknown. To investigate this issue, the work described here was directed to cloning and analysis of the gene and gene product for the *Xenopus laevis* homolog of MyoD.

An *X. laevis* gastrula cDNA library was screened with a mouse MyoD cDNA probe. The cDNA sequence deduced from overlaps of cDNAs obtained has extensive similarity to that of mouse MyoD, both within and without the basic and Myc-like regions which mediate *in vitro* myogenic transformation. Northern blot analysis showed transcripts of two sizes hybridizing to the isolated cDNA. The smaller transcript was found in unfertilized eggs, cleavage stages, and blastulas, while the larger one was first seen in stage 10 gastrulas, peaking in late neurulas, and then declining in later embryos. A careful analysis showed that MyoD transcripts appear about 2 h before cardiac actin transcripts accumulate.

Northern blot analysis and *in situ* hybridization showd that MyoD RNA was found in the early embryos only in the developing somites and in gastrula only in mesoderm. Maternal MyoD transcripts did not appear to be prelocalized in stage 8 1/2 blastulas. While *Xenopus* embryonic, but not adult heart, tissue expresses skeletal muscle α actin, no MyoD expression was detected in embryonic heart.

These findings provide evidence that expression of MyoD in *X. laevis* is activated by induction. Its transcription precedes that of cardiac actin, and is thus the earliest muscle-specific response to mesoderm induction described. Although it generally preceeds, it does not seem to be required for expression of skeletal muscle α actin. MyoD may act transiently in development to establish muscle gene expression.

The *Xenopus MyoD* Gene: An Unlocalised Maternal mRNA Predates Lineage-Restricted Expression in the Early Embryo
R. P. Harvey
Development, 108, 669—680, 1990   2-8

Members of the *MyoD* gene family seem to be involved in the process of commitment to the muscle lineage. Expression of one or more of these genes may be required for the development and maintenance of muscle tissue. In *Xenopus,* commitment to the muscle lineage begins during mesoderm induction by the vegetal pole cells towards more animal cells. This induction is thought to be mediated by growth-factor-like molecules of the TGFβ and FGF families. The work presented here was directed to assessing the role of MyoD genes in mesoderm induction in *Xenopus.*

A *Xenopus MyoD* gene was cloned by hybridization of a mouse MyoD clone to a *Xenopus* genomic library; two related *Xenopus MyoD* cDNAs were cloned by hybridization to a neurula cDNA library. The encoded proteins had sequences similar to murine MyoD across its entire length.

Northern blot analysis of polyadenylated RNA showed low level expression of a 1.5-kilobase molecule hybridizing to a *Xenopus MyoD* probe in oocytes and eggs. These RNAs seemed to have disappeared at the mid-blastula stage, but they reappeared at the gastrula stage, peaked in the late neurula, and diminished in tailbud embryos. RNAse protection assays suggested that zygotic expression of *MyoD* mRNAs began at the mid-blastula transition, increasing gradually through blastula and gastrula stages. Experiments with blastula explants provided evidence suggesting that maternal *MyoD* mRNA is found in all regions of the blastula. When animal pole explants of *Xenopus* mid-blastula embryos were treated with a TGFβ-like mesoderm-inducing factor secreted by a *Xenopus* cell line, dose-dependent expression of MyoD mRNA occurred. Weaker expression of *MyoD* mRNA was induced by bovine FGF.

This work provides evidence that in *Xenopus* the *MyoD* gene is first expressed in oogenesis and is found unlocalized throughout the early blastula. The *MyoD* gene seems also to be expressed in the zygote beginning at the mid-blastula transition, with transcripts localized to somites. *MyoD* gene expression is the earliest molecular indication of muscle commitment in embryos.

♦ In the past few years, the study of mesoderm induction in the frog embryo has tended to focus on muscle formation, since this tissue represents the most massive early mesodermal derivative, and owing to the easily obtained molecular markers for muscle, notably α-actin. However, it has become clear that myogenesis is not a direct consequence of mesoderm induction; there are one or more intervening steps (see Gurdon, 1988 and Symes et al., 1988; discussed in Yearbook 1990). Nevertheless, *Xenopus* is an excellent model system for studying vertebrate development, including muscle specification. Three groups have independently taken a fairly obvious step in this direction, i.e., cloning frog homologs of the myogenic control gene, MyoD1.

These are all relatively straightforward clone and characterize papers with few if any surprises in store. One curiosity is the presence of a small amount of MyoD1 mRNA in the oocyte. The function, if any, of maternal MyoD1 is completely obscure, but its existence is reminiscent of maternal mRNA encoding NCAM, an important protein limited in early development to the nervous system.

The availability of full-length *Xenopus* MyoD1 clones will make it possible to perform relatively simple experiments to test the capacity of this gene to alter cell identity in whole vertebrate embryos. Similar lines of investigation could be pursued as well with the various other myogenic genes that have been identified.*Thomas D. Sargent*

KTF-1, a Transcriptional Activator of *Xenopus* Embryonic Keratin Expression

A. M. Snape, E. A. Jonas, and T. D. Sargent
Development, 109, 257—165, 1990   2-9

The embryonic, epidermis-specific keratin gene of *Xenopus laevis,* XK81A1, is subject to spatial and temporal regulation of its expression. Previous work from this laboratory has suggested that both positive and negative transcription factors regulate the expression of XK81A1 by binding to sequence elements between −487 and +26 of the transcription initiation site. The work reported here sought to characterize the DNA binding activity or activities and the corresponding DNA binding sites involved in regulating transcription of this gene.

Nuclear extracts from early tailbud *Xenopus* embryos were found to contain an activity that altered the gel electrophoretic mobility of DNA from −274 to −140 of the XK81A1 gene transcription start site. The use of a series of 30-bp oligodeoxyribonucleotide competitors in the mobility shift assay gave results that implied that the sequence from −165 to −136 contained the binding site. This site contains the imperfect palindromic sequence ACCCTGAGGCT. DMS methylation interference and 1,10-phenanthroline-copper footprinting analyses confirmed the results of the mobility shift experiments.

A series of mutations in nucleotides −165 to −136 were constructed, placed into cloning vectors, and injected into fertilized *Xenopus* embryos.

When gene expression of tailbud stage embryos was assessed using RNAse protection assays, the results showed that mutations of the –165 to –136 site lowered by about eightfold the level of transcription from the injected keratin construct.

The keratin transcription factor (KTF-1) binding site was cloned into a plasmid 500 bp upstream from the initiation site of the *Xenopus* adult β-globin gene and injected into *Xenopus* embryos. Analysis of the resultant β-globin transcription showed no consistent effects when a single copy of the KTF-1 binding site was present on the construct, but constructs with two binding sites usually resulted in high levels of β-globin transcription. This transcription of β-globin seemed to occur in both epidermal and non-epidermal tissues.

These experiments identify a protein-binding sequence in the upstream promoter region of the *Xenopus* keratin gene XK81A1. This sequence binds a nuclear protein, named KTF-1, from *Xenopus* embryos *in vitro*. In *Xenopus* embryos, this sequence seems to act as a general activator of embryonic transcription. KTF-1 may act in concert with other factors to direct regulated transcription of the keratin gene.

♦ The past year has not seen a great deal of activity in the area of class II gene regulation in *Xenopus*. This paper from our laboratory describes Keratin Transcription Factor 1 (KTF-1) which was originally identified by bandshifting of a region of the XK81A1 keratin gene promoter. KTF-1 footprints a 19-base sequence containing an imperfect palindrome, ACCCTGAGGCT. Removal of this site drastically reduces the activity of keratin gene constructs introduced into embryos by microinjection. Likewise, adding KTF-1 sites to a *Xenopus* β globin gene greatly increases the expression of this construct. However, KTF-1 binding is apparently not sufficient for exclusive expression of injected constructs in epidermis, the tissue in which the endogenous keratin gene is specifically expressed.*Thomas D. Sargent*

A Family of Octamer-Specific Proteins Present During Mouse Embryogenesis: Evidence for Germline-Specific Expression of an Oct Factor

H. R. Schöler, A. K. Hatzopoulos, R. Balling, N. Suzuki, and P. Gruss
EMBO J., 8, 2543—2550, 1989   2-10

The octamer motif is a DNA sequence found in many promoters and enhancers. The octamer binding protein Oct1 has been found in all cell types tested; the octamer binding protein Oct2 has been found only in B lymphocytes. Both Oct1 and Oct2 contain homeobox sequences. Since a large family of developmentally regulated mammalian genes contain homeoboxes, this work was undertaken to look for additional octamer binding proteins, especially developmentally regulated ones, in the mouse.

Nuclear extracts were prepared from a variety of adult mouse tissues and from 12-day mouse embryos, placenta, and yolk sac. These extracts were assayed for the presence of octamer-binding proteins using an electrophoretic mobility shift assay with a radiolabeled fragment of the immunoglobulin heavy chain gene enhancer, which contains the octamer motif. Besides Oct1 and Oct2, 8 other binding proteins, named Oct3 through Oct10 were seen.

Oct2, previously thought to be found only in B lymphocytes, was found in brain, kidney, sperm, and embryos. The brain and the B lymphocyte Oct2 complexes had different heat stabilities and yielded different DNA binding domains after digestion with trypsin.

Oct3 was found in embryos and in the adult brain, always accompanied by Oct2 and Oct7. Different regions of the brain had different levels and ratios of these proteins, with the cerebellum having a notably lower abundance of all of these proteins than other brain regions.

Oct4 and Oct5 were found in unfertilized oocytes and embryonic stem cells. Although Oct4 was found in male and female primordial germs cells, it was not found in sperm or testes. Oct5 was found only in oocytes and embryonic stem cells.

These experiments suggest that a family of octamer-binding proteins is present in development in the mouse. These proteins seem to be differentially expressed during early embryogenesis.

A Novel Octamer Binding Transcription Factor Is Differentially Expressed in Mouse Embryonic Cells

K. Okamoto, H. Okazawa, A. Okuda, M. Sakai, M. Muramatsu, and H. Hamada

Cell, 60, 461—472, 1990 2-11

Transcription factors that appear in embryogenesis in a regulated fashion are likely candidates for factors that control mammalian development. The work described in this paper was a search for *trans*-acting factors found only in embryos. Proteins were sought from the mouse embryonal carcinoma cell line P19 that recognize enhancers previously found to be specific to embryonic stem cells.

DNAse 1 footprint assays were performed on nuclear extracts from differentiated and undifferentiated P19 cells. One protein detected was a novel octamer binding factor, Oct-3. This protein bound to both the typical octamer motif and to an AT-rich sequence found in the enhancer. Oct-3 was not found in P19 cells stimulated to differentiate by retinoic acid; instead, another binding activity, Oct-4, was found in these differentiated cells.

A cDNA encoding OCT-3 was cloned and analyzed. The encoded protein was predicted to comprise 377 amino acids and to contain a unique POU domain (a sequence of about 150 amino acids conserved among Pit-1, Oct-1, Oct-2, and unc-86). Northern blot analysis suggested that Oct-3 mRNA was present as a 1.5-kilobase species in P19 cells, but absent when those cells were induced to differentiate by retinoic acid. Oct-3 mRNA was not detected in various organs of adult mouse and mouse embryos at different stages.

To determine whether Oct-3 is a transcription factor, an expression vector was created in which the N-terminal, proline-rich half of Oct-3 was fused to a DNA binding domain of c-Jun. When transfected into P19 cells, this vector plus reporter plasmids resulted in stimulation of a target reporter gene. In other experiments, when the level of Oct-3 was increased in P19 cells, stimulation of enhancer activity resulted.

These findings provide evidence that Oct-3, a novel octamer binding protein found in mouse embryonal carcinoma cells, is a transcription factor. The expression of this protein appears to be developmentally regulated. In turn, Oct-3 itself may regulate differentiation of embryonic stem cells and be part of a hierarchy of regulatory genes.

A POU-Domain Transcription Factor in Early Stem Cells and Germ Cells of the Mammalian Embryo
M. H. Rosner, M. A. Vigano, K. Ozato, P. M. Timmons, F. Poirier,
P. W. J. Rigby, and L. M. Staudt
Nature, 345, 686—692, 1990   2-12

The work presented in this paper was intended to study transcription factors involved in early mammalian embryogenesis. To do this, the murine structural gene encoding Oct-3 (called Oct-4 in Schöler et al., 1989, above) was cloned and analyzed.

The predicted protein encoded a protein of 352 amino acids. Oct-3 contains a novel POU domain, a homeodomain, an N-terminal region rich in proline and glycine, and a C-terminal region rich in proline, glycine, serine, and threonine. When transcribed and translated *in vitro,* Oct-3 shifted bands in a mobility-shift DNA binding assay using an octamer motif sequence probe. When inserted into an expression vector and co-transfected with an octamer-dependent reporter plasmid into HeLa cells, Oct-3 stimulated correctly initiated mRNA synthesis from the reporter plasmid.

Northern blot analysis detected Oct-3 mRNA of 1.55 kilobases in undifferentiated embryonal carcinomal (EC) and embryonic stem cells. Retinoic acid differentiation of EC cells *in vitro* abolished detectable Oct-3 mRNA. *In situ* hybridization experiments detected Oct-3 mRNA in oocytes, fertilized ova, the 2.5-d.p.c. morula, and in both the trophoectoderm and the inner-cell mass cells of the 3.5-d.p.c. blastocyst. After 7 d.p.c., when mesodermal differentiation begins, Oct-3 mRNA was not detected in somatic cells, but was found in priomordial germ cells. Oct-3 mRNA was found in adult ovary and testis, but not in skeletal muscle, brain, heart, liver, spleen, kidney, or pancreas. Northern blot analysis suggested that Oct-3 mRNA was present in maturing and ovulated oocytes, but absent from resting oocytes.

These experiments suggest that Oct-3 is a transcription factor expressed in undifferentiated pluripotent cells of the early embryo and in germ cells. All these cells have the capability to differentiate along multiple lineages.

New Type of POU Domain in Germ Line-Specific Protein Oct-4
H. R. Schöler, S. Ruppert, N. Suzuki, K. Chowdhury, and P. Gruss
Nature, 344, 435—439, 1990   2-13

The Oct-4 protein is a prime candidate for an early developmental control gene since it is maternally expressed and present in the pre-implantation mouse embryo. This report describes the cloning of a cDNA for the mouse Oct-4 gene and characterization of the encoded protein.

A cDNA library from F9 stem cells was screened with a POU domain from the mouse Oct-2 gene. The inserts from positive recombinant phages were transcribed and translated *in vitro,* and the translation products were assayed for DNA binding in a gel mobility shift assay with the octamer motif probe. In this way, a clone was identified that resulted in identical shifts as did Oct-4 and Oct-5 of F9 nuclear extracts. The translation products had a proteolytic fingerprint identical to that of Oct-4. Two *in vitro* translation products resulted from this cDNA clone.

The isolated cDNA encodes a predicted protein of 324 amino acids. This protein contains a unique POU-specific domain and a POU homeodomain. The N-terminal region is rich in prolines interspersed with acidic amino acids. Use of deletions of the Oct-4 gene in *in vitro* experiments showed that sequences in the POU-specific domain contribute to the DNA binding capabilities of the Oct-4 protein.

Use of the Oct-4 probe in Northern blot analysis resulted in detection of a single transcript of 1.6 kilobases in F9 stem cells. No Oct-4 mRNA was found in 12.5-d embryos, or adult testis, liver, lung, intestine, brain, or kidney.

These studies provide evidence that Oct-4 is a novel POU protein. The expression patterns of this putative transcription factor suggest that it functions at the top of a hypothetical cascade of control events during murine embryogenesis.

♦ As described in the 1990 Year Book, a family of octamer-binding (Oct) proteins has been identified in various mouse tissues. The Oct proteins are known to interact specifically with the octamer motif, a *cis*-acting transcription regulatory element found in the promoter and enhance regions of many genes. In addition to the octamer-binding motif, sequence analysis of some of the *oct* genes has revealed a homeobox domain. A total of 10 Oct proteins have now been identified and their expression patterns vary. Some are found in adult mouse tissues whereas others are expressed in the developing embryo. Not all Oct proteins appear to be encoded by distinct genes. For example, Oct-5, may actually be an alternative form of Oct-4 resulting from RNA processing. Oct-1 appears to regulate ubiquitously expressed genes whereas Oct-2 regulates lymphoid-specific genes. Oct-2-like proteins have also been found in the embryo, brain, and kidney. Oct-3 has been detected in extracts of 12-d mouse embryos and in adult brain. Oct-4 and Oct-5 are two maternally expressed octamer-binding proteins that are also found during early embryogenesis. Both are present in unfertilized oocytes and ES/EC cells, whereas only Oct-4 is found in male and female primordial germ cells, maturing oocytes, and in the inner cell mass of the mouse blastocyst (note that Okamoto et al. and Rosner et al. refer to Oct-4 as Oct-3). Oct-6 has been found only in ES/EC cells. In contrast, Oct-7 to 10 have been found in a number of adult tissues.

Three of the papers cited above report the cDNA cloning and characterization of *Oct-4*. The reason so much attention has been placed on this specific gene is because of its expression pattern. *oct-4* is expressed in cells that retain the capacity to differentiate along multiple lineages. The expression is subsequently down-regulated as differentiation occurs. Primordial germ cells and maturing oocytes, however, continue to express the gene. This pattern is in contrast to other embryonic-specific genes such as homeobox genes which are expressed in post-gastrulation embryos in cells that have begun to differentiate into more committed cell types. Thus, the general conclusion is that Oct-4 may be distinct from other transcription factors so far described in that it is expressed exclusively in pluripotent cell types and may function by acting as regulator of the octamer motif thereby turning on or off genes involved in differentiation. The interesting questions now focus on those factors that down regulate *oct-4* expression as differentiation occurs and what genes are actually being modulated by Oct-4. *Terry Magnuson*

Cloning of the *T* Gene Required in Mesoderm Formation in the Mouse
B. G. Herrmann, S. Labeit, A. Poutska, T. R. King, and H. Lehrach
Nature, 343, 617—622, 1990   2-14

The *T (Brachyury)* gene of the mouse, highly conserved among vertebrates, is essential for mesoderm formation. Previous work seeking to isolate this gene, starting from the mapped position of a mutation, analyzed the fine structure of a 1400-kilobase (kb) chromosome piece containing 2 markers closely linked to *T*. In this paper, a combination of molecular and gentic techniques was employed to clone the *T* gene.

Chromosome walking and jumping resulted in the isolation of further probes in the region of the *T* gene. These probes were used to analyze several *T* alleles, and deletions were identified in mutant chromosomes. One cDNA clone, pme 75, was identified by the use of a likely probe to screen a DNA library prepared from RNA of 8.5-day-old embryos, the age at which mesoderm and notochord normally form. Hybridization with pme 75 identified alterations in *T*^wis, a gain-of-function allele of *T*. *T*^wis was found to have an insertion of 5.5 kb of a murine early transposon-like element DNA; sequence analysis showed that the insertion altered a splice site donor, and would result in the production of a modified protein.

The pme 75 DNA was found to be highly conserved among vertabrates. The predicted gene product consists of 436 amino acids and is not significantly homologous with proteins from the databases searched.

These findings provide evidence that the *T* gene has been cloned. This cloning was based only on the phenotype of the mutation, employing overlapping deletions of several mutant alleles and genetic methods. The modified protein predicted from the structure of the *T*^wis allele is consistent with the mutant phenotype, which is much less severe than that of the null allele. The use of this gene should assist study of murine mesoderm formation.

Expression Pattern of the Mouse *T* Gene and Its Role in Mesoderm Formation

D. G. Wilkinson, S. Bhatt, and B. G. Herrmann
Nature, 657—659, 1990   2-15

The *T* gene of the mouse is involved in mesoderm formation. Mutants homozygous for the null allele form insufficient quantities of mesoderm and have severely disrupted morphogenesis of notochord, allantois, and other mesoderm-derived structures, and die during gestation. Previous work has resulted in the cloning of the *T* gene. In the experiments reported here, this gene was used as probe to analyze the expression pattern of *T*.

Northern blot analysis of RNA from embryos and adult tissues detected a 2.1-kilobase transcript in the 8.5 and 9.5-d gastrulation-stage embryos, which was sharply diminished by 10.5 and 11.5 d. No expression of *T* RNA was detected in any adult tissues tested.

In situ hybridization experiments were next used to probe the spatial pattern of *T* gene expression in embryos. At 7, 8.5, and 9.5 d, mesoderm cells and primitive ectoderm adjacent to the primitive streak showed the presence of *T* RNA. *T* gene expression persisted in the notochordal plate of 8.5-d embryos and in the definitive notochord of 9.5-d embryos, although it had become undetectable in paraxial mesoderm and lateral mesoderm and extra-embryonic mesodermal tissues by that time. After gastrulation, *T* gene expression was detected only in notochord and in clusters of cells along the length of the spinal cord.

These findings provide evidence that the *T* gene is directly involved in mesoderm formation and notochord morphogenesis. The tissues that express the *T* gene are also those that are disrupted in mutants. Further study of *T* gene expression should provide insight into the molecular mechanisms of mesoderm formation.

♦ The *T* or *brachyury* gene in mouse was described almost 65 years ago with the discovery of a short-tailed mouse. Subsequent genetic analyses showed that the tail effect occurred in mice heterozygous for the mutant gene. Homozygous fetuses die at midgestation. Extensive phenotypic analyses have suggested that the wild-type form of the gene is necessary for mesoderm production and development of the notochord. Death of the embryo apparently results from poor development of the allantois. The tail shortening in heterozygotes results from abnormal associations of notochord and the gut and neural tube. In addition to these embryological defects, the *T* mutation interacts with the tailless-interaction factor of the *t*-complex resulting in tailless mice.

The papers cited above have identified a candidate *T* gene and represent an eloquent example of how reverse genetics can be used to clone genes. First, a long-range physical map of the area encompassing the *T* gene was established. Then, by examining different *T* alleles, deletions and duplications were discovered. The identification of these were instrumental in localizing the position of *T*. Chromosome walking and jumping were then used to locate a CpG island associated with a gene (*me75*) expressed in the primitive ectoderm and mesoderm next to the primitive streak and subsequently in the notochord. These are the cell types affected in the mutant embryos. One of the *T* alleles (T^{Wis}) showed an insertion in *me75*. Although the evidence is strong that *me75* and *T* are probably the same locus, final proof awaits genetic complementation in transgenic mice. Work will probably now focus on a characterization of the gene and production of an antibody for protein localization, as well as identification of genes such as the tailless-interaction factor that interact with *T* to produce tailless mice.

One of the leaders in mouse genetics in general and in the *t*-complex in specific was Dorothea Bennett. It was with great sadness to learn of her death this year. Dr. Bennett was an intellectual giant whose enthusiasm touched the lives of many. We will miss her. *Terry Magnuson*

Concerted Generation of Ig Isotype Diversity in Human Fetal Bone Marrow

H.-M. Dosch, P. Lam, M. F. Hui, and T. Hibi

J. Immunol., 143, 2464—2469, 1989   2-16

It is often assumed, but has not been proven, that a continuous developmental process results in the generation of B cells from stem cells. This work investigated this tissue, test whether, as in the T cell lineage, distinct phases exist in B cell ontogeny.

Human bone marrow B cells from 14- to 21-d-old fetuses were examined for surface immunoglobulin (sIg) heavy-chain expression by both flow cytometry and fluorescence miscoscopy. Parallel aliquots of cells were infected with Epstein-Barr virus and assayed by ELISA procedures for subsequent secretion of immunoglobulins.

In fetal marrow samples of cells from 14-week fetuses, B lineage cells positive for CD20/21 were present, as were the much rarer sIgM$^+$ cells. A burst in B cell development at 16- to 18-weeks gestational age resulted in nearly 1/4 of all bone marrow cells belonging to the B lineage. Surface expression of IgD increased rapidly at that time, and light-chain gene expression came to resemble the 6:4 κ:λ ratio seen adult serum. At 18 weeks and beyond, proportion s of lymphocytes positive for sIgG and sIgA approached the levels found in adult marrow. By 20 weeks of gestational age, the size of bone marrow B cell pools fell. Surface IgG expression was seen at very low levels on or after 16 weeks of gestation.

During the 16- to 18-week gestation period, 60% of marrow cells of the B lineage were transformable by EBV and secreted immunoglobulins. At 14 weeks, few B cells able to secrete immunoglobulins were found, and those were IgM. By 16 to 18 weeks, the larger B cell pools of bone marrow were seen by this assay to comprise about half pre-B cells and half lymphocytes capable of secreting immunoglobulins. These secreted immunoglobulins included IgM, G, D, A, and E, with the adult ratio of κ and λ light chains (Figure 2-16).

These findings provide evidence that B cell ontogeny of fetal human bone marrow consists of distinct, rapidly changing stages. The data suggest that inherent B cell programming may determine heavy-chain choice of B lymphocytes.

♦ Similar to the results reported with murine fetal liver cells in the previous paper, this study provides evidence for developmental bursts of activity at a discreet time during development. In this instance, human fetal bone marrow cells were isolated for study between weeks 14 and 21 of gestation. By week 14 of gestation, fully one fourth of all bone marrow cells were within the B lineage. At this time, 60% of the bone marrow B

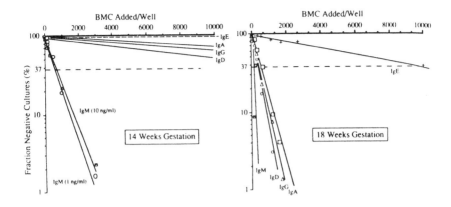

FIGURE 2-16. Frequency and isotype diversity of EBV-transformable cells in fetal bone marrow of 14- or 18-week gestational age. Fifth week limiting dilution cultures of 3 to 30,000 EBV-infected marrow cells were analyzed by particle concentration fluoroimmunoassay for the presence of Ig secretion. Positive wells were defined to contain ≥10 ng/ml of a given Ig isotype. Similar results were obtained by using 1 ng/ml as cutoff. Chi-square derived *p* values varied between 0.18 and 0.72, consistent with single hit kinetics of each regression fit. No valid frequency estimate was obtained for IgE-committed cells in 14-week samples (stipled line). Growing colonies were present in >90% of all Ig⁺ cultures and in considerable proportion of Ig⁻ wells. (From Dosch, H.-M., Lam, P., Hui, M. F., and Hibi, T., *J. Immunol.*, 143, 2464—2469, 1989. With permission.)

cells were transformable by EBV. This was exploited in this study to evaluate changes in B lineage cells during this time in development.

The EBV infected fetal bone marrow cells were analyzed under limiting dilution conditions for isotypes of antibody secreted by 5 weeks of culture. Virtually all of the week-14 transformed B cells secreted IgM. By week 18, however, transformed clones secreting all five isotypes were detectable. Therefore, there exists a significant lag between the appearance of IgM-secreting transformants and EBV-infected cells capable of secreting other isotypes. A third phase was identified, in which the fetal bone marrow B cell pool sizes decline. These data are consistent with a model in which the bone marrow develops competent precursor pools early during fetal development. *E. Charles Snow*

V$_H$ Gene Family Reportoire of Resting B Cells: Preferential Use of D-Proximal Families Early in Development May Be Due to Distinct B Cell Subsets
H. D. Jeong and J. M. Teale
J. Immunol., 143, 2752—2760, 1989   2-17

In the B cells of the adult mouse bone marrow, V$_H$ gene expression consists of near stochastic proportions of members of the 11 families of

V_H genes. In contrast, the functional B cell repertoire during fetal life is biased towards expression of the D-proximal V_H families Q52 and 7183. This preference by B cells early in ontogeny may reflect the microenvironment of the fetal liver or intrinsic genetically controlled factors. These investigations were designed to study this issue.

This work was based on the lymphocyte culture system. Identical microenvironments were provided by adult bone marrow stromal cells, which supported fetal or adult B cell development. Expression of V_H gene families was assessed in B and pre-B cells by sensitive *in situ* hybridization techniques using particular V_H family probes.

When adult bone marrow cell suspensions or fetal liver cells were grown on adult bone marrow stromal cell layers in the presence of lipopolysaccharide (LPS) for 5 to 6 d, the adult cells expressed the adult-like repertoire of V_H gene. In contrast, the fetal B and pre-B cells preferentially expressed the D-proximal V_H gene families. When fetal spleen cells were used, or when fetal cells unstimulated by LPS were tested in similar assays, they too showed preferences for D-proximal V_H gene family expression. After culture on adult bone marrow stromal cells for 8 weeks, fetal liver cells gave rise to B cells which had patterns of V_H expression similar to those of adult cells.

These results provide evidence that preferential use of D-proximal V_H gene families by fetal murine B cells is due to intrinsic factors more than to the fetal microenvironment.

♦ Antibody diversity is generated by the selection and rearrangement of gene segments which contribute the information necessary to synthesize the variable regions of immunoglobulin heavy and light chains. In the case of the heavy chain, the developing B cells select one variable (V_H), one diversity (D), and one joining (J_H) gene segment. Subsequent to the selection of the heavy-chain gene segments, the cells continue on to select one light chain variable region. At this time, the cell is capable of synthesizing immunoglobulin molecules, and, upon expressing some as receptors for antigen, develop into immunocompetent B cells capable of responding to antigen.

During B-cell development, there is an ordered expression of B cell reactivities (Klinman and Press, *J. Exp. Med.,* 141, 1133, 1975), which suggests that a mechanism exists for the ordered utilization of V_H gene segments. This was a puzzling observation until more was learned concerning the organization of V_H genes. It is now believed that there are between 100 and 200 V_H gene segments which are clustered into 11 homology gene families. Alt and colleagues first demonstrated that transformed pre B cells exhibited a preferential utilization of D-proximal V_H gene families (Yancopoulos et al., *Nature,* 311, 727, 1984). This result indicated that, at least during early B-cell development, those V_H gene segments closest to the D genes, with which they rearrange, are preferentially expressed. This has been subsequently shown to be the case with normal fetal and neonatal B cells (Jeong and Teale, *J. Exp. Med.,* 168, 589, 1988).

The preferential utilization of D-proximal V_H gene families could be regulated by the bone marrow microenvironment present during the early stages of B-cell development or by intrinsic, genetically controlled factors. This paper attempts to differentiate between these two possibilities. For this purpose, fetal liver cells (from day 11 to day 19) and bone marrow cells from 3- to 6-week-old mice were added to established, primary stromal cell layers. The fetal liver and adult bone marrow cells were passed over G-10 columns to remove intrinsic stromal cell components. B cells from such cultures were studied at both 4 to 5 d (short-term cultures) and 8 weeks (long-term cultures).

In a preliminary experiment, LPS-stimulated 18-d fetal liver cells and adult bone marrow cells were examined by *in situ* hybridization utilizing various V_H gene family probes, and a C_u probe to identify LPS-induced blasts. The results confirmed earlier studies showing a preferential utilization of D-proximal V_H gene families by fetal B cells. LPS-stimulation of fetal liver and adult bone marrow cells following short-term culture upon stromal cell cultures resulted in the same results seen when the cells were stimulated with LPS fresh from the animal. This pattern was seen when fetal livers from 11, 13, and 18 d of gestation were used as sources of developing B cells. However, the long-term cultivation of fetal liver cells on bone marrow stromal cell cultures resulted in the expression of an adult pattern of V_H gene usage.

Since the fetal and adult pre B cells were cultured in the same microenvironment, the relative contribution of environmental vs. genetic influences upon V_H gene usage could be determined. The results indicate that environmental factors may not be involved in regulating the preferential utilization of D-proximal gene families by fetal B cells, and are consistent with the possibility that the B-cell progenitors present early during development differ from those present in the adult animal. The results from the long-term cultures suggest that the predominant B-cell progenitors present in fetal liver are short lived compared to the B-cell progenitors expressed by the adult animal. These results favor the importance of inherent genetic factors controlling the expression of V_H gene segments during the early stages of the development of the B-cell repertoire. *E. Charles Snow*

Novel Post-Translational Regulation of TCR Expression in CD4⁺CD8⁺ Thymocytes Influenced by CD4

J. S. Bonifacino, S. A. McCarthy, J. E. Maguire, T. Nakayama, D. S. Singer, R. D. Klausner, and A. Singer

Nature, 344, 247—251, 1990   2-18

Most immature CD4+CD8+ and CD4−CD8− thymocytes express few or no T-cell antigen receptor (TCR) complexes on their surfaces. Coincident with their maturation to CD4+CD8− or CD4−CD8+ phenotypes, these cells acquire large numbers of surface TCR complexes. It is unknown what mechanisms regulate these changes during intrathymic development, although previous work has shown that injection of anti-CD4 monoclonal antibodies into mice results in dramatic increases in the number of TCRs on immature CD4+CD8+ thymocytes. The work presented here investigated the mechanisms for increased expression of TCR on these cells.

Anti-CD4 monoclonal antibodies were injected into mice, resulting in increased numbers of assembled TCR complexes on CD4+CD8+ thymocytes. Northern blot analysis showed that these thymocytes had lower levels of RNA encoding most of the TCR chains than did cells from uninjected mice. Pulse chase experiments using [35S]methionine showed that treatment with antibodies reduced or did not affect biosynthesis of β-, CD3-, γ-, δ-, ε-, and ζ-chains. When the survival of newly synthesized, CD3-γ subunits was examined, rapid degradation was found with control thymocytes labeled *in vitro,* although most protein from these cells was not rapidly degraded. Pulse chase experiments with thymic fragments showed that *in vivo* treatment with anti-CD4 resulted in up to 220% increases in relative survival of CD3-γ chains of thymic fragments. Biochemical experiments showed that degradation of these TCR subunits resembled the characteristic selectivity of the endoplasmic reticulum (ER) pathway.

These experiments provide evidence that the increase of surface expression of TCR in thymic development of CD4+CD8+cells occurs via decreased destruction of newly synthesized receptor within the ER. The underlying mechanism for this novel regulatory activity of the ER is not known.

♦ This study was based upon the observation that injection of anti-CD4 antibody into mice results in the increased expression of TCR upon CD4+CD8+ thymocytes. This was shown to be an absolute increase in surface-expressed TCR proteins (alpha, beta, and CD3 complex). This increase was not due to the increased transcription of the relevant TCR genes. Pulse-chase experiments revealed that thymocytes from anti-CD4-treated mice were synthesizing the proteins which comprise the TCR complex at the same rate as the nontreated controls. Therefore, this increase was not due to elevated RNA transcription, synthesis, or assembly of TCR proteins.

The protein components of the TCR complex in nontreated thymocytes were found to be rapidly degraded. In fact, anti-CD4 treatment resulted in up to 220% increase in survival of the TCR proteins. A careful examination of this degradation process revealed that the TCR proteins were being degraded within the endoplasmic reticulum and, therefore, never reached the surface membrane. This study indicates that the ligation of CD4 (and presumably CD8) on CD4$^+$CD8$^+$TCR$^-$ immature thymocytes initiates a biochemical process by which the degradation of TCR proteins is prevented. This allows the TCR complex to form and to be expressed at the surface membrane. These results are consistent with a featured role for CD4 or CD8 molecules as signaling proteins which participate during the continued differentiation of CD4$^+$CD8$^+$ immature thymocytes. *E. Charles Snow*

Developmental and Environmental Regulation of a Phenylalanine Ammonia-Lyase-b-Glucuronidase Gene Fusion in Transgenic Tobacco Plants
X. Liang, M. Dron, J. Schmid, R. A. Dixon, and C. J. Lamb
Proc. Natl. Acad. Sci. U.S.A., 86, 1989   2-19

The activity of phenylalanine ammonia-lyase (PAL) in plants is highly regulated in response to a variety of developmental and environmental cues. The work investigated the role of the promoter region of PAL2, one of three PAL genes. This was accomplished by transforming leaf disks with a gene fusion construction containing 1.1 kilobases of the sequences upstream to PAL, fused in-frame with the coding region of the reporter gene β-glucuronidase. The resulting distribution of β-glucuronidase activity in transgenic plants was used to assess the activity of the PAL2 promoter.

Transgenic plants had low levels of β-glucuronidase activity in leaves and high levels in roots and stems. Flowers had low levels of activity in sepals and ovaries and high levels in anthers, stigmas, and petals, especially the pigmented parts. This distribution of activity within the plant closely resembled the pattern of endogenous PAL2 gene expression.

Activity of β-glucuronidase in transgenic plants was induced by wounding of tissues from vegetative organs and by illuminating plants that had been previously maintained in darkness. This also resembled PAL2 gene regulation. *In situ* histochemical analysis showed β-glucuronidase activity located in a band of tissue corresponding to the region of cell proliferation at the apical tips of roots, in prexylem cells of stems, and at the vascular connections between leaf primordia and stems.

These results provide evidence that developmental spatial and temporal regulation of PAL2 activity is largely dependent on sequences 5¢ to the gene. This promoter apparently can transduce various cues to regulate gene expression.

♦ Plant developmental biology has progressed in parallel with the progress in animal development. Our insight into animal development has shown us that the promoter region of a given gene contains information necessary for the correct transcriptional regulation of that gene's expression. The current article shows that the promoter region of a plant gene also contains such information. The authors design a unique DNA fusion construct and use it to generate a transgenic plant. Upon investigation, they show that the fusion gene product is expressed temporally and spatially in a pattern predicted by the fusion gene promoter. This data suggests that the promoter region contains the appropriate signals necessary to direct the accurate developmental and environmental expression of this particular plant gene. This observation is important because it suggests that appropriately regulated expression of genes introduced into plants can be attained. *Joel M. Schindler*

Developmental Cell Biology 3

INTRODUCTION

Cell biology remains at the core of development. When one analyzes phenotype, it is ultimately the behavior of the constituent cells that inform us if development is either normal or abnormal. The cell is the ultimate developmental signal transducer — how a cell receives stimuli and responds to them has enormous development consequence.

Growth factors have become an increasingly important area of investigation. As their numbers increase and the membership in superfamilies continues to grow, the diversity of their functions suggests that as a class, they represent extremely important bioregulators. The cellular responses to growth factors are extensive and can include, but are not limited to, proliferation, differentiation, elongation, or relocation. In fact, their are few cellular responses not associated with some type of growth factor. The specificity of growth factor receptors and the relationship between receptors and their ligands also has developmental importance. Thus, as modulators of cell behavior, they play important developmental roles.

The developmental consequence of various cell behaviors is addressed in several of the articles included in this chapter. The extent to which cell movement impacts development is discussed, as is the developmental role of controlled meiotic and mitotic cell divisions. Second messengers, such as cAMP, and the effect phosphorylation of specific proteins have on development are examined. The biological nature of stem cells is discussed and the role cell death plays in development is addressed.

In aggregate, many different cell behaviors are presented, all exhibiting different developmental outcomes. In this regard, it is important to remember that all those behaviors, correctly integrated in space and time, result in normal biological development.

Structure and Function of the Cytoskeleton of a *Dictyostelium* Myosin-Defective Mutant
Y. Fukui, A. De Lozanne, and J. A. Spudich
J. Cell Biol., 110, 367—378, 1990   3-1

Myosin has been implicated in all aspects of non-muscle cell motility.

Previous work from this laboratory resulted in the creation of a myosin-defective *Dictyostelium* mutant whose cells contain only the truncated myosin fragment *hmm* instead of the normal myosin heavy chain. These *hmm* cells were created by insertional mutagenesis of the single genomic copy of the myosin heavy chain gene, and, while defective in development and cytokinesis, were viable and capable of chemotactic movements. Continuing their investigations into the role of myosin in non-muscle cell motility, these workers studied the organization of the truncated hmm protein, actin, and microtubules in mutant *hmm* cells.

These mutant cells showed diffuse cytoplasmic distribution of hmm protein, with no conventional myosin filaments visible. Actin distribution appeared normal during spreading, migration, cell-cell adhesion, and phagocytosis. Cytoplasmic microtubules were abnormal, as were microtubule networks. Microtubules often penetrated into cortical lamellipodial regions of F-actin accumulation.

Little behavioral or morphological polarity was exhibited by the mutant cells, with random extension of lamellipodia and pseudopodia. In response to cAMP, the cells formed surface blebs; this occurred more slowly than the "cringing" of wild-type cells. While wild-type cells round-up in response to cAMP, these mutants became irregular but did not round up when stimulated. Mutant cells were unable to cap Con-A receptors or to contract their cortical cytoskeleton. The cells were able to actively phagocytose bacteria.

Dictyostelium cells lacking normal myosin heavy chain are defective in microtubule organization, in contractile events leading to cytokinesis, and in responding to cAMP. Conventional myosin appears to play an important role in cortical motile activities of *Dictyostelium*.

Myosin I Is Located at the Leading Edges of Locomoting *Dictyostelium* Amoebae

Y. Fukui, T. J. Lynch, H. Brzeska, and E. D. Korn
Nature, 341, 328—331, 1989   3-2

Ameboid movements of eukaryotic cells occur by extension of lamellipodia and pseudopodia at the leading edges and retraction at the trailing edges. Much is unknown about the molecular and structural mechanisms of these movements. To investigate this matter, the work reported here studied the distribution of filamentous and nonfilamentous myosin in migrating *Dictyostelium* amoebae.

The subcellular distributions of the two myosin isozymes were assessed using immunofluorescence microscopy with specific antibodies. In normal migrating cells, filamentous myosin II was found concentrated in the posterior cortical region and was absent from lamellipodia and pseudopodia. Myosin filaments were scattered throughout the cytoplasm. Nonfilamentous myosin I was found concentrated at the leading edges of lamellipodia and pseudopodia, and diffusely distributed in the cytoplasm, especially the posterior. Myosin I was absent or nearly absent from regions behind the leading edge.

In cells undergoing cytokinesis, myosin II was found in the cleavage furrow and myosin I was found in the pseudopodia at the poles of the cells. Phagocytosizing cells showed strong staining for F-actin and myosin I in the "phagocytic cup", an early cortical projection.

These results provide evidence that myosin I is found in lamellipodia, pseudopodia, and phagocytic cups of normal *Dictyostelium* cells. In contract, myosin II appears concentrated in the posterior cortex of migrating cells and the contractile ring of dividing cells. It is possible that actomyosin-I-dependent forces occur at the leading edge and actomyosin-II-dependent forces at the trailing edge of migrating amoebae.

♦ Cell movement plays a central role in morphogenesis and is amenable to study in the amoebae of the cellular slime mold *Dictyostelium discoideum*. Major advances in the cloning of the genes encoding cytoskeletal proteins and the subsequent alteration of the genes by antisense mutagenesis or homologous recombination is allowing new insights into function of these molecules during development. Two papers by Fukui et al. present new insight into the organization and function of myosin in this organism.

The first paper takes advantage of a strain with a disrupted myosin II gene to demonstrate that myosin II is necessary for several important cell biological processes associated with the organization of the cell cortex. These include the rounding up of cells in response to cAMP, restriction of the cleavage furrow, capping of surface receptors and the establishment of cell polarity. The organization of the microtubules is particularly affected. Previous work has shown that these mutant cells were abnormal in development, but the corresponding defects at the cellular level were not known.

The related paper shows the distribution of myosin I and myosin II isoforms. They have distinct distributions in the cell with myosin I being in the lamellipodia and pseudopodia of moving cells as well as the phagocytic cup — all regions with well developed actin networks. In contrast, myosin II is localized in the posterior cortex of the migrating cell and the contractile ring. The data help to understand why mutants defective in myosin II are still able to extend cellular processes and undergo phagocytosis but have substantially impaired motility.

Continuation of this line of investigation will surely help to reveal the cell biological basis of development. *Stephen Alexander*

In Vivo Receptor-Mediated Phosphorylation of a G Protein in *Dictyostelium*

R. E. Gundersen and P. N. Devreotes

Science, 248, 591—593, 1990   3-3

Extracellular cyclic AMP functions in *Dictyostelium* development as a chemoattractant, a cell-cell signaling molecule, and an inducer of differentiation. Based on studies of mutants containing deletions or point mutations in the Gα2 gene, the G-protein α subunit of the cAMP receptor, Gα2, seems to be the major transducer of the cAMP signal. This work tested whether Gα2 undergoes covalent modification during cAMP stimulation.

A transient, time-dependent alteration in mobility during SDS-polyacrylamide gel electrophoresis was seen in Gα2 after incubation of cells with cAMP. This transition in mobility appeared after 20 s, peaked at 1 to 2 min, and returned to normal at 15 min. The transition showed adaptation and had a similar cAMP dose-response curve to other responses coupled to the cAMP receptor. The transition was not seen in cell lines transformed with cAMP receptor antisense DNA, which do not express cAMP receptors.

When cells were incubated with ^{32}P-labeled inorganic phosphate, specific immunoprecipitation of phosphorylated Gα2 was seen with a time course resembling the Gα2 electrophoretic mobility transition. Alkaline phosphatase treatment removed radioactive phosphorus from immunoprecipitated Gα2; this treatment restored electrophoretic mobility to normal. Electrophoresis on cellulose plates of partially hydrolyzed, phosporylated Gα2 resulted in detection of radioactive phosphoserine. When Gα2−-mutant cells were transformed with the Gα2 gene, the cAMP-induced phosphorylation of Gα2 was restored.

These experiments provide evidence that the Ga2 subunit of the cAMP receptor of *Dictyostelium* is phosphorylated at serine residues in response to cAMP. This modification appears to be reversible and may be involved in the regulation of the cAMP effector pathway.

♦ cAMP induces a number of important phenotypic changes during the development of *Dictyostelium discoideum*. These changes include chemotaxis and the induction and repression of specific gene expression. Overall, cAMP plays a significant role in the development of this organism. cAMP is bound by a specific receptor protein with seven transmembrane domains and thus appears to be a characteristic G protein-linked receptor. The paper by Gunderson and Devreotes supports this contention by showing that the Gα2 subunit is transiently phosphorylated (resulting in altered mobility on SDS gels) after addition of cAMP and that cell lines transformed with antisense sequences of the cAMP receptor are not phosphorylated. The authors suggest that the phosphorylation is an essential *in vivo* step in the activation of the G protein subunits, so that they can interact with their effector molecules. It will be interesting to test this hypothesis by constructing strains with Gα2 protein phosphorylation sites altered by *in vitro* mutagenesis. The altered genes can be transformed into the *frigid A* strain, which has its Gα2 protein deleted, to determine the phenotypic effects conferred by the mutagenesis. *Stephen Alexander*

The Exogastrula-Inducing Peptides in Embryos of the Sea Urchin *Anthocidaris crassispina* — Isolation and Determination of the Primary Structure

T. Suyemitsu, T. Asami-Yoshizumi , S. Noguchi, Y. Tonegawa, and K. Ishihara

Cell Diff. Dev., 26, 53—66, 1989   3-4

The molecular mechanisms for sea urchin gastrulation are unknown. Previous work from this laboratory demonstrated the existence of low molecular weight substances from the blastocoelic fluid of sea urchin embryos which induced extrogastrulation when present at very low concentrations. The resulting exogastrulas developed into pluteus larvae with extruded archenterons. The work described here was directed to the isolation and characterization of these extrogastrula-inducing peptides.

Embryo homogenates were chromatographed on DEAE-cellulose, resulting in three peaks of exogastrula-inducing activity. The homogenate contained a lethal factor that killed embryos, but this factor was separated from the active fraction by DEAE-cellulose chromatography. Three active fractions were obtained after several additional chromatographic steps. The apparent molecular weights based on SDS-polyacrylamide gel electrophoresis were 4800, 7200, and 5500/8300 (both obtained from a single fraction). Reversed-phase HPLC resulted in the purification of four peptides.

Two of the peptides were completely sequenced by manual Edman degradation. Both peptides contained six cysteine residues in common positions. When the disulfide bonds were reduced, the peptides lost their biological activity. The two peptides had similar numbers of amino acids and similar positions of cysteine residues as murine epidermal growth factor (EGF), but were otherwise of different sequence from EGF.

This work resulted in the isolation of four peptides from mesenchyme blastula of *Anthocidaris crassispina* and the sequencing of two of them. The normal role of these peptides during development is unknown.

Unusual Pattern of Accumulation of mRNA Encoding EGF-Related Protein in Sea Urchin Embryos

Q. Yang, L. M. Angerer, and R. C. Angerer
Science, 246, 806—808, 1989   3-5

Proteins with domains similar to epidermal growth factor (EGF) include many involved in specification of cell fate during embryonic development. Here, the messenger RNA encoding a new member of this family, expressed in embryos of the sea urchin *Strongylocentrotus purpuratus,* is characterized.

The encoded protein, SpEGF2, was predicted to comprise 325 amino acids, with a molecular mass of 36.9 kDa. The protein is primarily composed of four tandem repeats of an EGF-like domain, based on the positions of six cysteine residues. This simple molecule contains a hydrophobic leader, but lacks a potential transmembrane domain. It is predicted to be secreted into the blastocoel, in part because of its similarity to the exogastrulation-inducing peptides recently isolated from the sea urchin *Anthocidaris crassisipina* (see preceding paper by Suyemitsu et al.).

In situ hybridization experiments suggested that the SpEGF2 mRNA begins to accumulate at the blastula stage, only in ectoderm. In pluteus larvae, it was found in a spatial pattern that did not correspond to known histological borders. It was concentrated in the aboral ectoderm, especially at the pointed vertex opposite the mouth and at the border with the ciliary band. This distribution is similar to that of a mRNA encoding a putative transcription factor containing a homeodomain. The SpEGF2 mRNA is also expressed inhomogeneously among oral ectoderm cells, being concentrated in cells adjacent to the anal side of the mouth.

While the exact developmental role of SpEGF2 is unknown, accumulating evidence suggests that it functions in early gastrulation. Endodermal and secondary mesenchyme precursor cells may respond to SpEGF2.

♦ Work in these two laboratories converged on an interesting peptide found in sea urchin embryos that is related to epidermal growth factor. Suyemitsu et al. used a traditional biochemical approach to follow up observations made by several workers, that fluid from the blastocoel of embryos could cause early developmental arrest or the specific defect of exogastrulation. Although the peptides they purified were obtained from honogenates of whole *Anthocidaris crassispina* embryos, and thus have not been shown directly to reside in the blastocoel, they cause the same developmental defect as does the low molecular weight fraction from blastocoelic fluid. Two of the four peptides were purified sufficiently to determine their sequence, and to include them in the large family of EGF-related peptides.

Yang et al. selected the SpEGF2 cDNA for study because the mRNA was distributed in a pattern that did not correspond to recognized cell-type borders: At pluteus stage SpEGF2 message is nonuniformly distributed among cells of the aboral ectoderm, and demarcate different regions of the oral ectoderm. This phenomenon had previously been observed only for a mRNA encoding a homeobox protein (see this *Year Book 1990*, p. 245). Sequencing of this cDNA showed that it is very likely the homolog of the gene that encodes the *A. crassispina* peptides. Many proteins with very diverse functions include EGF-like domains. In many of these, which have mosaic structures including a variety of domains, the role of the EGF-like domain is not clear. In some of these cases (such as *notch,* and *delta* in *Drosophila,* and *lin 12* in *Caenorhabditis,* the protein is implicated in reception rather than origination of developmental signals. In other proteins, such as the blood clotting factors, it seems likely that the domain has a role unrelated to signaling. Thus, an important feature of the inferred SpEGF2 protein is that it contains only 4EGF-like domains and a putative hydrophobic leader region, thus implying that its function is executed exclusively via the EGF-like domains.

The immediate effect of blastocoel fluid from *A. crassispina* is to cause changes in the shape of cells in the vegetal plate (see Ishihara et al., *J. Exp. Zool.,* 220, 227—233, 1982), which appear to lose their apical-basal polarity. SpEGF2 mRNA accumulates during early blastula stage when this polarity is established. Thus, EGF2 protein may serve as a signal from differentiating ectoderm that is required for regulation of cell shape and cell division in the vegetal plate. Another suggestive observation is that the highest concentrations of SpEGF2 mRNA in the pluteus larva are in ectoderm cells adjacent to the growing ends of the spicules. *Robert C. Angerer*

Sequence of an Unusually Large Protein Implicated in Regulation of Myosin Activity in *C. elegans*

G. M. Benian, J. E. Kiff, N. Neckelmann, D. G. Moerman, and R. H. Waterston

Nature, 342, 45—50, 1989   3-6

The *unc-22* mutation of *Caenorhabditis elegans* results in a nearly constant twitching of body muscles that contract transiently but do not develop normal contractions. All known extragenic suppressors of this mutation are missense mutations of the myosin heavy-chain gene. The protein product of *unc-22* , named twitchin, has a predicted molecular mass of about 600,000 and is found in the myosin-containing regions of muscle cells. To better understand twitchin, and perhaps other very large proteins found in vertebrate muscle, such as titin, nebulin, and dystrophin, the sequence of the *unc-22* gene was determined.

Over 47,000 kilobases (kb) of DNA were sequenced, as was a complementary DNA. The sequence was predicted to contain 14 exons, with the largest over 9 kb in length. The predicted protein is largely composed of two repeated motifs. The first motif is about 100 residues, found in 31 copies, all rich in proline; the second is about 93 residues and found in 26 copies. One region of twitchin devoid of either motif was found to be homologous to the catalytic domain of protein kinases, especially chicken smooth-muscle myosin light-chain kinase. The latter protein also contained both twitchin motifs.

The first twitchin motif resembles sequences in the neural cell adhesion molecule (N-CAM); the second twitchin motif resembles the immunoglobulin superfamily, especially the subset including N-CAM, myelin-associated protein, and the Fc receptor. Both motifs were found in titin, a protein of the myofilament lattice of vertebrate striated muscle probably bound to thick filaments.

Twitchin may have evolved by fusion of primordial gene motifs, followed by duplication and divergence. The strong homology of twitchin to the catalytic domains of protein kinases suggests that twitchin functions to phosphorylate a muscle protein, presumably the regulatory myosin light chain. The large size of the protein may imply that it has other functions as well.

Functions of the Myosin ATP and Actin Binding Sites Are Required for *C. elegans* Thick Filament Assembly
A. Bejsovec and P. Anderson
Cell, 60, 133—140, 1990   3-7

An understanding of macromolecular assembly is often aided by genetic analysis. To study muscle assembly and function, these workers previously described EMS-induced mutations in the *C. elegans unc-54* myosin heavy chain gene. These mutations resulted in altered myosin heavy chains and disruptions of assembly of thick filaments and sarcomeres. The *unc-54(d)* mutations resulted in dominant, muscle-defective phenotypes. In the work described here, 31 alleles of *unc-54(d)* were sequenced and implications from these sequences were drawn about the process of thick filament assembly.

Some 31 *unc-54(d)* mutations were localized using single base mismatch analysis and sequenced using the polymerase chain reaction. All these alleles contained single base substitutions within the globular head region of myosin (Figure 3-7). The 10 most strongly dominant alleles, those that most disrupt assembly of myosin heavy chains A and B, all contained mutations in a glycine-rich region highly conserved among nucleotide binding proteins. This region is essential for nucleotide binding and/or hydrolysis. Five other alleles of *unc-54(d)* contain altered

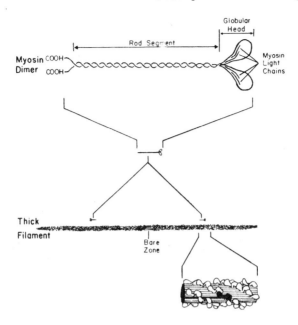

FIGURE 3-7. Myosin and thick filament structure. Vertebrate thick filaments are about 1.5 μm in length and consist of several hundred myosin dimers. *C. elegans* body-wall thick filaments are approximately 10 μm in length. *C. elegans* thick filaments contain, in addition to myosin, paramyosin (Waterston et al., 1974) and core protein (Epstein et al., 1985). (From Benian, G. M. Kiff, J. E., Neckelmann, N., Moerman, D. G., and Waterston, R. H., *Nature*, 342, 45—50, 1989. With permission.)

residues in a region that contributes to the ATPase of myosin. Five additional *unc-54(d)* alleles contained mutations at the major site of actin binding.

The *unc-54(d)* appear to be defective in some aspect of head function required for filament assembly. These results suggest that functions of the myosin ATPase are important for the assembly of thick filaments and sarcomeres. Actin-myosin interactions may also be involved in the assembly process.

♦ These two papers present some rather unexpected findings on myosin filament assembly and regulation in *C. elegans* muscle. Bejsovec and Anderson found that all of the 31 dominant mutations that disrupt myosin heavy chain assembly into thick filaments map to the globular head of myosin. The strongest alleles map to the ATP-binding site, and several other alleles map to the site that binds actin. These results are surprising because *in vitro* studies have demonstrated that the rod segments of myosin are both necessary and sufficient for assembly *in vitro*. The results of Bejsovec and Anderson clearly point to a critical role of the myosin head *in vivo*. Their results further suggest that actin-myosin interactions are an important part of the assembly process and that the myosin ATPase must be functional. The paper by Benian et al. presents the sequence of the *unc-22* gene, which encodes another abundant component of thick filaments. The *unc-22* gene product, called twitchin after the phenotype of *unc-22* mutants (constant twitching of the body muscles), is composed mainly of multiple copies of two different 100 amino acid motifs, one of which appears to belong to the immunoglobulin superfamily. It is the similarity of the C-terminal region to the catalytic domain of myosin light-chain kinases that suggests that twitchin functions by regulating myosin. Previous genetic, phenotypic, and cytological results had suggested that twitchin associates with thick filaments and participates in the con-traction-relaxation cycle. The sequence of *unc-22* further suggests that twitchin operates at the level of myosin light chain phosphorylation. *Susan Strome*

---

daf-1, a *C. elegans* Gene Controlling Dauer Larva Development, Encodes a Novel Receptor Protein Kinase
L. L. Georgi, P. S. Albert, and D. L. Riddle
Cell, 61, 635—645, 1990   3-8

When *C. elegans* is exposed to overcrowding and limited food supply, it may develop into an arrested, non-feeding stage called the dauer larva. The formation of dauer larva involves specific, reversible changes in virtually every tissue, resulting in a unique cuticle and an occluded mouth that render the dauer larva resistant to many chemicals. Animals bearing mutations in the *daf-1* gene form dauer larvae constitutively. The gene *daf-1* occupies an intermediate position in the hierarchy of genes believed to be involved in the pathway for neural transduction of environmental cues involved in dauer larva development (Figure 3-8). Since mutations in this gene do not result in defects in chemosensory behavior, nor in ab-normalities in morphogenesis of the dauer larva, and since the gene has been hypothesized to affect a cell or cells that receive neuronal signals, *daf-1* may play a role in signal transduction. To find out more about *daf-1*, these workers sought to clone and analyze the gene at the molecular level.

Insertions of transposon Tc1 resulted in tagged mutations in *daf-1;* the excision of these insertions correlated with reversions of the mutations. Once established, these tagged mutations and their flanking sequences were isolated. A fragment of these sequences was used to identify a low-abundance, 2.5-kilobase transcript on Northern blots of total RNA from wild-type worms. This same fragment was used to screen a genomic library, and the resulting clone was analyzed and sequenced.

The predicted protein contains 669 amino acid residues, and has a

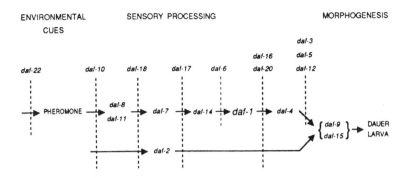

FIGURE 3-8.   A genetic pathway for dauer larva development. Genes are ordered based on epistatic relationships between dauer-constitutive mutations (drawn in the pathway) and dauer-defective mutations (represented as blocking the pathway). The pathway parallels events in dauer-inducing pheromone) and implementation of the appropriate developmental program. Mutations in *del-22* affect pheromone biosynthesis (Golden and Riddle, 1985); mutations in *daf-10* and *daf-6* result in abnormal chemosensory behavior and neuroanatomy (Albert et al., 1981), as well as failure to respond to the dauer-inducing pheromone (Golden and Riddle, 1984b); and mutations in *daf-9* and *daf-15* result in abnormal dauer larva morphogenesis (Albert and Riddle, 1988). (From Georgi, L. L., Albert, P. S., and Riddle, D. L., *Cell,* 61, 635—645, 1990. With permission.)

putative transmembrane domain. The C-terminal region resembles both tyrosine- and serine/threonine-specific protein kinases, especially that of the *raf* proto-oncogene family. Three *daf-1* mutants examined had alterations in the kinase domain.

This work provides evidence that *daf-1* encodes a membrane-bound protein kinase. This is consistent with the role of this gene product in signal transduction.

♦ More than 25 genes participate in dauer larva formation, which is *C. elegans'* response to overcrowding and starvation. Mutations in these *daf* genes result in either the inability to form dauers or constitutive dauer formation. Based on its position in the dauer larva genetic pathway, the dauer-constitutive gene *daf-1* is thought to affect cells that receive signals from amphidial neurons. Cloning and sequencing of *daf-1* by Georgi *et al.* has revealed that the *daf-1* gene encodes a transmembrane protein whose C terminal cytoplasmic domain probably functions as a protein kinase. The kinase domain most closely resembles (although the percent identical residues is low) the *raf* proto-oncogene family of serine/threonine protein kinases. Molecular analysis of other *daf* genes, in combination with the existing genetic and phenotypic information about *daf* mutants, ought to elucidate how environmental cues are sensed and transduced into the morphogenetic events of dauer larva development. *Susan Strome*

Mitotic Domains Reveal Early Commitment of Cells in *Drosophila* Embryos

V. E. Foe

Development, 107, 1—22, 1989   3-9

In *Drosophila* embryogenesis, all the nuclei of the syncytial egg divide synchronously for 13 cell cycles; but when cellularization occurs, this global synchrony ends. The work presented here describes the "mitotic domains" which undergo locally synchronous mitosis beginning with cycle 14. These domains were visualized using microscopic techniques, including time-lapse movies of live embryos and staining of fixed embryos with antibodies against tubulin and the *engrailed* protein.

Observation of many embryos showed that a highly reproducible pattern of mitosis occurs beginning with the 14th nuclear cycle and into gastrulation. At least 25 domains exist, with some consisting of a single midline cell cluster, most consisting of a pair of bilaterally symmetric cell clusters, and others consisting of metameric repeats of bilateral pairs of cell clusters.

In all embryos, the temporal sequence of the occurrence of mitoses in each domain was reproducible, as were the spatial boundaries of each domain. The boundaries of many domains coincided precisely with the boundaries of *engrailed* protein stripes. Cells of at least some domains shared specific morphogenetic traits distinct from cells in neighboring domains. In some cases, mitotic domains marked embryonic primordia. Gastrulation seemed to be effected by the specialized behavior of the different domains.

These findings suggest that the development of mitotic domains in the newly cellularized *Drosophila* embryos is an early expression of commitment to specific developmental fates. It is possible that the presence of mitotic domains may be a general feature of development.

♦ This is an exquisitely well crafted study of patterns of cell division during early embryonic development of *Drosophila*. It took Victoria Foe 4 years of meticulous observations to describe the 3 h of embryonic development that follow cellularization of the blastoderm. Spatial patterns of cell division in precisely staged embryos were visualized by immunofluorescent staining of tubulin. *Mitotic domains* can be identified as patches of dividing cells (mitotic spindles are clearly visible) surrounded by interphase cells. Precise mapping of each domain onto the surface of the embryo was achieved by measuring their position relative to zones of *engrailed* expression and relative to the ventral midline. Altogether, 29 domains are described. Two of these (domains A and B) do not divide during the developmental period covered by the study and cells in two others (domains M and N) divide less synchronously. The mitotic domains form in precise and reproducible chronological order and are numbered in order of appearance. Cells in domain 1 divide 70 min after the last global embryonic division (mitosis 13) and cells in domain 25 divide after 115 min. Thus, interphase 14 varies by as much as 45 min. The number of cells comprising each domain is constant and characteristic of each domain. Domain 10 consists of about 800 cells, domain 25 consists of metamerically arranged and bilaterally symmetrical single cells. The order of appearance of the domains and their position in the embryo are invariant. Certain mitotic domains are characterized by spindles that are either parallel or perpendicular to the surface of the embryo, in others the spindles are perpendicular to the long axis of the embryo, and in yet others, the direction of the spindle is random. Together the domains form a patchwork that covers the entire surface of the embryo.

There is strong suggestive evidence that the mitotic domains represent units of cell determination and this observation adds a great deal of interest to the work. In several cases the correspondence between a given domain and the fate of its constituent cells is striking. For instance, the boundaries of the domains frequently coincide with the boundaries of engrailed expression. In addition, all cells of a domain often share specific properties that distinguish them from cells of neighboring domains. For instance, all cells of a given domain invaginate into the interior of the embryo while their neighbors do not. The mitotic domains provide the earliest, most accurate and detailed landmarks for deduction of cell fate. They are by far more precise and easy to identify than the fate maps that were established previously by use of cell ablations and transplantations. Mitotic domains are likely to occur in other insect species (cf. Foe and Odell, *Am. Zool.*, 29, 617—652, 1989), thus providing an important tool for evolutionary studies.

It should be noted that there is a one-to-one correspondence between the mitotic domains an patterns of expression of the *string* gene (Edgar, B. A. and O'Farrell, P. H., *Cell*, 57, 177—187, 1989; discussed in last year's volume). The results of the present paper generate many extremely interesting and challenging questions concerning the molecular mechanisms that underlie the formation of the mitotic domains. These can best be summarized by quoting from the concluding remarks: 1. What mechanism orchestrates mitotic synchrony in each domain? 2. What specific combination of gene products among all those expressed within a domain, uniquely defines that domain? 3. What combinations of prior gene expression initiates expression of those domain-specific genes? 4. What cellular actions do those domain-specific genes orchestrate? *Marcelo Jacobs-Lorena*

Drosophila Ribosomal RNA Genes Function As an X-Y Pairing Site During Male Meiosis

B. D. McKee and G. H. Karpen

Cell, 61, 61—72, 1990

3-10

Despite cytological and genetical analysis, the molecular mechanisms responsible for homologous pairing of chromosomes during meiosis are poorly understood. Pairing of sex chromosomes during meiosis in *Drosophila* males has been a useful model system for studying this problem because of the absence of synaptonemal complexes and chiasmata. During meiosis, the centric X heterochromatin pairs with the base of the short arm of the Y; both of these sites of pairing are closely linked genetically to the nucleolus organizers. The work reported here provided a molecular test of whether ribosomal DNA is responsible for X-Y pairing in *Drosophila.*

A heterochromatically deficient X chromosome which disjoins randomly from the Y had a single X-linked ribosomal RNA (rRNA) gene inserted into it by P element-mediated transformation. This insert conferred upon the chromosome a significantly increased frequency of pairing and disjunction from the Y chromosome; pairing occurred at the site of the insertion. Excision of the single copy of rDNA restored the defects in pairing and disjunction. When single copies of the same gene were inserted at three autosomal sites, the resulting chromosomes did not stimulate disjunction or pairing of the heterochromatin-deficient X chromosome. Cells bearing X chromosomes containing two inserted copies of rDNA had further increases in frequency of pairing and disjunction compared to cells with a single copy of the insert.

These results show that rRNA genes can promote X-Y pairing and disjunction in *Drosophila.* This suggests that nucleolus organizers are the sites of X-Y pairing in wild-type flies.

♦ Proper chromosome segregation during meiosis is fundamental for the survival of the eukaryote. Many of the components involved in this process, such as centrioles, microtubules, and kinetochores, are beginning to be understood from a molecular as well as mechanistic point of view. Another essential process, however, the recognition and pairing of chromosomes, remains poorly understood. The complexity of the process and the probable involvement of numerous recognition sites contribute to the difficulties in devising proper experimental approaches. Meiosis in the *Drosophila* male has an important advantage for the study of chromosome segregation. This is because unlike what occurs in most other organisms, chromosome segregation during spermatogenesis does not depend on the formation of synaptonemal complexes and chiasmata (meiotic recombination does not occur in males). This property has allowed the partial mapping of chromosome regions required for proper pairing and segregation (disjunction) of chromosomes. For instance, pairing of chromosome 2 appears to be restricted to euchromatic regions while pairing of the X appears to require centric heterochromatin. In particular, pairing between the X and Y chromosomes was known to involve the region surrounding the nucleolus organizer of which the ribosomal RNA (rDNA) genes are part. Thus, it seemed possible that rDNA is involved in X-Y pairing. Taking into account that the X and the Y have each about 200 rRNA genes, the finding by McKee and Karpen that one rDNA unit is sufficient to promote chromosome pairing and disjunction is striking. Two rDNA units work better than one, and four to eight units are probably sufficient to promote complete disjunction. The relative location of the rDNA sequences on each of the chromosomes is apparently not important for proper pairing or segregation. However, the rDNA sequences clearly act in *cis,* since the presence of these sequences in other chromosomes has no effect on X-Y disjunction. Another question raised by these results is whether the presence of any homologous sequences on both chromosomes is sufficient to confer proper pairing. The answer is clearly "no". X-to-Y translocations involving hundreds of kilobases of either euchromatic or repetitive DNA have no influence on chromosome pairing. In other words, the effect is specific for rDNA. Having an assay system, it is now possible to ask which specific rDNA sequences are involved in the functional X-Y pairing and, with some luck, it may be possible to identify proteins that mediate chromosome-chromosome interactions. *Marcelo Jacobs-Lorena*

Cell Division in Malpighian Tubule Development in *D. melanogaster* Is Regulated by a Single Tip Cell
H. Skaer
Nature, 342, 566—569, 1989   3-11

Malpighian tubule development in *Drosophila* begins with 4 individual primordia of 20 to 25 cells each, outpouchings of the prospective hindgut first seen at 6.25 h after egg-laying, which proliferate to 484 cells per tubule well before the end of embryogenesis. Further development of the Malpighian tubules is achieved by an increase in cell size, accompanied by an increase in C number of the cells. This study examined cell division during the organogenesis of Malpighian tubules.

Incorporation of 5'-bromodeoxyuridine was used to identify replicating cells. At first, the tubules showed labeled cells scattered along their lengths, but this changed to a pattern in which only distal cells were labeled; after proliferation of these cells, and generation of the mature cell number by 10.25 h after egg-laying, the tubules increased in length by cell rearrangement, and all nuclei underwent several endomitotic cycles.

Early into the cell division period, a large cell at the distal tip of each

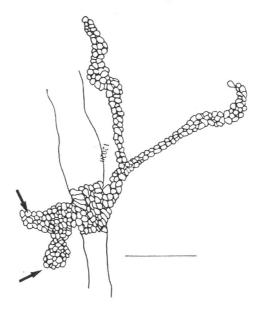

FIGURE 3-11. Camera lucida drawing of Malpighian tubules from an embryo in which two of the four tip cells were ablated. The tubulues without tip cells are arrowed. Both cell proliferation and tubule elongation are affected. Scale bar, 50 μm. Methods: Ablations were performed at the earliest time that the tip cell becomes prominent (7.75 h) and the embryo was cultured for a further 6 h in M3 medium at 25°C before dissection and staining with toluidine blue. Ablation of the tip cell was carried out by removing it, by suction, with a glass microplate. (From Skaer, H., *Nature,* 342, 566—569, 1989. With permission.)

Malpighian tubule became prominent. These tip cells did not cycle and divide; their position remained unchanged, while those of the other cells were changing during tubule elongation. When the tip cells of two of four tubules were ablated, the operated tubules arrested their development while the unoperated tubules developed normally (Figure 3-11).

The tip cells of *Drosophila* Malpighian tubules seem required for the normal pattern of cell division. In several ways, these cells resemble the distal tip cells of the gonad of *Caenorhabditis elegans,* whose functions seem to involve the *glp-1* gene product, a putative transmembrane protein that contains epidermal growth factor-like domains. Models of the mechanism of action of the tip cells should include the previously reported observation that the tip cells, rich in endoplasmic reticulum and Golgi apparatus, appear to be actively involved in secretion.

♦ The Malpighian tubules (MT) of insects are excretory organs akin to the mammalian kidney. They remove waste products from the body cavity, discarding them into the hindgut. In *Drosophila,* the MT originate during embryogenesis from four outpocketings in the prospective hindgut. Tubule primordia consist of about 20 to 25 cells and can be first detected at the extended germ band stage (6.25 h after egg laying). These cells divide to reach the mature number of 484 cells by 10.25 h. Thereafter, the cells enlarge, elongate, and increase their DNA content through a series of endomitotic cycles without further cell division. DNA synthesis (S phase) can be assayed in individual cells by allowing the tubules to incorporate bromodeoxyuridine (BUdR) into their DNA followed by detection of the incorporated BUdR by immunocytochemical techniques. Using this assay Skaer determined that at early developmental times cells along the entire length of the tubule incorporate DNA, while later (e.g., at 8.5 h of development) only cells in the distal quarter of the tubule incorporate DNA. About 1.5 h after the primordia are first detected, a single cell at the distal-most position of the tubule acquires morphological characteristics very different from all others: this cell is larger, protrudes from the distal tip, and never incorporates DNA or divides during the time covered by this analysis. Ablation of this tip cell (but not of other more "internal" cells) has dramatic consequences to the development of the tubules: no further growth or cell division is observed. Thus, Skaer clearly established that the tubule tip cell plays an important role in the regulation of Malpighian tubule cell division and morphogenesis. Cell-cell communication is certainly involved in this process. We should now hope that mutants that affect the appearance of the tip cell or the development of the Malpighian tubules become available to gain more insight into this fascinating process. A role analogous to the tubule distal cell appears to be played by the so-called distal tip cell of the *C. elegans* gonad, whose ablation arrests mitotic division of germ cell precursors. It is conceivable that similar regulatory mechanisms operate in both cases. *Marcelo Jacobs-Lorena*

Ectopic Expression of the Proto-Oncogene *int-1* in Xenopus Embryos Leads to Duplication of the Embryonic Axis
A. P. McMahon and R. T. Moon
Cell, 58, 1075—1084, 1989   3-12

The well-conserved *int-1* gene is a proto-oncogene activated by mouse mammary tumor virus, whose expression in early murine development is restricted to neural cells located at the dorsal midline of the neural tube after closure. The *Drosophila* homolog of *int-1* is the segment polarity gene, *wingless (wg)*, which is required in each segment for the establishment of posterior pattern elements. The protein product of the *wg* gene, possibly a secreted protein, may maintain expression of the *engrailed* gene. To further investigate the role of *int-1* in vertebrate development, normal *int-1* expression in *Xenopus* development was deregulated.

When *Xenopus* eggs were injected with either murine wild-type *int-1* RNA or with RNA encoding the *int-1* gene product fused with an epitope of the human *c-myc* protein, most of the resulting embryos survived and formed apparently normal gastrulas. However, the resulting neurulas reproducibly developed neural plates with bifurcated anterior portions and enlarged posterior portions. Eggs injected with an RNA that encoded a truncated *int-1* polypeptide developed normally. Whole-mount immunochemistry for the *myc* epitope showed that embryos injected with *int-1-myc* RNA had widespread distribution of *int-1-myc* protein, with concentrations in the neural folds of the neurula stage. When embryos were injected with RNA that encoded an *int-1* protein mutated at a single conserved C-terminal cysteine, they developed normally. When embryos were injected with RNA that encoded an *int-1* protein lacking the putative signal sequence, they developed normally; but when they were injected with an RNA that encoded an *int-1* protein fused to an unrelated signal peptide, they developed bifurcated neural plates. Examination of sections from neurulas injected with *int-1* RNA showed that notochords and somites had been duplicated.

These findings provide evidence that ectopic expression of murine *int-1* in *Xenopus* embryos results in dual axis formation. This effect seems to depend on translation of a functional *int-1* protein. It is possible that ectopic *int-1* expression splits the axis organizer region first described by Spemann and Mangold.

The Anterior Extent of Dorsal Development of the *Xenopus* Embryonic Axis Depends on the Quantity of Organizer in the Late Blastula
R. M. Stewart and J. C. Gerhart
Development, 109, 363—372, 1990   3-13

In amphibians, gastrulation is established by the organizer region of the marginal zone between the hemispheres. The organizer occupies the dorsal lip of the blastopore, the sector that begins the gastrulation process. It has long been known that transplantations of exogenous organizers into early gastrulas results in the development of a complete secondary body axis, and it is assumed that the endogenous organizer functions similarly in the normal development of the primary axis. While many studies have analyzed the qualitative inductive properties of the organizer, few have studied the effects on the anterior-posterior pattern of varying the quantity of the organizer. The latter question was the subject of the investigation presented here.

These experiments involved a novel technique. Late blastula embryos were bisected along the animal-vegetal axis at various angles to the bilateral plane; these halves were recombined to form full-sized embryos with normal-sized marginal zones (Figure 3-13). Some recombinants were formed from bisections of "ventralized" late blastulas, derived from UV-

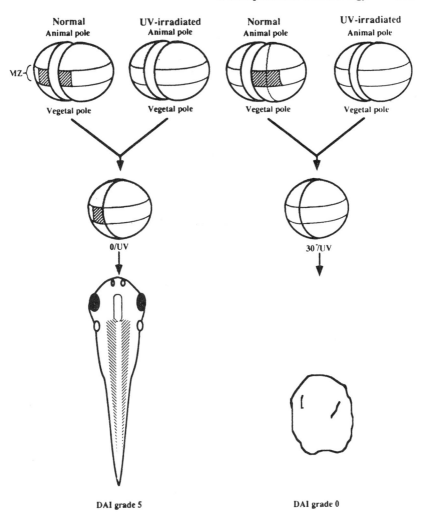

FIGURE 3-13. The surgical operations to produce lateral half/UV (0/UV) and 30˚/UV recombinants. Late blastula embryos are cut in half through the animal-vegetal axis and recombined in the correct animal-vegetal orientation. The organizer region of the marginal zone (MZ) is cross-hatched. On the far left, the normal embryo is cut along the dorsal midline. Third from the left, the normal embryo is cut 30° away from the dorsal midline, but still through the animal-vegetal axis (position of dorsal midline indicated by dotted line). Recombinant embryos are incubated to stage 34¹ and scored for the anterior extent of development of the body axis. (From Stewart, R. M. and Gerhart, J. C., *Development,* 109, 363—372, 1990. With permission.)

irradiated eggs, that differentiate only into ventral cell types and thus seem to contain no organizer tissue.

Recombinants lacking regions within 30° of the dorsal midline became ventralized embryos, almost completely lacking dorsal structures. Whenever some dorsal structures were missing, they were deleted from the anterior end. When recombinant embryos containing less than sufficient organizer amounts were analyzed, it was found that the extent of anterior development of dorsal structure correlated with the amount of organizer.

These experiments provide evidence that the organizer region of the late blastula of *Xenopus* is 60° wide, centered on the dorsal midline. The lateral width of the organizer seems to determine the anterior extent of development of dorsal structures.

♦ The concept of an "organizer" region in amphibian embryos has been around since Spemann and Mangold demonstrated its existence in the 1930s. We know the organizer is located more or less in the dorsal quadrant of the marginal zone, and that transplanting this region into a host blastocoel can result in the induction of a Siamese twin embryo on the host ventral side. We also know (Cooke et al., 1987; see Yearbook 1989) that ectoderm treated with XTC-derived mesoderm inducer (activin A) can act as an organizer in the same context. There is, however, not much known about how the organizer works. The papers by McMahon and Moon and by Stewart and Gerhart provide some insight into this question from two very different perspectives.

The first paper is one of most dramatic and most convincing examples of "artificial genetics" in the *Xenopus* system (see figure). The *Drosophila* homolog of *int*-1, *wingless,* is involved in pattern formation in fly embryos, and it is clear from these results that *int*-1 does something important in vertebrates. Since injection of mouse *int*-1 mRNA into fertilized eggs duplicates the anterior and enlarges the posterior components of the dorsal axis, and considering that a superficially similar effect on embryos can be elicited by transplanting a second organizer close to the endogenous one, it is reasonable to conclude that *int*-1 is involved in organizer establishment or function. On the other hand, *Xenopus int*-1 gene expression in normal embryogenesis is not detectable by RNA blots until midneurula, after the primary axis specification events have taken place. More data will be needed before this puzzle can be solved.

Stewart and Gerhart take a more classical approach to studying the organizer. Their experimental embryos with reduced organizer volume look quite similar to the ventralized embryos generated by ultraviolet irradiation of fertilized eggs, i.e., reduced anterior and dorsal structures. This fits with a general model in which the extent of cortical rotation determines the dimensions of the organizer, which in turn determines the extent of dorsoanterior development. Unfortunately, these investigators did not continue the series to include embryos with greater than normal organizer volume: it would be interesting to see what kind of embryos would result, i.e., hyperdorsalized embryos such as those that result from treatment of blastulae with lithium salts, or bifurcated embryos similar to those obtained by *int*-1 RNA injection. *Thomas D. Sargent*

Identification of a Potent *Xenopus* Mesoderm-Inducing Factor as a Homologue of Activin A

J. C. Smith, B. M. J. Prive, K. Van Nimmen, and D. Huylebroeck   3-14

The mesoderm-inducing factors (MIF) of *Xenopus* include members of the fibroblast growth factor family and of the transforming growth factor type β family. The most potent of those from the latter family is XTC-MIF, produced by *Xenopus* XTC cells. The work here characterized purified XTC-MIF.

This factor was purified by previously published methods which resulted in a single peak on reverse-phase HPLC and as a single band on polyacrylamide gel electrophoresis under both reducing and non-reducing conditions. Once purified, XTC-MIF induced expression of muscle-specific actin in animal caps at concentrations as low as 0.2 ng/ml, less than 1/10 the concentration of TGF-β required for the same effect.

N-terminal amino acid sequencing of XTC-MIF showed a single sequence identical in at least eight of the first ten residues to mammalian activin A. In a bioassay for growth inhibition of mink lung epithelial cells, TGF-β was at least 300 times more potent than purified XTC-MIF. In stimulating the release of follicle-stimulating hormone from pituitary cells, XTC-MIF had an ED_{50} similar to that of activin A. Like recombinant activin A, but unlike TGF-β, XTC-MIF stimulated the differentiation of K562 cells into hemoglobin-synthesizing cells.

These findings provide structural and functional evidence that the mesoderm-inducing factor from *Xenopus* XTC cells is a homologue of activin, and not TGF-β. The authors hypothesize that it may be common for organisms to use the same factors transiently in development and again in regulating or maintaining the adult.

Mesodermal Induction in Early Amphibian Embryos by Activin A (Erythroid Differentiation Factor)

M. Asashima, H. Nakano, K. Shimada, K. Kinoshita, K. Ishii, H. Shibai, and N. Ueno

Roux's Arch. Dev. Biol., 198, 330—335, 1990   3-15

In all vertebrates, mesoderm development is believed to be controlled by mesoderm-inducing factors. Recently, it has been suggested that some members of the transforming growth factor β family of proteins have mesoderm-inducing activity. To investigate this issue, the present study tested the mesoderm-inducing effects of two of these proteins: activin A (erythroid differentiation factor) and inhibin A.

Explants of ectoderm from the animal pole region of stage 9 *Xenopus laevis* blastulae were cultured in media with or without various exogenous factors. Subsequent formation of mesodermal tissues was assayed grossly, microscopically, and by indirect immunofluorescence, using a polyclonal antibody specific for muscle myosin.

Most explants treated with human recombinant activin A swelled, formed spheres, elongated, and developed various mesodermal derivatives. Explants treated with low doses of activin A developed blood cells, mesenchyme, mesoblasts within coelomic epithelia, and muscle cells; those treated with high doses often produced notochords and muscle blocks. Cultured explants varied greatly in the particular mesodermal structures they produced.

Bovine inhibin A, even at high doses, had no major morphological or histological effects on the explants, although it did result in swelling and in some dose-dependent formation of mesenchyme and epidermis. When explants were incubated in the presence of both inhibin A and activin A, inhibin A did not seem to reduce the mesoderm-forming activity of activin A.

The results of these experiments provide evidence that activin A, but not inhibin A, has mesoderm-inducing activity. Inhibin A does not seem to affect the activity of activin A. The lack of uniformity in mesodermal tissues produced by explants treated with activin A may imply problems with the assay system.

Activin-Like Factor from a *Xenopus laevis* Cell Line Responsible for Mesoderm Induction
A. J. M. van den Eijnden-Van Raaij, E. J. J. van Zoelent,
K. van Nimmen, C. H. Koster, G. T. Snoek, A. J. Durston, and
D. Huylebroeck
Nature, 345, 732—734, 1990   3-16

Previous work has shown that cultured *Xenopus laevis* XTC cells produce a factor that induces mesoderm from isolated ectodermal explants. When the heterogeneous XTC cell line was subcloned, the different cell lines all produced a number of known growth factors, including transforming growth factor β, but only the XTC-GTX-11 line also produced mesoderm-inducing factor (MIF). This factor has been shown to be unrelated to TGF-β and fibroblast growth factor. However, activin A has recently been found to be a potent MIF. The work presented here tested the hypothesis that XTC-MIF is activin.

These experiments compared the activin-like activity produced by some of these cell lines with their MIF activity. Recombinant activin A induced morphological changes in cultured stage 10 *Xenopus* ectoderm explants, and histological examination showed that differentiation of notochord, muscle, and neural tissue had occurred. These were similar to the mesodermal derivatives induced by MIF from XTC and XTC-GTX-11 cells.

Since activin A is identical to erythroid differentiation factor (EDF) and to follicle-stimulating hormone-releasing protein, further experiments tested the activity of conditioned media from the XTC cell lines in these two assays. Media from XTC-GTX-11 cells stimulated release of follicle-stimulating hormone, and heat-treated media from these cells showed an EDF activity. Media from other cloned XTC cell lines, which did not contain MIF activity, had no significant effects in these assays.

These findings provide evidence that the presence of activin correlates with the MIF activity in conditioned media from cloned XTC cells. This suggests that XTC-MIF is the *Xenopus* homolog of mammalian activin.

The Biological Effects of XTC-MIF: Quantitative Comparison with *Xenopus* bFGF

J. B. A. Green, G. Hower, K. Symes, J. Cooke, and J. C. Smith
Development, 108, 173—183, 1990   3-17

Recent work has identified 2 mesoderm-inducing factors (MIF) from *Xenopus* XTC-MIF, produced by the *Xenopus* XTC cell line, and basic fibroblast growth factor (bFGF), present in the egg and early embryo. Results from most experiments, performed with partially purified factors, have suggested that XTC-MIF produces dorsal, intermediate, and ventral mesoderm, while bFGF produces only intermediate and ventral cell types. To test this possibility, the work described here involved a direct quantitative comparison of the effects of the two factors in their pure, endogenous forms on animal pole explants.

Both factors had similar dose-response curves for the induction of mesoderm from explants, as assayed by external appearance. However, XTC-MIF induced more dramatic gastrulation-like movements in the explants than did *Xenopus* bFGF (XbFGF). Histological analysis and RNAse protection studies for muscle-specific actin and N-CAM messengers showed that XTC-MIF was a potent inducer of muscle and nervous tissue, while XbFGF was nearly ineffective at similar concentrations. XTC-MIF frequently induced notochord, but XbFGF did not. Explants treated with XbFGF developed normal epidermis, mesenchyme, and mesothelium. Animal pole cells at stage 10 had lost their competence to respond to XbFGF. In contrast, competence to XTC-MIF was retained until stage 11.

These findings provide evidence that bFGF is not the sole MIF *in vivo,* and that the two factors have different activities. The qualitative and quantitative differences shown here between XbFGF and XTC-MIF are consistent with a model in which the latter acts as a dorsoanterior inducer and the former as a ventroposterior inducer, both interacting to form the body pattern.

♦ It is generally acknowledged that Jim Smith's discovery that the Xenopus XTC cell line secretes a potent, soluble inducer revolutionized the mesoderm induction field. This inducer, called XTC-MIF (mesoderm inducing factor) has become an indispensable tool to investigators interested in this problem, and much work has been carried out utilizing crude XTC-MIF. However, a key question has gone unanswered until recently: what is XTC-MIF?

Rosa and co-workers reported in 1988 (1989 Yearbook, p. 64) that the inducing activity of XTC-MIF could be neutralized, at least partially, by antibody highly specific for transforming growth factor β2 (TGFβ2), and concluded on this and other grounds that XTC-MIF was a *Xenopus* homolog of TGFβ2. This turns out to have been insightful, but not quite correct. Smith and colleagues have purified XTC-MIF and obtained sufficient amino terminal sequence data to unequivocally identify this protein as Activin A. The papers by van den Eijnden-Van Raaij et al. and by Asashima et al. present indirect but strong evidence supporting this conclusion. Activin A is a distant (35% homology) relative of TGFβ identified by its ability to stimulate pituitary cells to release follicle stimulating hormone. The active form of this growth factor is a homodimer of two β_A chains. There exists a related polypeptide, termed β_B, and activin B is a homodimer of this. The heterodimeric form also exists, and all have similar activities. When one of these β polypeptides associates with another chain, called α, the resulting dimer is inhibin A or inhibin B, respectively, and as the name suggests, these have an opposite effect on FSH release.

It must be kept in mind that XTC-MIF/activin was isolated from tissue culture cells, and there is no evidence yet that activin is present at significant levels in embryos. A different mesoderm inducer, fibroblast growth factor (FGF) is, however, found in frog embryos at biologically active concentrations (Slack and Isaacs, 1989; see 1990 Year Book). The paper by Green et al. compares the effects on ectodermal explants of purified FGF and activin, and generally confirms the notion that FGF induces a ventral, posterior mesoderm and activin a dorsal, anterior mesoderm. This qualitative difference has lead to models in which a nonuniform distribution of FGF and activin (or its equivalent in embryos) results in the establishment of the dorsoanterior-ventroposterior (DV) axis. As shown in the diagram below, this asymmetry could be achieved by segregating either the inducers (A) or the receptors (AR). Whatever its molecular basis, this polarization presumably occurs during the cortical rotation prior to first cleavage. It should be possible in the near future to test these ideas. *Thomas D. Sargent*

Binding of Brain-Derived Neurotrophic Factor to the Nerve Growth Factor Receptor

A. Rodrigues-Tébar, G. Dechant, and Y.-A. Barde

Neuron, 4, 487—492, 1990   3-18

Brain-derived neurophophic factor (BDNF) is a protein produced in the central nervous system (CNS) that seems to promote neuronal survival during development in the same way that the target-derived, neurotrophic protein nerve growth factor (NGF) is believed to act. This activity begins by binding to specific receptors on neuronal surfaces. Both BDNF and NGF have been shown to bind to high- affinity ($K_D \sim 10^{-11}$) and low-affinity ($K_D \sim 10^{-9}$) receptors. The recent cloning of the gene encoding BDNF showed that NGF and BDNF have similar amino acid sequences, including six cysteine residues located at identical positions. The present work was designed to study the binding characteristics of the two related factors to their receptors in the presence of the heterologous ligand.

A 1000-fold excess of NGF was needed to prevent the binding of BDNF to its high-affinity receptors on dorsal root ganglion neurons from 9-d-old chick embryos; conversely, a 1000-fold excess of BDNF was needed to prevent the binding of NGF to its high-affinity receptors on the same cells. When cultures of BDNF-responsive, non-NGF-responsive neurons from the placode-derived nodose gangion were assayed, NGF, in pharmacological concentrations, acted as a specific BDNF agonist. NGF and BDNF competed equivalently for binding to the same low-affinity receptors of sensory neurons isolated from dorsal root ganglia. Experiments with a fibroblast cell line stably transfected with the NGF receptor gene showed that the cells had acquired binding capabilities to both NGF and BDNF.

These results provide evidence that the low-affinity NGF receptor also binds BDNF with low affinity. The alteration that results in high-affinity binding is accompanied by a dramatic ability to discriminate between the two factors. It is possible that the receptors for both NGF and BDNF share a binding component, with an additional component involved in high-affinity binding.

♦ Brain-derived neurotrophic factor (BDNF) and nerve growth factor (NGF) are proteins that support survival of embryonic neurons. Molecular cloning of the two growth factors has revealed significant similarities; the molecules are structurally related and share 50% amino acid identity. The populations of cells affected by these growth factors are somewhat overlapping in the case of young neurons, though later in development, distinct subpopulations appear to be affected by BDNF and NGF, respectively. Both factors appear to bind to cells by means of a high-affinity receptor, as well as a low-affinity receptor. The gene encoding the low affinity NGF receptor has been cloned and is a single copy gene.

The similarities in ligand structure prompted Rodriguez-Tebar and colleagues to study the binding properties of BDNF and NGF to their receptors in response to the heterologous ligand. They found a major difference between the high- and low-affinity receptors when challenged with the other ligand. For both the high affinity NGF and BDNF receptors, 1000-fold excess of the heterologous ligand was necessary to achieve 50% reduction in binding. In contrast, essentially identical concentrations of the heterologous ligand caused 50% reduction of binding to the low-affinity receptor. Cells transfected with the cDNA for the low-affinity NGF receptor bind NGF and BDNF equally well.

Their major conclusion is that the low-affinity NGF receptor is also a low-affinity BDNF receptor. It is likely that the high-affinity NGF receptor is derived from the same gene product as the low-affinity NGF receptor. Although the mechanisms that lead to high-affinity binding are not yet clear, there appears to be a dramatic increase in the ability of the high-affinity receptors to discriminate between NGF and BDNF. Thus, related growth factors may share common receptors which are later modified to discriminate between their ligands. *Marianne Bronner-Fraser*

A v-*myc*-Immortalized Sympathoadrenal Progenitor Cell Line in which Neuronal Differentiation Is Initiated by FGF but Not NGF

S. J. Birren and D. J. Anderson
Neuron, 4, 189—201, 1990   3-19

A growing body of evidence has led to the hypothesis that a common developmentally restricted progenitor cell gives rise to adrenal medullary chromaffin cells and sympathetic neurons. A progenitor cell has been isolated from primary cultures of embryonic rat adrenal gland, using fluorescence-activated cell sorting (FACS) with specific cell surface monoclonal antibodies, that becomes a chromaffin cell in the presence of glucocorticoids, and can give rise to sympathetic neurons in the absence of these hormones. Much is still unclear on what is required for the neuronal differentiation of these cells. To study this issue, the present work was designed to establish and examine immortalized cell lines from these progenitor cells.

Sympathoadrenal progenitor cells were isolated from rat sympathetic ganglion priomordia by FACS and labeling with HNK-1, a cell surface antibody specific for tyrosine hydroxylase-positive cells. These cells were immortalized by infection with a v-*myc*-containing recombinant retrovirus, resulting in the establishment of ten separate c-*myc*-infected, adrenal-derived HNK-1-positive (MAH) cell lines. In these lines, all cells expressed a panel of markers characteristic of normal progenitor cells, including L1, N-CAM, and Thy-1.

Nerve growth factor (NGF) did not rescue MAH cells deprived of glucocorticoids, nor did it induce neurite outgrowth. Rather, basic fibroblast growth factor (bFGF) stimulated proliferation and morphological differentiation of the cells. The response to bFGF was inhibited by dexamethasone. After 1 week of exposure to bFGF, the cells died. MAH cells cultured with or without dexamethasone did not contain NGF receptor mRNA detectable by Northern blotting.

When MAH cells were grown in FGF plus NGF, or were grown first in FGF and then switched to NGF, a very small percentage of cells survived and became stable, post-mitotic neurons. This survival and differentiation was blocked by antibodies to NGF. Induction of NGF receptor mRNA was detected in two MAH cell lines exposed to FGF.

These findings provide evidence that sympathoadrenal progenitor cells can be immortalized and cultured with the retention of many of their developmental properties and potentials. The first stage of neuronal differentiation of these bipotential cells may require FGF, which induces NGF receptors and NGF responsiveness (Figure 3-19). The resultant cells may then mature into post-mitotic neurons in the presence of NGF.

♦ Sympathetic neurons and adrenal chromaffin cells both differentiate from a common progenitor cell derived from the neural crest. This sympathoadrenal precursor is a partially restricted cell that gives rise to adrenal chromaffin cells in the presence of glucocorticoids and can give rise to sympathetic neurons in the absence of glucocorticoids. After differentiation, the survival of sympathetic neurons is dependent on nerve growth factor (NGF). The factors that trigger neuronal differentiation of the sympathoadrenal sublineage are not clear, largely because of difficulties in obtaining pure populations of precursors in sufficient numbers to study.

To circumvent this problem, Birren and Anderson have produced immortalized sympathoadrenal progenitor cells lines. Sympathoadrenal progenitors were isolated using the HNK-1 antibody and fluorescence-activated cell sorting. The cells then were immortalized using v-myc-containing retrovirus. These lines can initiate neuronal differentiation in the presence of exogenous FGF, but not other growth factors including NGF. Furthermore, they possess no NGF-receptor mRNA. A small number of these cells survive for long periods of time, when they differentiate into sympathetic-like neurons responsive to NGF.

These results suggest that the acquisition of NGF responsiveness and dependence may occur sequentially in the sympathoadrenal lineage. FGF appears to initiate neuronal differentiation and induce the appearance of

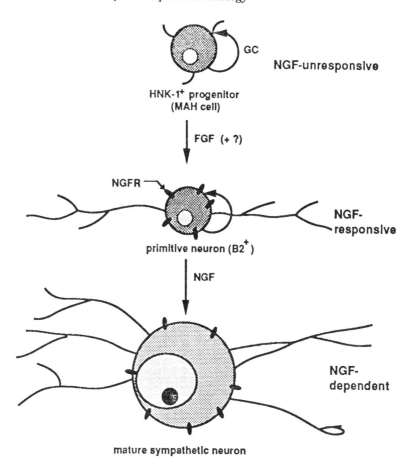

FIGURE 3-19. Schematic illustrating the stages of neuronal differentiation that may be controlled by FGF and NGF. The sympathoadrenal progenitors represented by MAH cells (top) extend neurites in response to FGF , but do not yet express NGF receptors. In the presence of glucocorticoid, these cells divide (circular arrow) but do not differentiate. Following treatment with FGF, a primitive neuron can be observed, which bears processes and which is still dividing. This cell type has been described in primary cultures previously (Anderson and Axel, 1986; Anderson, 1988). At least a subpopulation of these primitive neurons begins to express the NGF receptor. In the presence of NGF, these cells mature into postmitotic neurons with large cell bodies, prominent nucleoli, and extensive networks of neurites. Such neurons are then dependent on NGF for survival. (From Birren, S. J. and Anderson, D. J., *Neuron,* 4, 189—201, 1990. With permission.)

NGF receptors. This is followed by dependence of the neurons on NGF. Thus, FGF either may trigger or be required continuously for transition to an NGF-dependent state. Although the factors that function *in vivo* are not yet known, FGF appears to be a good candidate molecule for biasing the sympathoadrenal lineage along neuronal lines. *Marianne Bronner-Fraser*

Dynamics of Early B Lymphocyte Precursor Cells in Mouse Bone Marrow: Proliferation of Cells Containing Terminal Deoxynucleotidyl Transferase

Y.-H. Park and D. G. Osmond

Eur. J. Immunol., 19, 2139—2144, 1989   3-20

In B cell development, it has been proposed that three phenotypically distinct populations of cells that do not express μ heavy chains exist: those containing the nuclear enzyme terminal deoxynucleotidyl transferase but not expressing the B220 cell surface glycoprotein detected by monoclonal antibody 14.8 (TdT$^+$14.8$^-$ cells), those that are TdT$^+$14.8$^+$, and those that are TdT$^-$14.8$^+$. These populations may represent successive stages in the development of early B lymphocyte precursors. To help understand the genesis of primary B cells in murine bone marrow, this study analyzed the frequency, size distribution, proliferative properties, and production rates of TdT$^+$ cells.

TdT$^+$ cells from mouse bone marrow totaled 1.6% of all bone marrow cells, and were mainly medium-sized cells. Some were normally seen in metaphase, when TdT labeling was dispersed throughout the cytoplasm. When treated with vincristine sulfate to stop cells in metaphase, many of the larger cells accumulated in metaphase, and the cell size of the entire population increased.

About half the TdT$^+$ cells co-expressed the B220 surface glycoprotein. These cells were slightly larger than TdT$^+$14.8$^-$ cells, but both cell types included large mitotic cells whose frequency was increased by treatment with vincristine. Both populations were passing through cell cycles, but the TdT$^+$14.8$^+$ cells had a more rapid rate of entry into mitosis, a shorter apparent average cell cycle time, and about twice the cell production rate as the TdT$^+$14.8$^-$ cells. The total turnover of cells from murine bone marrow was 25×10^5 cells per day for TdT$^+$14.8$^-$ cells and 50×10^5 cells per day for TdT$^+$14.8$^+$ cells.

Together with previous findings, these results permitted the quantitation and description of the proliferative kinetics of three m$^-$ cell populations. The data provide evidence that early precursor cells committed to the B cell lineage express intranuclear TdT, that later B220 expression is added, and that subsequently TdT expression is abolished. During this period of development, the average apparent cell cycle time shortens, while a progressive expansion of cell production occurs. This study also provides estimates of the numbers of cell cycles undergone at each stage and the time required for these stages of development.

♦ The precursors of B cells have recently been divided into three subgroups which may represent successive stages of B-cell development. These subgroups are all negative for cytoplasmic mu heavy chain, and differentially express terminal deoxynucleotidyl transferase (TdT) and a 220-kDa cell surface glycoprotein referred to as B220. The order of appearance is apparently TdT+B220−, TdT+B220+, and Tdt−B220+. It is the TdT−B220+ cells which become mu positive, indicating that heavy chain gene rearrangement has occurred. The present study was conducted to determine the population dynamics of these three precursor B-cell subpopulations.

TdT+ bone marrow cells were shown by indirect immunofluorescence to be mainly medium-sized cells of 6 to 15 μm in diameter, representing 1.6% of all bone marrow cells. Fifty percent of these TdT+ bone marrow cells also expressed B220. The double positive cells were slightly larger and there was more of them in metaphase than the TdT+B220− cells. This was most obvious following the administration of vincristine, where the double-positive cells more rapidly accumulated in metaphase. These double-positive cells also displayed a more rapid passage through the entire cell cycle, resulting in twice the actual cell production with unit time compared to the TdT+B220− cells. The results of this study indicate that there is at least one mitotic division in each of the three phenotypic stages of precursor B-cell differentiation. Each succeeding subpopulation demonstrates a progressive increase in cell numbers, showing that there exists a substantial increase in numbers of precursor B cells prior to the rearrangement of the immunoglobulin genes. B-cell genesis, therefore, consists of at least 6 rounds of division over a 62-h period prior to mu chain synthesis. *E. Charles Snow*

Analysis of Thymic Stromal Cell Subpopulations Grown *In Vitro* on Extracellular Matrix in Defined Medium: II. Cytokine Activities in Murine Thymic Epithelial and Mesenchymal Cell Culture Supernatants

I. Eshel, N. Savion, and J. Shoham
J. Immunol., 144, 1563—1570, 1990   3-21

These authors have established two morphologically distinct primary cultures of mesenchymal (MC) and medullary epithelial cells (EC) from murine thymic stroma. These cultures can both be maintained on extracellular matrix and grown in defined media. The study reported here capitalized on the ability of these cultures to preserve the functional capabilities of the cells and permit an analysis of cell interactions between thymocytes and stromal cells. The culture supernatants from these cells were therefore assayed for cytokine production and effects on T cell function.

Neither culture supernatant contained interleukin-1 (IL-1), IL-2, IL-4, or interferon activity. Both culture supernatants contained IL-3 activity, with supernatant from epithelial cultures containing much more than that from mesenchymal cultures. Supernatants from both cultures contained granulocyte/macrophage colony stimulating factor. Thymic epithelial cells, but not mesenchymal cells, produced prostaglandin E_2.

Epithelial cell supernatant stimulated spontaneous $[^3H]TdR$ incorporation, Con A-induced proliferation and lymphokine production by thymocytes; mesenchymal cell supernatant had significant levels of activity only in stimulating spontaneous thymidine incorporation.

These findings suggest that the cell lines studied here may better preserve physiologic function than established cell lines. Different stromal cell types seem to have different effects on T cell maturation. Thymic stromal cells may act on thymocytes at different stages of their development by either cytokine secretion or by direct contact.

♦ This paper investigates some of the characteristics exhibited by soluble mediators secreted by primary cultures of murine thymic epithelial cells. These cells were MHC class I and II positive, and displayed many of the physical attributes previously described for epithelial cells in general, and thymic epithelial cells in particular. For these studies, culture supernatant (CS) was collected at various times after cultivation of the epithelial cells, and the CS were analyzed for growth-inducing activities on thymocytes and T cells.

The CS did not contain detectable levels of IL-1, IL-2, IL-4, or antiviral activity (ruling out interferons and IL-6). The CS did contain high levels of IL-3, with low levels of CSF (presumably GM-CSF). Also, the supernatants contained high levels of PGE_2. The CS from thymic epithelial cells was found to enhance the spontaneous proliferation of thymocytes and the Con A-induced proliferation of T cells. Also, the CS enhanced the secretion of IL-2 and IL-3 by Con A-stimulated thymocytes, and this occurred with the peanut agglutinate-negative population. This study provides evidence for thymic epithelial cell-derived soluble mediators functioning as positive growth stimulators of preactivated thymocytes. *E. Charles Snow*

Activation of a Suicide Process of Thymocytes through DNA Fragmentation by Calcium Ionophores and Phorbol Esters

H. Kizaki, T. Tadakuma, C. Odaka, J. Muramatsu, and Y. Ishimura
J. Immunol., 143, 1790—1794, 1989   3-22

It is not known why the majority of thymocytes die during their development while a minority proliferate and mature in the thymic cortex, but this process appears to be regulated. Some evidence suggests that activation of protein kinase C and elevation of cytoplasmic Ca^{2+} may be involved in T cell proliferation, but other work has implicated the calcium iono-phore A23187 in the suicide process, or apoptosis, of rat thymocytes. This apoptosis involves DNA fragmentation at linker regions. Recent work from this laboratory implied that the phorbol ester TPA induced DNA fragmentation of mouse thymocytes by activation of protein kinase C. The work presented here reflects further studies on the effects of calcium ionophores and phorbol esters on immature murine thymocytes.

The ionophore A23187-induced DNA fragmentation of thymocytes in a dose-dependent fashion, with corresponding effects on cell death. The fragmentation induced by A23187 was inhibited by cycloheximide, an inhibitor of protein synthesis, H-7, an inhibitor of protein phosphorylation, and actinomycin D, and an inhibitor of mRNA synthesis. The phorbol ester PDB induced DNA fragmentation in mouse thymocytes. When TPA was added to thymocytes in the presence of A23187, the DNA fragmen-tation induced by A23178 was dramatically reduced, and the incorpora-tion of thymidine was dramatically increased. Immature thymocytes seemed more susceptible to fragmentation by either A23187 or TPA than mature thymocytes.

These experiments provide evidence that either calcium ionophores or phorbol esters can induce DNA fragmentation in mouse immature lym-phocytes. Small amounts of TPA inhibited DNA fragmentation induced by A23187. DNA fragmentation induced by A23187 seems to require protein phosphorylation and protein and mRNA synthesis. It is possible that the intracellular signals of protein kinases and calcium ions may regulate the fate of developing thymocytes by their interactions.

♦ Probably one of the most exciting areas of work over the past year has centered upon understanding the process of apoptosis. Apoptosis refers to the process of programmed cell death and is manifested by condensation of the cell. During this process, there appears within the cells numerous membrane-bound vesicles (apoptotic bodies) representing the destruction of the cell from within. This process may represent a means by which the immune system permanently depletes some of its self-reactive clones by a process referred to as negative selection. This process has been best analyzed during the depletion of self-reactive thymocytes in the thymus and may occur as a sequence of the delivery of incomplete growth signals to thymocytes reacting to self antigens. This means that the elicitation of TCR-mediated signal pathways can lead both to cell death or cell activa-tion, depending upon the differentiative state of the cell. This concept is compatible with two-signal models of cellular activation in which the delivery of only signal one results in cellular paralysis or death. The more stringent requirement for multiple signals to activate cellular proliferation might help to minimize the accidental movement of inappropriate cells into the cell cycle. The work over the past year has begun to focus upon the mechanisms which mediate receptor-initiated programmed cell death.

Thymocytes were prepared from young adults and some were separated into peanut agglutinin (PNA) positive and negative fractions. The basic experimental design was to follow the levels of DNA fragmentation in the cells polyclonally stimulated with phorbol esters and calcium ionophores. DNA fragmentation increased over background levels by 4 h after the addition of a calcium ionophore (A23187). The extent of plasma membrane damage was assessed by following the release of lactate dehydrogenase (LDH). A23187 caused a dose-dependent increase in DNA fragmentation up to 1 μM, when it also elicited an increase in LDH release. Since thymocytes which die in nonstimulated cultures do not demonstrate increased DNA fragmentation, this change in DNA fragmentation following addition of the calcium ionophore is due to a cellular response to biochemical signals. Inhibitors of protein phosphorylation (H-7), RNA synthesis (actinomycin D), and protein synthesis (cycloheximide) all reversed the increase in DNA fragmentation induced by A23187.

The induction of DNA fragmentation by phorbol esters displayed a different kinetic pattern, with fragmentation detectable by 6 h and optimal fragmentation requiring the presence of the phorbol ester for 12 to 24 h. The addition of the phorbol ester to A23187-stimulated cells reduced the level of DNA fragmentation to background levels, while increasing the level of thymocytes entering the cell cycle. The induction of DNA fragmentation by calcium ionophore or phorbol ester was highest in the most immature (PNA$^+$) thymocyte populations.

These results suggest that the delivery of incomplete activation signals to thymocytes, particularly the most immature, results in the initiation of a cellular program which elicits the fragmentation of the cell's DNA. This cellular process requires RNA and DNA synthesis, as well as some level of kinase activity. Therefore, the inappropriate activation of thymocytes initiates a process of rapid cell death and may be responsible for some of the massive cellular turnover which occurs within the thymus. *E. Charles Snow*

Erythropoietin Retards DNA Breakdown and Prevents Programmed Death in Erythroid Progenitor Cells

M. J. Koury and M. C. Bondurant
Science, 248, 378—381, 1990   3-23

Erythropoietin (Epo) acts directly on erythroid progenitor cells to stimulate erythrocyte production, but the mechanism of its action is unknown. Previous work has shown that erythroid progenitor cells have different rates of DNA breakdown depending on whether they are cultured with or without Epo. The experiments described here studied the effects of Epo on the DNA of erythroid progenitor cells.

A homogeneous population of erythroid progenitor cells for study was isolated from spleens of mice infected with Friend leukemia virus (FVA cells). These cells were similar in their Epo dependence and in their rate of DNA synthesis to cells of the colony-forming unit erythroid stage of normal mice. FVA cells contained broken DNA whether or not they were cultured in the presence of Epo; but an analysis of the fate of [^{3}H]thymidine-labeled DNA showed that Epo retarded DNA fragmentation by a factor of 2.6. The fragmented DNA from FVA cells appeared to be double-stranded and of periodic sizes, consistent with the length of DNA associated with nucleosomes. When cells were incubated in the presence of thymidine concentrations permissive for DNA synthesis, the addition of Epo retarded the rate of DNA cleavage enough to maintain most DNA as high molecular weight molecules. The absence of Epo in similar culture conditions resulted in net DNA breakdown. DNA breakdown in the absence of Epo led to cell death, but FVA cells cultured with Epo survived and differentiated into reticulocytes.

These experiments provide evidence that erythropoietin controls erythrocyte production by inhibiting DNA breakdown that leads to cell death. Normal erythropoiesis seems to involve the process of apoptosis, or programmed cell death, which is retarded in a dose-dependent fashion by erythropoietin.

♦ For these experiments, an erythroid progenitor cell line (FVA cells) isolated from mice infected with Friend leukemia virus and exhibiting an erythropoietin (Epo)-dependent stage was used. The culture of FVA cells in the absence of Epo results in detectable DNA fragmentation by 8 h. This effect of Epo was shown to be upon the cleavage rate of DNA, not upon its repair rate. In fact, in the absence of Epo, the cleavage rate was increased by a factor of 2.6. The resultant fragments were double-stranded DNA fragments. The cells cultured with Epo were found to start dying by 12 to 16 h, although most of the cells did survive out to 24 h. In the presence of Epo these differentiate into reticulocytes by 48 h.

These results indicate that erythroid progenitors go through an Epo-dependent phase. This allows the overall rate of red cell production to be regulated by continuously altering the survival rate of the erythroid progenitors by regulating the available levels of Epo. *E. Charles Snow*

Anti-Immunoglobulin Antibodies Induce Apoptosis in Immature B Cell Lymphomas

J. Hasbold and G. G. B. Klaus
Eur. J. Immunol., 20, 1685—1690, 1990   3-24

Several phenotypically immature B cell lymphomas have served as model systems for the study of the inhibitory effects of antigen on B cells. The lines WEHI-231 and CH31 are sIgM$^+$ lymphomas whose growth is inhibited by anti-Ig antibodies; the mechanism of this effect is unknown.

In the experiments described here, the effects of anti-Ig antibodies was studied.

Culturing in the presence of anti-μ antibodies specifically resulted in about 90% inhibition of tritiated thymidine uptake in both cell lines. Many of the cells were arrested in the G1 phase of the cell cycle. Staining with propidium iodide showed that treated cells of both lines contained small, brightly stained bodies (apoptotic bodies) characteristic of cells undergoing apoptosis or programmed cell death. When the DNA from treated cells of both lines was run on an agarose gel, the classic ladder pattern of oligonucleosomes was seen. Lipopolysaccharide reversed the DNA fragmentation induced by anti-μ in both cell lines, just as lipopolysaccharide reverses the anti-Ig-induced growth inhibition of these cells.

These findings provide evidence that growth inhibition of both CH31 and WEHI-231 lymphomas by anti-m is accompanied by DNA fragmentation and formation of apoptotic bodies characteristic of programmed cell death. Apoptosis may be involved in the development of B cell tolerance to self-antigens.

♦ The treatment of two immature B lymphomas (WEHI-231 and CH31) with greater than 1 μg/ml anti-IgM inhibited greater than 90% of [3H]thymidine incorporation. Many of the cells appeared to arrest in the G1 stage of the cycle. By 24 h following the addition of anti-IgM to CD31 cells, 60% of the cells were dead and 90% contained apoptotic vesicles. For similar results, WEHI-231 cells required exposure to anti-IgM for at least 48 h. The anti-IgM-treatment was formally shown to induce the appearance of DNA degradation. The simultaneous addition of lipopolysaccharide reversed both growth inhibition and apoptosis induction in anti-IgM-stimulated lymphoma cells. The delivery of a partial growth stimulus through the sIg receptors displayed upon immature B cells results in the same induction of programmed cell death seen following a similar action with immature thymocytes. This process may represent, therefore, a uniformed mechanism by which the immune system depletes potentially harmful self-reactive clones during lymphocyte ontogeny. *E. Charles Snow*

Antigen-Induced Apoptosis in Developing T Cells: A Mechanism for Negative Selection of the T Cell Receptor Repertoire

E. J. Jenkinson, R. Kingston, C. A. Smith, G. T. Williams, and J. J. T. Owen

Eur. J. Immunol., 19, 2175—2177, 1989   3-25

Previous work from this laboratory has shown that apoptosis occurs *in vitro* when antibodies to the CD3 component of the T cell receptor complex are added to fetal thymus culture. This finding suggested that autoreactive T cells become deleted during development by programmed cell death, or apoptosis. If it could be shown that apoptosis can be triggered by antigen binding as well as by direct engagement of CD3, this hypothesis would be strengthened. The experiments reported here tested this possibility.

The bacterial "superantigen" staphylococcal enterotoxin B (SEB), which binds to all T cells, was added to 7- to 11-d-old thymus organ cultures. When DNA was extracted from treated thymocytes and electrophoretically separated on agarose gels, a distinct ladder of oligonucleosomal fragments was visible. Fluorescence-activated cell sorting profiles showed that exposure to SEB resulted in significant reductions in overall cell yield per lobe and in significant reductions in the number of $V_\beta 8^+$ cells. The number of $V_\beta 6^+$ cells and $V_\beta 11^+$ cells was unaffected by SEB treatment.

These results show that SEB induces apoptosis in T cells in thymus organ cultures. This effect is specific to $V_\beta 8^+$ cells. The model system used here should be useful for further study of antigen-induced apoptosis in T cell development.

♦ For these experiments, thymocytes were recovered from day 14 mouse embryos and exposed to *Staphylococcus aureus* enterotoxin B (SEB) *in vitro*. SEB is a superantigen which stimulates all T cells expressing $V_\beta 3$ or $V_\beta 8$ elements (as high as 20% of all T cells in some strains of mice). This allows for an experimental means for testing the effect of antigen, rather than anti-receptor antibody, occupancy of TCR expressed on immature thymocytes. The addition of SEB to 7 to 11 d thymus organ cultures induced the appearance of DNA fragmentation. In addition, such cultures had 25% fewer cells recovered at 18 h after SEB addition. The recovered cells exhibited a 40 to 90% reduction in $V_\beta 8^+$ thymocytes without changes in $V_\beta 6^+$ and $V_\beta 11^+$ thymocytes, which cannot bind SEB. It should be emphasized at this time that superantigens bind directly to the TCR and are not presented by APC. Therefore, the stimulation of T cells with superantigens results in the occupancy of TCR similarly to the way the receptors are occupied by antireceptor antibodies. These results show that this means of TCR occupancy with immature populations of thymocytes also results in the initiation of programmed cell death. *E. Charles Snow*

Haemopoietic Colony Stimulating Factors Promote Cell Survival by Suppressing Apoptosis

G. T. Williams, C. A. Smith, E. Spooncer, T. M. Dexter, and D. R. Taylor

Nature, 343, 76—79, 1990

3-26

Colony stimulating factors (CSFs) enhance the cell survival of hemopoietic precursor cells in addition to stimulating their proliferation. When hemopoietic cells are deprived of the appropriate CSF, they rapidly die. Much cell death that occurs during embryogenesis occurs by the active process of apoptosis. These experiments tested whether hemopoietic precursor cells undergo apoptosis on withdrawal of the appropriate CSF.

Three hemopoietic precursor cell lines derived from long-term mouse bone marrow culture were studied. These lines were dependent for growth on CSFs, and died rapidly in their absence. When cultured in the presence of interleukin-3 (IL-3) and horse serum, they exhibited >95% viability and >98% primitive blast cell morphology. When cultures were deprived of IL-3, extensive cell death was observed. After gel electrophoresis, the DNA extracted from cell cultured in the absence of IL-3 formed a ladder pattern of oligonucleosomal fragments characteristic of apoptosis. Electron microscopy of IL-3-deprived cells revealed morphological features of apoptosis, such as condensed chromatin and loss of plasma membrane microvilli. Incubation of IL-3-deprived cells in the presence of the protein synthesis inhibitor cycloheximide inhibited the rate of cell death. When two related cell lines that were dependent on GM-CSF and G-CSF were cultured in the absence of these factors, the specific DNA degradation patterns and morphological changes characteristic of apoptosis were seen.

These results provide evidence that IL-3, GM-CSF, and G-CSF suppress apoptosis of hemopoietic precursor cell lines. Cytokines and growth factors may be important for regulating the sizes of hemopoietic precursor populations.

♦ This study utilized two hematopoietic precursor cell lines (FDCP-1 and FDCP-Mix) which were originally derived from long-term murine bone marrow cultures. FDCP-1 is a granulocyte progenitor, while two FDCP-Mix clones (A4 and 1) display characteristics of hematopoietic stem cells. These cells grow in cultures only when various colony-stimulating factors (CSF, such as IL-3, GM-CSF, and G-CSF) are included within the culture system. In the absence of one of these CSF, the cells die rapidly. When FDCP-1, FDCP-Mix A4, and FDCP-Mix 1 were cultured for 19 h in the absence of a CSF source they exhibited the type of DNA fragmentation which is characteristic of cells undergoing programmed cell death. DNA fragmentation began as early as 6 h, and the cells also displayed other indications of apoptosis, such as condensed chromatin and appearance of apoptotic vesicles. Also, the addition of cycloheximide to cultures deprived of a CSF source displayed a significantly reduced level of cell death.

The ability to suppress the development of apoptosis was shown to be shared by IL-3, GM-CSF, and G-CSF. This indicates that such an effect upon hematopoietic stem survival may represent a common functional property of CSFs. *E. Charles Snow*

Cellular and Developmental Properties of Fetal Hematopoietic Stem Cells

C. T. Jordan, J. P. McKearn, and I. R. Lemischka

Cell, 61, 953—963, 1990   3-27

A complete description of the totipotent hematopoietic stem cells, which contribute to all the hematopoietic lineages, has not yet been possible because of deficiencies in the present methods for studying the properties of enriched cell populations. In this paper, a new approach for studying the fetal totitpotent hematopoietic stem cells is presented.

The first step was the infection of fetal liver with two discrete retrovital markers. Next, murine liver cells from 14-d embryos were fractionated using the monoclonal antibody AA4.1. This antibody has previously been shown to recognize less than 1% of mid-gestation fetal liver cells; the AA4.1 antigen is similar to the CD34 surface antigen of humans, both marking pre-B cells. Differentially marked sAg$^+$ and sAg$^-$ subpopulations were mixed and engrafted into irradiated recipients. Analysis of the fate, or *in vivo* potential, of the engrafted cells involved periodic Southern blot analysis of peripheral blood cell DNA, or autopsy for analysis of proviral distribution.

Use of this strategy resulted in the demonstration that the AA4.1$^+$ cells included multipotential cells that are both necessary and sufficient for complete and permanent reconstitution of irradiated hosts. These cells did not adhere to fibronectin, and had a density of 1.065 to 1.070 g/ml. These totipotent stem cells displayed low levels of antibody staining with one or more monoclonal antibodies directed against epitopes found on differentiating hematopoietic lineages.

These experiments provide evidence that the techniques employed can identify totipotent hematopoietic stem cells. They also suggest that fetal liver cells expressing the cell surface marker AA4.1 define the stem cells. The use of these techniques should facilitate the assessment of the properties of stem cells without requiring homogeneous preparations of cells or engraftment of limiting cell numbers.

♦ Over the past several years, the search for the totipotential stem cell has brought us tantalizingly close to identifying this self-renewing stem cell. The search for this cell has taken two paths. In the first, monoclonal antibodies, specific for minor antigens found upon bone marrow cells, have been utilized for the purification of potential candidates for the totipotential stem cell (see Spangrude et al., *Science,* 241, 58, 1988). In a second approach, the function of potential stem cell clones is followed through the use of genetic (retroviral) markers (first demonstrated by Snodgrass and Keller, *EMBO J.,* 6, 3955, 1987). This approach has been used to show that these stem cell clones can reconstitute lethally irradiated hosts (Jordan and Lemischka, *Genes Dev.*, 4, 220, 1990). This method has recently been modified to allow the identification of physical characteristics of stem cells at the same time one is determining the cloning behavior of the cells.

In this approach, fetal liver cells are divided into two populations, each genetically "marked" with provirus A or B (each of these proviruses are distinguishable by Southern blot analysis). Next, the two populations are subdivided based upon differentially expressed surface proteins (so, for example, surface protein X positive cells are marked with both virus A and virus B). Therefore, the surface protein expressing and nonexpressing subpopulations, each marked by a different virus, can be mixed together and the mixture utilized to reconstitute lethally irradiated animals. Peripheral blood cells are used to identify provirally marked animals, and the distribution of the virus marker in the four major peripheral blood cell lineages (monocytes, granulocytes, T lymphocytes, and B lymphocytes) is determined. This approach allows for an evaluation of the long-term contribution of surface protein X positive or negative cells towards the repopulation of these peripheral blood cells. In addition, the marker can be exploited to study the location of stem cells over long periods of time. Probably the greatest value of this approach is that any physically distinguishable subpopulation of hematopoietic cells can be studied within the context of the *complete* hematopoietic system.

Fetal liver cells were fractionated into two populations based upon the expression of antigen AA4.1 (approximately 0.5 to 1.0% of fetal liver cells display this surface protein). The results of these experiments are consistent with this cell surface protein being expressed on all totipotential stem cells present at day 14 of gestation. This surface protein also identifies day 12 to 14 CFU-S. Therefore, the most primitive stem/progenitor cells are included within the 0.5 to 1.0% AA4.1-positive cells. A further characterization revealed that the most primitive hematopoietic stem cells display a density of 1.065 to 1.070 g/ml and do not adhere to fibronectin-coated plates. In conjunction with a collection of monoclonal antibodies detecting epitopes present upon more differentiated hematopoietic cells, the totipotential stem cell was purified an estimated 500 to 1000-fold.

The regulation of cellular growth within the hematopoietic system has received much attention over the past several years. It has become increasingly obvious that soluble mediators released from elements of the mature immune system (macrophages and lymphocytes) play a role in this regulation. Such a situation allows for the coupling of the rate of hematopoiesis to the ever-changing requirements of the peripheral tissues for hematopoietic cells. For example, during a bacterial siege at some site distal to the bone marrow, the need for reinforcements must be conveyed back to the relevant stem cells within the bone marrow. It makes sense that the cells participating during the defense against the foreign challenge should be the cells assigned the task to notify the cellular factory concerning the needs of the defense system. In addition, there are times in which the allocation of hematopoietic cells into the periphery meets the needs of these distal sites, and the factory must be told to reduce its output of blood cells. Such a stem cell inhibitor was originally described in the human system to be released by normal human bone marrow cells (Wright et al., *Leukaemia Res.,* 4, 309, 1980). Recently, a similar factor, derived from murine bone marrow macrophages, has been identified, providing the opportunity to characterize this class of negative regulators of stem cell proliferation. *E. Charles Snow*

Identification and Characterization of an Inhibitor of Haemopoietic Stem Cell Proliferation
G. J. Graham, E. G. Wright, R. Hewick, S. D. Wolpe, N. M. Wilkie, D. Donaldson, S. Lorimore, and I. B. Pragnell
Nature, 344, 442—444, 1990   3-28

Little is known about what controls hemopoietic stem cell proliferation, although locally acting regulatory elements of the stromal microenvironment are thought to be involved. Identification of these elements has been hampered by lack of suitable culture systems. Recent work from this laboratory has resulted in a novel *in vitro* colony assay that detects a primitive cell, CFU-A, that is similar to hemopoietic stem cells of bone marrow as defined by the spleen colony assay. These CFU-A cells respond to CFU-S proliferation regulators. The present work was directed to the isolation and characterization of an inhibitor of hemopoietic stem cell proliferation.

Capitalizing on the previously described CFU-A assay, involving growth of bone marrow cells in the presence of CSF-1and GM-CSF into colonies containing mainly macrophages, it was found that culture supernatant from normal marrow cells inhibited the formation of CFU-A colonies. A murine macrophage line, J774.2, gave rise to a culture supernatant that was strongly inhibitory in both CFU-A and CFU-S assays. Chromatographic purification of the inhibitory factor resulted in an active fraction that migrated as a doublet with an apparent relative molecular mass of 8000 on sodium dodecyl sulfate polyacrylamide gel electrophoresis; the active fraction seemed to have a molecular mass of 50 to 100 K during gel filtration. The purified factor had no affect on the proliferation of more mature lineage-restricted progenitors such as granulocyte-macrophage colony-forming cells.

Partial N-terminal sequencing of the two components of the doublet showed sequences corresponding to the cytokine macrophage inflammatory proteins (MIP) 1α and 1β. The J774.2 cells contained mRNAs for both these cytokines. When the corresponding DNAs were transcribed and translated in COS cells, it was found that MIP-1α had identical activity as the purified stem cell inhibitor (SCI). Furthermore, blocking antibodies to native MIP-1 neutralized the activity of both MIP-1α and SCI. MIP-1β had no detectable activity on CFU-A cells.

These findings provide evidence that a specific inhibitor of CFU-S proliferation is released by normal bone marrow cells. This inhibitor seems identical to MIP-1a.

♦ The hematopoietic system can be regulated at many levels. The three most obvious are the totipotential stem cell, the intermediate precursor cells which feed directly into select lineages, and the mature blood cells. The most difficult level to study, of course, is the regulation of stem cell proliferation. An experimental approach is described which allows the analysis for the *in vitro* growth characteristics of a primitive colony forming cell (CFU-A). The addition of conditioned medium from cultures of J774.2 cells, a bone marrow-derived macrophage cell line, to this *in vitro* assay system demonstrated that a component present within this condition medium selectively inhibited the proliferation of CFU-A cells. This factor was referred to as a stem cell inhibitor (SCI).

SCI was purified by fractionating concentrated conditioned medium over ion-exchange, heparin-Sepharose, and blue-Sepharose columns. The purified SCI was monitored by SDS-PAGE and the functional assay described above. SCI was purified 50,000-fold and found as a doublet possessing a molecular mass of 8000 (8K). The purified SCI was found to inhibit CFU-A proliferation while not affecting the growth of more mature, lineage-restricted progenitors such as GM-CFC cells. HPLC was employed to separate the doublet into two peptides which were subjected to N-terminal sequencing. The major sequence of SCI corresponded to a previously described cytokine referred to as macrophage inflammatory protein-1α (MIP-1α). Recombinant MIP-1α was found to also inhibit CFU-A proliferation without affecting the growth of GM-CFC cells. In addition, evidence was provided for SCI existing as a high molecular mass complex (50 to 100K) under physiological conditions. It is possible that treatment of patients with SCI prior to their receiving cytotoxic therapies for cancer might serve to protect the integrity of the totipotential cell compartment. *E. Charles Snow*

Mouse *Mos* Protooncogene Produce Is Present and Functions During Oogenesis

R. S. Paules, R. Buccione, R. C. Moschel, G. F. Vande Woude, and J. J. Eppig

Proc. Natl. Acad. Sci. U.S.A., 86, 5395—5399, 1989

3-29

The cellular version of the homolog of the Moloney murine sarcoma virus transforming gene *v-mos* has not been found expressed in normal tissue until recently and is not well understood. Recent work has shown that the *c-mos* of *Xenopus* is expressed during oocyte development and is required for oocyte maturation. In the work presented here, the homologous murine protooncogene *Mos* was studied for its presence and function in mouse oocytes.

Protein extracts from mouse oocytes metabolically labeled *in vitro* showed the presence of immunoprecipitable Mos protein based on recognition with three Mos-specific antisera; immunoprecipitation was blocked by competing antigen. The immunoprecipitated protein appeared identical to that from transformed NIH 3T3 cells. The p39^{mos} was detected in fully grown arrested oocytes, in oocytes labeled during meiotic maturation, and in ovulated eggs, but not in growing oocytes or pronuclear-stage embryos. When cumulus-enclosed, fully grown, germinal vesicle-arrested oocytes were injected with *c-mos* antisense oligonucleotides, the first meiotic division and first polar body production were completely blocked. Microinjection of control oligonucleotides had only minor inhibitory effects. Microscopic examination of injected, inhibited oocytes showed that the block in maturation occurred after germinal vesicle breakdown.

These experiments provide evidence that the *Mos*-encoded protein product is synthesized in fully grown and in maturing mouse oocytes and in ovulated ova. The microinjection of *Mos* antisense oligonucleotides into oocytes blocks the first meiotic division and inhibits the formation of the first polar body. The *Mos* gene product may regulate meiosis in vertebrate oocytes.

Microinjection of Antisense c-*mos* Oligonucleotides Prevents Meiosis II in the Maturing Mouse Egg
S. J. O'Keefe, H. Wolfes, A. A. Kiessling, and G. M. Cooper
Proc. Natl. Acad. Sci. U.S.A., 86, 7038—7042, 1989   3-30

The fact that activated oncogenes can induce neoplastic transformation suggests that their progenitors, the protooncogenes, may also be important in regulating cell growth and differentiation. The protooncogene c-*mos* is unique in that its major expression is limited to the male and female germ cells of the mouse and several other species, indicating a specific function in meiotic cell types. This study utilized injection of antisense oligonucleotides to delineate the function of c-*mos* in the murine oocyte, since meiosis is readily followed *in vitro* in this model.

Oocytes injected with antisense c-*mos* oligonucleotides completed the initial meiotic division, but failed to initiate meiosis II. Loss of c-*mos* resulted in chromosomal decondensation and reformation of a nucleus following meiosis I. Cleavage to two cells ensued.

These findings indicate that the c-*mos* protooncogene acts as a maternal message in normal meiotic maturation of the murine oocyte. The sequelae of loss of c-*mos* are events that also would follow the loss of maturation-promoting factor (MPF) activity. The c-*mos* protein may interact directly with MPF or other components of the MPF pathway, or it may act indirectly on MPF through regulating protein synthesis.

♦ Male and female germ cells of mice, as well as several other species, are the major sites for expression of the c-*mos* proto-oncogene. This expression pattern is consistent with a specific function for c-*mos* in meiotic cell types. The fully grown mouse oocyte accumulates large amounts of c-*mos* transcripts which lack detectable poly (A) tails. Subsequent polyadenylation occurs during maturation which is indicative of recruitment of maternal mRNAs for translation. The bulk of the maternally derived c-*mos* RNA is then degraded by the two-cell stage which is consistent with the fate of other maternal mRNAs. To address the functional significance of c-*mos* expression in the developing mouse oocyte, the goal of the work reported in the studies cited above was to block the appearance of the c-*mos* protein by injecting antisense oligonucleotides. Similar work has been done with amphibian oocytes, resulting in a block to germinal vesicle breakdown (Sagata et al., *Nature*, 335, 519—525, 1988). The results from the reports using mouse oocytes indicate c-*mos* expression is required for completion of meiosis but at slightly later times during oocyte maturation than what was observed for amphibians. The differences between amphibians and mouse may not be unexpected given the differences in utilization of maternal RNA between the two species. For example, in amphibians, new protein synthesis is required for germinal vesicle breakdown, whereas in mouse it is not required until sometime around metaphase I. Thus, although the phenotypes produced by antisense oligonucleotides are different between the two species, each occurs around the time that requires new protein synthesis. The confusing aspect of the two mouse reports is that they differ from one another as to the exact timing of the block in oogenesis. For example, in the paper by Paules et al., microinjection of three different c-*mos* antisense oligonucleotides into fully grown oocytes prevented first polar-body emission, thereby blocking meiotic maturation. In contrast, the results reported by O'Keefe et al. showed that fully grown oocytes injected with antisense c-*mos* oligonucleotides were able to complete the first meiotic division but failed to initiate meiosis II. Instead, chromosome decondensation occurred followed by nucleus formation and division to two cells. The different phenotypes reported in the two mouse papers are difficult to reconcile. The timing of the injection is similar and genetic background does not appear to be a factor. The major difference between the two reports is the specificity of the antisense oligonucleotides. The oligonucleotides used in the experiments reported by Paules et al. were complementary to the first 20 nucleotides surrounding the first ATG codon. In contrast, the oligonucleotides used by O'Keefe et al. were complementary to sequences near the middle and 3' terminus of the c-*mos* coding sequence. A detailed quantitative study of message and protein levels is needed to determine absolute levels of reduction in protein function produced by the two sets of oligonucleotides. Alternatively, the experiment could be repeated using targeted mutagenesis techniques to introduce a mutation that would result in complete loss of protein function. In either case, both papers indicate normal c-*mos* expression is required for completion of meiosis. The c-*mos* proto-oncogene encodes a serine/threonine protein kinase. It is already known that a number of proteins are specifically modified by phosphorylation during oocyte maturation, and it is possible that c-*mos* may play a regulatory role in meiosis through mechanisms of phosphorylation. Thus, in addition to mutagenesis experiments, future work on regulation of meiosis in mice is likely to concentrate on identifying targets of c-*mos* phosphorylation in the oocyte. *Terry Magnuson*

---

c-*kit* mRNA Expression in Human and Murine Hematopoietic Cell Lines

C. Andre, L. d'Auriol, C. Lacombe, S. Gisselbrecht, and
F. Galibert
Oncogene, 4, 1047—1049, 1989   3-31

The human c-*kit* mRNA present in brain tissue and hemopoietic cells encodes a polypeptide sharing major structural features with the PDGF receptor. It is likely that c-*kit* encodes a tyrosine kinase receptor, but its ligand remains unknown. The authors examined c-*kit* expression in hemopoietic cell lines if the erythroid, myeloid, and lymphoid lineages and in hemopoietic organs of both humans and mice.

Expression of c-*kit* mRNA was observed at early stages of erythroid and myeloid differentiation, but no such expression was associated with the lymphoid lineage. It would seem that c-*kit* expression is not limited to a single hemopoietic linage, and that the oncogene is expressed during fetal development but not in normal adult hemopoietic tissues.

The protooncogene c-*fms* is one of the many hemopoietic growth factors having specific patterns of expression. In view of the structural homology between c-*kit* and c-*fms* and the specific hemopoietic pattern of c-*kit* hemopoietic expression, the c-*kit* ligand may have a role in the proliferation of immature stem cells and/or their differentiation from the erythroid/myeloid lineage.

W Mutant Mice with Mild or Severe Developmental Defects Contain Distinct Point Mutations in the Kinase Domain of the c-*kit* Receptor

A. D. Reith, R. Rottapel, E. Giddens, C. Brady, L. Forrester, and A. Bernstein

Genes Dev., 390—400,19   3-32

Developmental abnormalities result from mutations of receptor tyrosine kinases. Mutations at the murine *W/c-kit* locus produce intrinsic defects in stem cells of the melanocytic, hemopoietic, and germ-cell lineages. *W* alleles differ in the severity of phenotypes which they confer; some alleles exhibit an independence of pleiotropic effects. The goal of this study was to analyze the structure and activity of c-*kit* in a number of independent *W* alleles conferring a wide range of mutant phenotypes. Autophosphorylation activity associated with c-*kit* was studied in five different *W* mutants.

Mast-cell cultures derived from mice or embryos homozygous for each of the *W* alleles were deficient in c-*kit* autophosphorylation activity. The degree of deficiency paralled the severity of the phenotype conferred by a given *W* allele both *in vivo* and in a mast-cell co-culture assay. Two mildly dominant homozygous viable alleles expressed reduced levels of an apparently normal c-*kit* protein. The c-*kit* defects conferred by two moderately dominant homozygous viable alleles and by a homozygous lethal allele were ascribed to single-point mutations within the kinase domain of the c-*kit* polypeptide. These mutations result in point substitutions of amino acid residues which are highly conserved in protein tyrosine kinases.

Molecular characterization of additional *W* mutations will help define the structural elements required for signal transduction by transmembrane protein kinases *in vivo*. It will also provide insight into the role played by c-*kit* in mammalian embryogenesis and cellular differentiation.

The Dominant W[42] *spotting* Phenotype Results From a Missense Mutation in the c-*kit* Receptor Kinase

J. C. Tanb, K. Nocka, P. Ray, P. Traktman, and P. Besmer

Science, 247, 209—2212, 1990   3-33

The murine *white spotting* locus *(W)* is allelic with the protooncogene c-*kit,* which encodes a transmembrane tyrosine protein kinase receptor for a ligand that remains unknown. Mutations at the *W* locus influence hematopoiesis as well as the proliferation and migration of primordial germ cells and melanoblasts during development. The *W*[42] mutation is particularly severe in both the homozygous and heterozygous states. This mutation influences pigmentation, gametogenesis, and hematopoiesis. Mice homzygous for W^{42} have severe macrocytic anemia and small gonads, and lack virtually all coat pigment.

Mast cells were cultured from the liver of 2-week-old mouse fetuses in order to investigate c-*kit* protein products. c-*kit* protein products were expressed normally in hoozygous mutant mast cells, but exhibited defective tyrosine kinase activity *in vitro*. Nucleotide sequence analysis of mutant cDNAs revealed a missense mutation replacing aspartic acid with asparagine at position 790 in the c-*kit* protein product. Aspartic acid-790 is a conserved residue in all protein kinases.

The c-*kit* W^{42} mutation has the characteristics of dominant loss-of-function mutation with regard to the known pleoitropic functions of c-*kit*. Studies with other dominant receptor mutations and with transgenic or chimeric mice may help define the role of kinase receptor systems in mammalian development.

◆ Earlier reports (Chabot et al., *Nature,* 335, 88, 1988; Geissler et al., *Cell,* 55, 185, 1988) presented convincing evidence that the protein encoded by the c-*kit* protooncogene is the gene product of the *W* locus. Mice homozygous for the *W* mutation are sterile, show extensive white spotting and have severe anemia that results in perinatal death. Based on these observations, the *W* locus was thought to be involved in stem cell proliferation and survival during embryogenesis. Many independent mutations are known at the *W* locus and the different alleles vary in their degree of severity. c-*kit* is known to encode a cell surface glycoprotein with tyrosine kinase activity, and thus, its role in stimulating proliferation of stem cells would make sense. The reports cited above extend the earlier observations by showing (1) expression of c-*kit* in cell types and lineages known to be affected by the *W* mutation and (2) impaired kinase activity in mast cell cultures from *W* mice. The kinase deficiency was found to be due to mutations that affect c-*kit* expression or, alternatively, the kinase domain. The nature of the mutation correlated with the severity of phenotype conferred by the alleles examined. These results support the suggestion that c-*kit* and *W* are allelic and suggest that phenotypic effects associated with the heterozygous state of the mutation may be due to the formation of receptor oligomers compromised by the presence of a mutant monomer. Although c-*kit* is known to have a broad range of expression, the developmental defects associated with the different alleles are specific. To explain this apparent discrepancy, the suggestion is made that c-*kit* interacts in a synergistic manner with other growth and differentiation factors which then might have specific affects at various stages of differentiation. Clearly, the work with c-*kit* and *W* indicates the importance of receptor protein tyrosine kinases in mammalian development and much effort will be devoted to these proteins over the next few years.
Terry Magnuson

Molecular Cloning of Mast Cell Growth Factor, a Hematopoietin that is Active in Both Membrane Bound and Soluble Forms

D. M. Anderson, S. D. Lyman, A. Baird, J. M. Wignall, J. Wisenman, C. Rauch, C. J. March, H. S. Boswell, S. D. Gimpel, D. Cosman, and D. E. Williams

Cell, 63, 235—243, 1990   3-34

A mast cell growth factor (MGF) has been identified as a ligand for the *c-kit* protooncogene. The gene encoding MGF maps to chromosome 10 near the *steel (Sl)* locus, implicating MGF as a putative product of this locus. A reduction in mast cells is one of the primary defects is *Sl* mice. The authors have now cloned cDNAs encoding the MGF protein.

The MGF protein encoded by the cloned cDNA can be expressed in biologically active form either as a membrane-bound protein or as a soluble factor. The soluble protein was found to promote the proliferation of MGF-responsive cell lines. In the presence of erythropoietin, this protein stimulated the formation of macrocytic erythroid and multi-lineage hematopoietic colonies. MGF appeared to act at an early stage of hematopoiesis, in accord with the known phenotype of the *Sl* defect.

MGF may be phsyiologically active as a membrane protein, a soluble protein, or both. Both cell-cell and soluble factor-membrane receptor interactions may take place between membrane-associated MGF and its receptor, *c-kit*. The ability of MGF to function as a cell-surface molecule may help explain the dominant nature of the *Sl* defect.

Mast Cell Growth Factor Maps Near the Steel Locus on Mouse Chromosome 10 and is Deleted in a Number of Steel Alleles

N. G. Copeland, D. J. Gilbert, B. C. Cho, P. J. Donovan, N. A. Jenkins, D. Cosman, D. Anderson, S. D. Lyman, and D. E. Williams

Cell, 63, 175—183, 1990   3-35

Mutations at the mouse steel *(Sl)* locus identify a gene required for the development of primordial germ cells, melanocytes, hemopoietic stem cells, and cells of the erythroid and mast-cell lineages. The *Sl* gene acts extracellularly rather than in the affected cells themselves. Dominant white spotting *(W)* mutations parallel those of the *Sl* series and affect the same cell lineages, but there is evidence that *W* acts within affected cells. Mutations in *W* and *Sl* have comparable phenotypic effects, including deficiencies of pigment cells, germ cells, and blood cells.

The recent finding that *W* encodes the *c-kit* protooncogene, a tyrosine kinase membrane receptor, suggests that *Sl* encodes a ligand for *c-kit*. The authors now report having identified and purified mast-cell growth factor (MGF), a *c-kit* ligand. Sequences encoding MGF were cloned, and

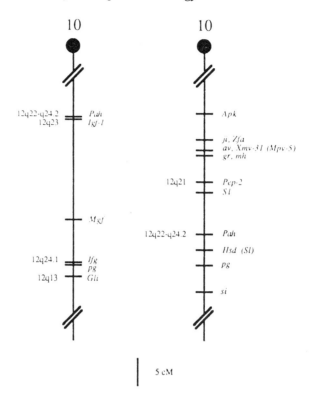

FIGURE 3-35. Linkage maps of mouse chromosome 10. The chromosome on the left shows the interspecific backcross map, including the position of *Mgf*. The chromosome on the right shows the June 1990 version of the chromosome 10 linkage map compiled by M. T. Davisson, T. H. Roderick, A. L. Hillyard, and Doolittle as provided from GBASE, a computerized database maintained at The Jackson Laboratory, Bar Harbor, ME. This map is based on data from genetic crosses among laboratory strains, recombinant inbred strains, and interspecific backcrosses. The two maps are aligned at the pg locus. Several of the genes located on mouse chromosome 10 have been mapped in human chromosomes, and the positions are indicated. (From Copeland, N. G., Gilbert, D. J., Cho, B. C., Donovan, P. J., Jenkins, N. A., Cosman, D., Anderson, D., Lyman, S. D., and Williams, D. E., *Cell,* 63, 175—183, 1990. With permission.)

linkage maps constructed (Figure 3-35) which show that *Mgh* maps near *Sl* in the distal region of chromosome 10. *Mgf* is structurally altered in a number of *Sl* alleles.

It seems very likely that *Sl* encodes the mast-cell growth factor. Many related issues can be resolved by cloning full-length MGF cDNAs, delineating the genomic structure of MGF, and comparing mutant and wild-type forms of MGF.

The *kit* Ligand: A Cell Surface Molecule Altered in Steel Mutant Fibroblasts

J. G. Flanagan and P. Leder

Cell, 63, 185—194, 1990   3-36

The c-*kit* protooncogene, the gene at the mouse *W* developmental locus, encodes a cell-surface protein having sequence homology with receptors for polypeptide growth factors. Studies showing that c-*kit* is affected by mutations at the dominant white spotting *(W)* locus indicate that it has important roles in normal development. The authors now have identified and characterized the *kit* ligand using a method based on the specificity and high affinity of the ligand-receptor interaction.

A genetic construct encoding only the extracellular domain of the receptor yielded a soluble protein which served as an affinity reagent. The intact receptor is an integral membrane protein containing a hydrophobic transmembrane region (Figure 3-36). The construct was made by insert-

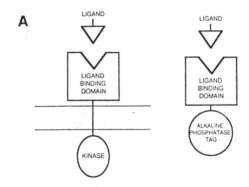

FIGURE 3-36. A soluble receptor affinity reagent with an enzyme tag. (A) In the native state (left-hand side) a typical tyrosine kinase cell surface receptor consists of an extracellular domain that can bind a ligand, a single hydrophobic transmembrane domain,. and an intracellular domain with protein tyrosine kinase activity. Ligand binding results in transduction of a signal to the interior of the cell. To produce a soluble affinity reagent (right-hand side) to detect and characterize the ligand of a receptor tyrosine kinase, a construct was made encoding the receptor extracellular domain without the transmembrane and intracellular domains. The C-terminal end of the extracellular domain was instead genetically fused to a domain of human placental alkaline phosphatase, providing a tag that binds to available monoclonal antibodies and also has an intrinsic enzyme activity that can be used to trace the fusion protein. (B)Mammalian expression vector APtag-1. The plasmid carries high level transcription control elements from the Moloney murine leukemia virus LTR and 3' splice and polyadenylation signals from rat insulin gene. Cloning sites in APtag-1 allow fusion of polypeptides of interest to the AP tag. In this case, the 5' end of a c-*kit* DNA encoding the extracellular domain was inserted in APtag-1m to give plasmid APtag-1. (C) Expression of the APtag-KIT fusion protein or unfused SEAP. Transfected J558l clones J558.AK and J558.AP were metabolically labeled with [^{35}S]methionine; the supernatants were immunoprecipitated with a monoclonal antibody to human placental alkaline phosphatase and analyzed by SDS-PAGE with a 10% gel. (From Flanagan, J. G. and Leder, P., *Cell*, 63, 185—194, 1990. With permission.)

ing the 5′ end of a murine c-*kit* cDNA— encoding the extracellular domain — into a new mammalian expression factor, APtag-1. The APtag-KIT fusion protein was used to directly identify the *kit* ligand at the molecular level. The protein demonstrated a specific binding interaction with a ligand on 3T3 fibroblasts. Staining demonstrated labeling over the entire surface of these cells, but not in nearby non-expressing cells. Binding was absent in 3T3 fibroblasts carrying the steel *(Sl)* mutation.

It now is apparent that the *kit* ligand is a cell-surface molecule, although it is possible that some of the ligand is released in soluble form. Mutations at the *Sl* locus influence the expression or structure of the *kit* ligand. Methods like those used may help in identifying and characterizing the partners of other receptors tyrosine kinases, in addition to different types of cell surface receptors and other proteins whose ligands are unknown.

The Hematopoietic Growth Factor KL is Encoded by the *Sl* Locus and Is the Ligand of the c-*kit* Receptor, the Gene Product of the *W* Locus

E. Huang, K. Nocka, D. R. Beier, T. Chu, J. Buck, H. Lahm, D. Wellner, P. Leder, and P. Besmer
Cell, 63, 225—233, 1990   3-37

Mutations at the steel *(Sl)* locus on mouse chromosome 10 produce phenotypic features very similar to those of mice carrying *W* mutations, including alterations in hemopoiesis, gametogenesis, and melanogenesis. The *W* locus is allelic with the c-*kit* protooncogene. The authors recently purified a hemopoietic growth factor, KL, from conditioned medium of BALB/c 3T3 fibroblasts, which exhibits the biologic properties of the c-*kit* ligand. The present binding and cross-linkage studies demonstrated that KL is the ligand of c-*kit.*

KL stimulates the proliferation of both mast cells and early erythroid progenitors, and it binds specifically to the c-*kit* receptor. The predicted amino acid sequence of isolated KL-specific cDNA clones indicates that KL is synthesized as an integral transmembrane protein. Linkage analysis mapped the KL gene to the *Sl* locus on mouse chromosome 10. KL sequences were deleted in the genome of the *Sl* mouse.

These findings show that the *Sl* locus encodes the ligand of KL, the c-*kit* receptor. Alterations in more mature cell populations often are found in *Sl* and *W* mutations where c-*kit*/KL function is only partially impaired. Many weak *Sl* alleles are known, and study of their phenotypes will be very helpful in delineating the pleiotropic functions of the c-*kit* receptor system.

Primary Structure and Functional Expression of Rat and Human Stem Cell Factor DNAs

F. H. Martin, S. V. Suggs, K. E. Langley, H. S. Lu, J. Ting, K. H. Okino, C. F. Morris, I. K. McNiece, F. W. Jacobsen, E. A. Mendiaz, N. C. Birkettt, K. A. Smith, M. J. Johnson, V. P. Parker, J. C. Flores, A. C. Patel, E. F. Fisher, H. O. Erjavec, C. J. Herrera, J. Wypych, R. K. Sachdev, J. A. Pope, I. Leslie, D. Wen, C. Lin, R. L. Cupples, and K. M. Zsebo

Cell, 63, 203—211, 1990   3-38

An early-acting hemopoietic growth factor, the stem cell factor (SCF), has been identified from its ability to stimulate marrow cells from mice treated with 5-fluorouracil. Partial cDNA and genomic clones of rat SCF have been isolated. The authors have now utilized probes based on the rat sequence to isolate partial and full-length cDNA and genomic clones of human SCF.

The two bioassays used in purifying rat SCF measure the stimulation of a colony-forming cell having high proliferative potential (HPP-CFC), and stimulation of the proliferation of a murine mast-call subclone. Based on the primary structure of the 164-amino acid protein purified from BRL-3A cells, truncated forms of both the rat and human proteins were expressed in *E. coli* and mammalian cells, and found to possess biologic activity. SCF augments the proliferation of both myeloid and lymphoid progenitors in marrow culture. In addition, it exhibits strong synergistic activity in combination with colony-stimulating factors.

A growth factor that stimulates multilineage hemopoietic cells could have therapeutic potential in a number of hemopoietic disorders. It also could enhance gene and cell replacement treatments which are dependant on marrow transplantation.

Candidate Ligand for the c-*kit* Transmembrane Kinase Receptor: KL, a Fibroblast Derived Growth Factor Stimulates Mast Cells and Erythroid Progenitors

K. Nocka, J. Buck, E. Levi, and P. Besmer

EMBO J., 9, 3287—3294, 1990   3-39

The c-*kit* protooncogene, which is allelic with the murine *white-spotting* locus *(W)*, encodes a transmembrane tyrosine kinase receptor for a ligand that has not been definitively identified. The receptor is one of a number of the family of colony-stimulating factor-1/platelet-derived growth factor/*kit* receptor family. It is important to identify the ligand for c-*kit* because of the pleiotropic effects it may have on the cell types expressing c-*kit* which are effected by *W* mutations. Targets of these mutations include erythroid and mast-cell lineages, as well as stem cells.

In the absence of interleukin-3 (IL-3), mast cells from normal mice but not those from *W* mutant mice are maintained by co-culture with 3T3 fibroblasts. The authors used a mast-cell proliferation assay to purify a 30-kDa protein, KL, from conditioned medium of Balb/3T3 fibroblasts. KL stimulated the proliferation of normal marrow-derived mast cells, but not the proliferation of mast cells from *W* mice. In contrast, normal and mutant mast cells responded similarly to IL-3. Connective tissue mast cells from the peritoneal cavity of normal mice expressed a high level of c-*kit* protein on their surfaces, and they proliferated in response to KL. KL combined with erythropoietin stimulated the formation of early erythroid progenitors from fetal liver and spleen cells, but not from marrow cells of adult cells or the fetal liver cells of *W/W* mice.

The biologic properties of KL accord with what is expected of the c-*kit* ligand. If KL is, in fact, the ligand of the c-*kit* receptor, administration of KL mice carrying *S1* mutations should eliminate at least some of the effects of mutations at this locus.

Identification of a Ligand for the c-*kit* Proto-Oncogene

D. E. Williams, J. Eisenman, A. Baird, C. Rauch, K. Van Ness, C. J. March, L. S. Park, U. Martin, D. Y. Mochizuki, H. S. Boswell, G. S. Burgess, D. Cosman, and S. D. Lyman

Cell, 63, 167—174, 1990   3-40

The *W* locus on murine chromosome 5 encodes the c-*kit* protooncogene, which in turn encodes a tyrosine kinase receptor. A pair of unique stromal cell clones derived from cultures of *Sl*-mutant bone marrow and normal littermate bone marrow have been described. The cell line from mutant mice was unable to maintain the viability of Il-3-dependent mast-cell lines in co-culture. A growth factor for these mast-cell lines was identified in supernates from the normal stromal cell line.

The mast-cell growth factor, termed MGF, has now been purified and sequenced. A panel of Il-3-dependent cell lines was screened for responsiveness to partially purified MGF in tritiated thymidine incorporation assays. Proliferative stimulation of the cells in response to MGF correlated with expression of mRNA for c-*kit*. Cross-linking labeled MGF to c-*kit*-expressing cells and immunoprecipitation of the complex with antiserum specific for the C-terminus of c-*kit* demonstrated MGF to be a ligand for protooncogene.

These findings support a role for MGF as a ligand for the c-*kit* protein. It remains possible that there are multiple ligands capable of binding to c-*kit*.

Identification, Purification, and Biological Characterization of Hematopoietic Stem Cell Factor from Buffalo Rat Liver-Conditioned Medium

K. M. Zsebo, J. Wypych, I. K. McNiece, H. S. Lu, K. A. Smith,
S. B. Karkare, R. K. Sachdev, V. N. Yuschenkoff, N. C. Birkett,
L. R. Williams, V. N. Satyagal, W. Tung, R. A. Bosselman,
E. A. Mendiaz, and K. E. Langley
Cell, 63, 195—201, 1990   3-41

The authors have identified a novel growth factor for primitive hemopoietic progenitors, termed stem cell factor (SCF). The protein was isolated from medium conditioned by Buffalo rat liver cells, and its activity tested on bone marrow cells derived from mice treated with 5-flourouracil.

The protein was found to be heavily glycosylated, containing both N-linked and O-linked carbohydrate. The N-terminal amino acid sequence of purified SCF was revealed 248 amino acids subsequent to the putative leader sequence. Both isolated natural rat SCF and recombinant-derived rat SCF were biologically active. The protein exhibited strong synergism with colony-stimulating factors in cultures of semi-solid bone marrow. In addition, it promoted the growth of mast cells.

Defects in stem cell support and survival are implicated in the cause of such disorders as aplastic anemia and diseases producing pancytopenia. A growth factor for stem cells could improve the outcome of patients having marrow transplantation or receiving stem cell-mediated gene replacement treatments.

Stem Cell Factor is Encoded at the *Sl* Locus of the Mouse and is the Ligand for the c-*kit* Tyrosine Kinase Receptor

K. M. Zsebo, D. A. Williams, E. N. Geissler, V. C. Broudy, F. H. Martin,
H. L. Atkins, R. Hsu, N. C. Birkett, K. H. Okino, D. C. Murdock,
F. W. Jacobsen, K. E. Langley, K. A. Smith, T. Takeishi,
B. M. Cattanach, S. J. Galli, and S. V. Suggs
Cell, 63, 213—224, 1990   3-42

The authors have cloned a partial cDNA encoding murine stem call factor (SCF) from Buffalo rat liver cells. The gene for SCF is syntenic with the *Sl* locus murine chromosome 10. The spectrum of biologic activities of SCF correlates with the phenotypic abnormalities present in *W* or *Sl* mutant mice. The present study confirmed that SCF maps to chromosome 10, and showed that three independent mutations at *Sl* are associated with alterations or deletions of SCF genomic sequences.

SCF genomic sequences were found to be deleted when retroviral vectors were used to immortalize fetal liver stromal cell lines from mice carrying lethal mutations at the S1 locus. Two other mutations at *Sl* (*Sl^d* and *Sl^{12N}*) were associated with deletions or alterations of SCF genomic sequences. Administration of exogenous SCF[164] to compound heterozygote steel (*Sl/Sl^d*) mice reduced the severity of macrocytic anemia and locally repaired their mast-cell deficiency. In addition, SCF[164] bound to recombinant c-*kit* expressed in COS-1 cells.

These findings add to the evidence that the phenotypic abnormalities expressed by *W* or *Sl* mutant mice result from primary defects in c-*kit* receptor-ligand interactions, which are critical for the normal development of many different cell types.

♦ The *dominant white spotting (W)* and *steel (Sl)* mutations show similar phenotypic effects which include absence of migrating melanocytes leading to alterations of coat color, varying degrees of anemia due to an absence of hematopoietic stem cells, and sterility due to an absence of primordial germ cells populating the genital ridges. *In vivo* transplantation as well as co-culture experiments indicate that the defect in *W* is intrinsic to the hematopoietic stem cell whereas the defect in *Sl* is in the stromal cell environment of the bone marrow. As described elsewhere in this yearbook and in the 1990 Year Book, c-*kit*, a membrane-bound tyrosine kinase, is actually the product of *W*. The more severe *W* alleles are due to large deletions whereas mutations within the kinase domain of the receptor lead to more subtle phenotypes. Once the c-*kit/W* correlation was made, the search was on for identifying the c-*kit* ligand. Four groups have now defined a new hematopoietic growth factor whose gene maps to a position coincident with *Sl*. The mapping data include a *M. spretus/ M. musculus* backcross panel, deletions of *Sl* associated with the more severe alleles, and a RFLP coupled with an altered mRNA associated with the *Sl^d* allele. The phenotypic variations observed with some of the deletions suggests that other closely linked genes may also be contributing to the phenotype. The new growth factor is heavily glycosylated, probably exists as a dimer, and has a broad range of biological activities. The purified factor was shown to bind to cells that express c-*kit* and can be cross-linked and co-immunoprecipitated with the transmembrane receptor. Sequence differences between the published reports indicate alternative splicing may be responsible for different forms of the factor. Taken together, these data support the idea that the new growth factor is a ligand for c-*kit*. Future work is likely to concentrate on the intracellular response of the tyrosine kinase domain elicited by ligand/receptor binding. In addition, the role of this factor in melanocyte and germ cell migration has yet to be established. Possibilities that have not yet been excluded include alternative forms of the ligand and/or receptor or the existence of other ligands capable of binding the receptor. Either of these possibilities could explain the broad range of biological effects associated with *Sl* and *W. Terry Magnuson*

Antibody Which Defines a Subset of Bone Marrow Cells That Can Migrate to Thymus

H. C. O'Neill

Immunology, 68, 59—65, 1989   3-43

The cell line 16C1 is a T-cell lineage of immature phenotype exhibiting the ability to migrate to, and localize in the thymus of irradiated mice. It has been proposed that anti-PgP-1 antibodies may bind to T-cell precursors in the thymus and inhibit their capacity to home back to thymus. In the present study, several antibodies specific for Pgp-1 were tested for their ability to inhibit the migration of 16C1 and of bone marrow to thymus in a 3-h double-label migration assay.

Analysis of Pgp-1 expression showed that 16C1 derived from C57BL mice typically binds antibody specific for the Pgp-1.1, but not the Pgp-1.2 allelic determinant. This unusual epitope expression has also been noted in several other cell lines having an origin similar to that of 16C1. Some 5% of C57BL bone marrow cells bind both allele-specific antibodies; these cells are found among the class I+, Thy-1+, T200+ subpopulation of marrow cells. Anti-Pgp-1 antibody inhibits the thymus-homing capacity of both 16C1 and bone marrow. These antibodies serve to deplete bone marrow of cells able to reconstitute the T-cell compartment of irradiated mice.

At least two distinct Pgp-1 determinants can be expressed by cells of C57BL mice. Antibody specific for either of them is capable of inhibit thymus homing capacity. The mechanism by which precursor cells from bone marrow enter the thymus remains to be established.

♦ The author identifies a T-cell line which displays many of the characteristics of immature T cells. Consistent with this connotation was the demonstration that this cell line preferentially homes into the thymus. A monoclonal antibody specific for Pgp-1 was found to inhibit the ability of this immature T-cell line to migrate into the thymus. In addition, the same antibody interfered with the ability of normal thymic progenitor cells from the bone marrow to migrate to the thymus. The pretreatment of bone marrow with anti-Pgp-1 and complement depleted the bone marrow of cells capable of homing to the thymus. Therefore, T-cell progenitor cells ready to leave the bone marrow and migrate to the thymus are Pgp-1 positive. In addition, the Pgp-1 molecule apparently participates during the process of pre T-cell localization within the thymus. It is possible that Pgp-1 interacts with a ligand displayed by thymic stromal cells. *E. Charles Snow*

Calcium-Dependant Killing of Immature Thymocytes by Stimulation via the CD3/T Cell Receptor Complex

D. J. McConkey, P. Hartzell, J. F. Amador-Perez, S. Orrenius, and M. Jondal

J. Immunol., 143, 1801—1806, 1989   3-44

Anti-CD3 antibody promotes a cell suicidal process known as programmed cell death (PCD) in immature thymocytes. This observation suggests a potential role for T-cell receptor (TCR)-mediated PCD in the induction of tolerance. Glycocorticoid-induced thymocyte DNA fragmentation and cell killing depend on an early and sustained rise in cytosolic calcium ion concentration. From what is known of CD3-mediated signal transduction in thymocytes, it seemed likely that anti-CD3-induced thymocyte PCD could involve a similar Ca-dependant mechanism.

Anti-CD3 antibody was found to promote DNA fragmentation in suspensions of human thymocytes. Both activation of endonuclease and actual cell killing were dependant on an early, sustained increase in cytosolic calcium, mostly of extracellular origin. The rise in calcium was similar to that induced by concanavalin A, but this mitogen failed to stimulate DNA fragmentation or cell death. Phorbol ester prevented calcium-dependant cell death in response to anti-CD3, suggesting that activation of protein kinase C prevented cell death. All agents that stimulated PCD tended to lead to the disappearance of immature $CD4^+CD8^+$ thymocytes. Mature peripheral blood lymphocytes were, in contrast, insensitive to attempted stimulation of PCD.

It appears that antibody-mediated stimulation of immature thymocytes via the TCR complex leads to calcium-dependent, endonuclease-mediated cell killing. It remains to identify the mechanisms providing positive signals to thymocytes. Expression of interleukin-2 may be necessary for the intrathymic generation of cells expressing mature, functional TCR.

♦ Thymocytes were obtained from patients up to 2 years of age undergoing corrective surgery, and prepared by standard protocols. The treatment of cells with anti-CD3 elicited a dose-dependent increase in cellular levels of calcium and levels of DNA fragmentation. The optimal dose of anti-CD3 elicited a similar level of DNA fragmentation as seen following the addition of a calcium ionophore. This induction of DNA fragmentation by anti-CD3 stimulation was blocked by buffering increases in cellular levels of calcium with quin-2. The anti-CD3-induced increase of calcium was due to both intracellular mobilization and extracellular influx of calcium. However, this increase in calcium levels does not, by itself, explain programmed cell death. This is underscored by the demonstration that treatment of thymocytes with Con A, which elicits an almost identical calcium response as seen by anti-CD3 treatment, does not induce programmed cell death. The addition of a phorbol ester to anti-CD3 stimulated thymocytes also reduced the level of DNA fragmentation. Finally, the anti-CD3-treatment induced cell death selectively within the immature CD4$^+$CD8$^+$ thymocytes. As T cells mature, they become increasingly resistant to programmed cell death.

The induction of enhanced cellular levels of calcium in immature thymocytes is a contributory factor for the initiation of programmed cell death. This must occur in the absence of a complete growth stimulus to these thymocytes. This is highlighted by the results seen when either Con A or calcium ionophore plus phorbol esters are added to thymocytes. In both of these situations, a prominent calcium response is observed, but in the context of a complete growth signal. Once again, clonal deletion apparently occurs as a consequence of partial delivery of growth signals to immature thymocytes. *E. Charles Snow*

Maternal Controls, Cytoplasmic Determinants, and Imprinting 4

INTRODUCTION

While each new generation brings with it an element of diversity and uniqueness, the parental generation does not immediately let go biologically. In fact, both genetic and epigenetic factors from the parental generation play a developmental role in the offspring.

Because the cytoplasmic contribution of the mother is far more extensive than the father, maternal effects resulting from molecules contained in the cytoplasm are well documented. Recent interest has focused on the molecular nature of those signals and the mechanism that localizes those signals to regions of the egg prior to fertilization. Correlations have been seen between the localization of known maternally contributed molecules and elements of the cytoskeleton, suggesting that the cytoskeleton plays a role in anchoring the effector molecules in the egg.

While examples of maternal effect genes are extensive, especially in *Drosophila* where they play such a central role in pattern formation (see Chapter 9), fewer paternal effect genes have been identified. However, an example of such a gene is included as a possible model for the transmission of specific paternal effects.

An area of particular excitement to investigators studying the genetic basis of human disease is the recent observation that some genetic diseases may be the result of genomic imprinting in the parents. Thus, understanding the mechanism of imprinting, and its developmental role, may have important clinical ramifications with regard to some genetic disorders.

A Sperm-Supplied Product Essential for Initiation of Normal Embryogenesis in *Caenorhabditis elegans* Is Encoded by the Paternal-Effect Embryonic-Lethal Gene, *spe-11*
D. P. Hill, D. C. Shakes, S. Ward, and S. Strome
Dev. Biol., 136, 154—166, 1989   4-1

While control of early embryogenesis is chiefly maternal, the sperm serves several critical functions, including activating the oocyte, contributing a vital component of the diploid zygote, and, in many organisms, contributing a microtubule-organizing center. Further, sperm entry occasionally has a role in establishing embryonic axes; this appears to be the case in *C. elegans*. Mutations in the *spe-11* gene of *C. elegans* result in a paternal-effect embryonic-lethal phenotype.

Fertilization of wild-type oocytes by sperm from homozygous *spe-11* mutant males led to abnormal zygotic development. In contrast, oocytes from homozygous *spe-11* hermaphrodites, when fertilized by wild-type sperm, developed normally. Embryos fertilized by sperm from homozygous *spe-11* worms failed to complete meiosis and exhibited defects in eggshell formation, orientation of mitotic spindles, and cytokinesis. Genetic analysis suggested that *spe-11* is expressed before spermatogenesis is completed, and that the wild-type locus encodes a product present in sperm which contributes to initiating correct early events in *C. elegans* embryos.

Some sperm factors or functions apparently are necessary for proper execution of the maternal instructions that guide early embryogenesis in *C. elegans*. An ontogenetic role of the *spe-11* gene product in early development distinguishes mutations of this gene from the paternal-affect mutations identified in *Drosophila*, which chiefly affect chromosomal behavior.

◆ Analysis of the large collection of maternal-effect embryonic-lethal mutations in fruit flies and nematodes has provided a wealth of information about the roles of oocyte-supplied gene products in early embryonic development. The participation of sperm-supplied components in embryogenesis is less well understood. This paper describes genetic and phenotypic analysis of *spe-11*, a paternal-effect embryonic-lethal gene whose product is present in sperm and is essential for normal zygote development. The earliest defects observed in eggs fertilized by mutant *spe-11* sperm are what may be considered activation defects, followed by defects in spindle placement and cytokinesis. Molecular analysis of the *C. elegans spe-11* gene and of the two paternal-effect lethal genes identified in *Drosophila, pal* and *ms(3)K81,* should contribute to our understanding of the events of fertilization and the roles of sperm factors in early development. *Susan Strome*

Structure of the *Drosophila BicaudalD* Protein and its Role in Localizing the Posterior Determinant *nanos*
R. P. Wharton and G. Struhl
Cell, 59, 881—892, 1989   4-2

The anteroposterior body pattern of *Drosophila* is specified early in development under the control of three maternal determinants: *bicoid (bcd)*, *nanos*, and *torso*. There is evidence that the *BicaudalD (BicD)* gene product may have a role in transporting nanos to the posterior pole or stabilizing it there. Mutations in *BicD* lead to global reorganization of the *Drosophila* body pattern in which the head, thoracic, and anterior abdominal segments are replaced by posterior abdominal segments and terminals.

The way in which ectopic *nanos* activity in embryos from *BicD* mutant mothers might promote abdominal development and suppress head and thoracic development was examined by studying the expression of some segmentation genes controlled by *nanos* or *bcd* (Figure 2). The primary cause of the phenotype appeared to be the inhibition, by incorrectly localized *nanos* activity, of two anterior factors, *bicoid* and *hunchback*. The *BicD* gene was isolated and found to encode a coiled-coil protein resembling the carboxy-terminal part of the myosin heavy chain. *BicD* protein was uniformly distributed throughout wild-type oocytes, but in *BicD* mutant oocytes was concentrated at the anterior pole along with ectopic nanos activity.

It appears that *BicD* encodes a cytoskeleton-like protein which plays a part in transporting or anchoring the *nanos* morphogen within the cytoplasm of the oocyte. A role for *BicD* in localizing *nanos* activity suggests a parallel with the localization of Vg1 mRNA in *Xenopus*.

♦ The initial steps in the organization of the embryonic body plan of *Drosophila* are under control of genes expressed during oogenesis. Products of several key maternal genes are localized within the egg at the position where genetic evidence shows they are required. For example, *bicoid* is required for the formation of anterior structures and *bicoid* mRNA is restricted to an anterior cap of the egg. *nanos* is essential for correct differentiation of posterior structures and its mRNA is localized to the posterior end. The correct localization of these gene products appears to be important. Mutants that alter the localization of *bicoid* mRNA cause serious disruption of embryonic development, even though these embryos contain normal amounts of wild-type *bicoid* mRNA. Little is known about how these mRNAs become localized.

Mutations in at least three maternally expressed genes, *bicaudal*, *Bicaudal-C*, and *Bicaudal-D (Bic-D)* can lead to a dramatic reorganization of the embryonic body plan to generate the so-called double-abdomen phenotype. The *Bic-D* gene is the subject of the two reports. Enhancing the interest of these studies is the suggestion that *Bic-D* might function in the localization of determinants of embryonic development. The evidence is derived from a combination of genetic and molecular evidence. The *Bic-D* phenotype shows that the fate of the region of the embryo that normally develops into head and thorax is changed toward the formation of abdominal structures. Using antibodies to stain whole-mount embryos, Wharton and Struhl show that *Bic-D* mutations result in the repression of genes such as *hunchback* and *Krüppel,* that are normally expressed in the middle-anterior part of the embryo. Other genes that are normally expressed in more posterior

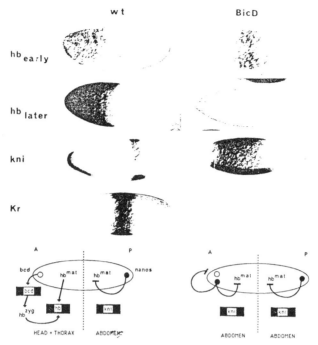

FIGURE 4-2. Origin of the bicaudal phenotype. Shown are typical embryos from wild-type (wt) and *BicD¹/ BicD²* (BicD) females stained with antisera specific for *hunchback* (hb), *knirps* (kni), or *Krüppel* (Kr) proteins, as indicated at the left. At the bottom are schematic summaries of the regulatory interactions among some of these proteins that play key roles in generating the wild-type and bicaudal body patterns. The anterior end of each embryo is at the left and the dorsal surface is at the top of each photograph in this and all subsequent figures. The *hb* protein in early embryos (before nuclear division cycle 10, first row) is derived from translation of maternal transcripts, whereas most of the *hb* protein in the later wild-type embryos (nuclear cycles 12 to 14, second row) comes from translation of mRNA generated zygotically under the control of *bicoid*. Not all of the early bicaudal embryos exhibit the central zone of weak hb protein shown; some have no detectable protein. We note that in early bicaudal embryos like the one shown, the *hb*-free region in the anterior is invariably larger than the corresponding region in the posterior; as if the anterior *nanos* activity is less tightly localized than the posterior *nanos* activity. Later bicaudal embryos do not express detectable *hb* protein through the beginning of nuclear cycle 14 (second row). After this stage, a *torso*-dependent stripe of *hb* protein appears at the posterior end of wild-type embryos (Tautz, 1988) and at both ends of bicaudal embryos (data not shown). In wild-type embryos, *kni* protein (third row) is expressed in an anterior ventral domain as well as in a band in the prospective abdominal region near the middle of the embryo. In bicaudal embryos, the anterior expression is eliminated, presumably a result of the absence of *bcd* function (data not shown), and the remaining domain of expression extends far anteriorly. The central domain of *Kr* protein expression seen in wild-type embryos (fourth row) is absent in bicaudal embryos. However, as for *hb* protein, *Kr* protein is expressed in a small posterior domain in somewhat older wild-type embryos (Gaul et al., 1987) and at both ends of bicaudal embryos (data not shown). At the bottom of the figure are schematic summaries of some of the initial steps in the genesis of wild-type and bicaudal embryos. A broken line divides each embryo into anterior (A) and posterior (P) halves. Wild-type embryos initially contain *bcd* mRNA (open circle) and *nanos* activity (closed circle) localized at the A and P poles, respectively, as well as maternal *hb* transcripts (*hb*mat) distributed uniformly throughout. *hb* protein (boxed) is produced in the anterior by translation of both *hb*mat transcripts and zygotic transcripts (*hb*zyg), the latter synthesized under the control of *bcd* protein (boxed). The *bcd* and *hb* proteins together direct the development of head and thoracic segments (see Nüsslein-Volhard and Roth, 1989). In the posterior, *nanos* activity blocks expression from *hb*mat transcripts, leading, by default, to expression of *kni* protein (boxed), which directs abdominal development. In bicaudal embryos, the ectopic *nanos* activity at the anterior pole blocks the expression of both *bcd* and *hb*mat transcripts, thereby blocking head and thoracic development. In the absence of any *hb* protein, *kni* protein expression occurs in both the anterior and posterior halves, promoting the development of mirror-symmetric abdomens. (From Wharton, R. P. and Struhl, G., *Cell*, 59, 881—892, 1989. With permission.)

parts of the embryo, such as *knirps,* are expressed over a broad anterior region in *Bic-D* embryos. Based on indirect evidence, Wharton and Struhl propose an attractive model whereby the repression of *hunchback* and *Krüppel,* and the ectopic activation of *knirps* (these are all em bryonically expressed genes) are a consequence of a single primary event, namely, the failure of correctly localizing the maternal gene *nanos.* The model proposes that the *nanos* gene product specifically promotes the degradation of the *bicoid* and *hunchback* mRNAs, and circumstantial evidence is presented to support this contention. It should be pointed out that loss-of-function *Bic-D* mutants have a phenotype that is quite different from the double-abdomen phenotype of dominant mutants. In homozygous loss-of-function mutants, the egg chambers contain 16 nurse cells, rather than the usual 15 nurse cells and 1 oocyte. Since all of the 16 germ cells of the egg chamber are interconnected by cytoplasmic bridges, *Bic-D* might function early in oogenesis in localizing or activating "oocyte differentiation factor(s)" in 1 of the 16 cells (the future oocyte) of the cluster. *Marcelo Jacobs-Lorena*

A Two-Step Model for the Localization of Maternal mRNA in *Xenopus* Oocytes: Involvement of Microtubules and Microfilaments in the Translocation and Anchoring of Vg1 mRNA

J. K. Yisraeli, S. Sokol, and D. A. Melton
Development, 108, 289—298, 1990   4-3

The animal-vegetal axis of *Xenopus* oocytes becomes the primary embryonic axis following fertilization. Among the several localized maternal mRNAs that have been isolated is Vg1, a vegetally localized message whose protein product is part of the TGF-β family. Vg1 mRNA is translocated in early stage IV oocytes to a shell along the vegetal cortex, a region including the plasma membrane and associated internal material.

The authors have analyzed the process by which Vg1 is localized in an attempt to learn how polarity is established in *Xenopus* oocytes. In fully grown oocytes, Vg1 mRNA was tightly localized at the vegetal cortex. Biochemical fractionation showed it to be associated with a detergent-insoluble subcellular fraction, which is thought to contain most of the cytoskeletal elements of the cell. Studies with cytoskeletal inhibitors such as cytochalasin B showed that microtubules are involved in translocation of the message to the vegetal hemisphere, and that microfilaments are important in anchoring the message at the cortex. Immunohistochemical studies demonstrated the presence during translocation of a cytoplasmic microtubule array.

These findings indicate that Bg1 mRNA is associated with microtubules in middle-stage oocytes, and that the association is required for the mRNA to translocate to the vegetal hemisphere (Figure 4-3). The cytoskeleton apparently has a role in localizing information in the oocyte. In addition,

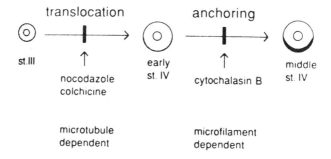

FIGURE 4-3. A two-step model for the localization of Vg1 mRNA in oocytes. The diagram represents the simplest interpretation of the data presented in the paper. Vg1 mRNA distribution (indicated by shading) becomes progressively restricted to the vegetal cortex during normal oogenesis (indicated by the horizontal arrows). Microtubule inhibitors, such as nocodazole or colchicine, block the first part of this process, which we have termed translocation, and result in the prevention of any noticeable migration of Vg1 mRNA. Microfilament inhibitors, such as cytochalasin B, prevent the second step of the localization, which we have called anchoring, and result in the release, in late stage oocytes, of the Vg1 mRNA from its tight cortical shell into a broad band along the cortex. (In middle stage oocytes when the translocation process is occurring, cytochalasin B prevents anchoring at the cortex as well and results in the ectopic accumulation of Vg1 mRNA in the cytoplasm above the cortex. (From Yisraeli, J. K., Sokol, S., and Melton, D. A., *Development,* 108, 289—298, 1990. With permission.)

recognition by microtubules of *cis*-acting sequences probably requires the intervention of factors specific for the RNA.

Localized Synthesis of the Vg1 Protein during Early *Xenopus*
Development
D. Tannahill and D. A. Melton
Development, 106, 775—785, 1989   4-4

The Vg1 gene of *Xenopus* encodes a maternal mRNA which is localized to the vegetal hemisphere of both oocytes and embryos. The mRNA encodes a protein related to the TGF-β family of small-sized, secreted growth factors. The role of Vg1 protein in early development is unclear, but it appears to be a candidate for a natural inducer of mesoderm. The putative Vg1 protein closely resembles molecules known to be active in mesoderm-inducing assays, and Vg1 mRNA is present in cells of the early embryo which emit inductive signals.

Antibodies to recombinant Vg1 protein were used to demonstrate that the protein is first present in stage IV oocytes, and reaches peak levels in stage VI oocytes and eggs. During embryogenesis, the protein was synthesized until the gastrula stage. Embryonically synthesized Vg1 protein was present only in vegetal cells of the early blastula. The protein was found to be glycosylated and to be associated with membranes in the early embryo.

A small proportion of full-length Vg1 protein was shown to be cleaved to yield a small peptide.

The Vg1 protein seems to be an endogenous growth factor-like molecule that is involved in inducing mesoderm in the *Xenopus* embryo. There is preliminary evidence that Vg1 can act synergistically with basic fibroblast growth factor to induce mesoderm.

MPF-Induced Breakdown of Cytokeratin Filament Organization in the Maturing *Xenopus* Oocyte Depends upon the Translation of Maternal mRNAs
M. W. Klymkowsky and L. A. Mynell
Dev. Biol., 134, 479—485, 1989   4-5

One of the most dramatic events during oocyte maturation in *Xenopus* is breakdown of the asymetrically organized system of cytokeratin-type intermediate filaments. The animal-vegetal gradient characterizing this breakdown process could reflect a gradient in the maturational signal or an indirect signal, such as the diffusion of nuclear proteins following nuclear breakdown. Active maturation-promoting factor (MPF) appears within the oocyte in conjunction with new protein synthesis but, once activated, MPF appears to act autocatalytically on maternal stores of inactive MPF precursor to generate further active factor.

Monitoring of oocyte maturation *in vitro* confirmed that cytokeratin filament organization breaks down in an animal-to-vegetal direction. The breakdown process takes place in enucleated oocytes. Injection of MPF led to the breakdown of cytokeratin filaments independently of new protein synthesis. The effect of MPF did, however, require the translation of maternal mRNAs. Breakdown of the cytokeratin filament system was not observed when oocytes were treated with cyclohemimide before injection of MPF.

The factors controlling cytokeratin reorganization in the maturing *Xenopus* oocyte are distinct from those involved in breakdown of the nuclear envelope.

♦ One of the more sensational discoveries in the *Xenopus* field was the identification of Vg1 as a mRNA tightly localized in the vegetal cortex of the oocyte and encoding a TGFβ-related polypeptide (Weeks and Melton, 1987; see 1989 Year Book). These properties lead in an appealingly obvious way to a model in which localized Vg1 is the molecular basis for endodermal capacity to induce mesoderm. Unfortunately, this mRNA seems reluctant to accept such a glamorous role; the only published evidence that Vg1 protein does something important in development is a rather guarded reference by Tannahill and Melton to preliminary results showing a synergistic effect with fibroblast growth factor. To repeat my comment from the 1989 Year Book, "it would be astounding if this growth-factor-like polypeptide were not involved in mesoderm induction in some important way."

While the function of Vg1 may be elusive, it is clearly a localized RNA, and the mechanism of this localization is an interesting question in biology. Yisraeli et al., using various inhibitors, conclude that transport of Vg1 depends on microtubules, while microfilaments anchor the RNA in the vegetal cortex. Intermediate filaments do not appear very important in this scheme, although it is not clear from this paper what fraction of Vg1 is released when microfilaments are dissociated by cytochalasin B.

One circumstantial reason to suspect a role for intermediate filaments in Vg1 localization is the disruption of oocyte keratin filaments that coincides with release of this RNA from the cortical shell. The paper by Klymkowsky and Maynell reveals some new information about this cytoskeletal breakdown and compares it to other events taking place during oocyte maturation. *Thomas D. Sargent*

Epigenetic Control of Transgene Expression and Imprinting by Genotype-Specific Modifiers

N. D. Allen, M. L. Norris, and M. A. Surani
Cell, 61, 853—861, 1990   4-6

Expression and DNA methylation of the murine transgene locus TKZ751 are controlled by genotype-specific modifier genes. The authors have developed a method for studying the control of transgene expression utilizing a readily detected *lacZ* reporter gene linked with an HSVtk cryptic promoter. Expression of each of these transgenes is very prone to a range of positional effects imposed by the *cis*-acting factors of the chromosomal domains where they reside, and by *trans*-acting factors that may produce epigenetic changes on the integration loci.

The DBA/2 and 129 genetic backgrounds enhanced expression of the transgene locus, while the BALB/c background suppressed its expression — though only after maternal inheritance of the BALB/c modifier. Epigenetic change in the transgene locus occurred cumulatively over successive generations. Irreversible methylation was noted after three consecutive germ line passages.

The authors conclude that genotype-specific modifier genes regulate the penetrance and expressivity, as well as parental imprinting of the TKZ751 locus through epigenetic modification. The influence of strain-specific modifier genes on the imprinting of transgenes may be a general phenomenon.

♦ The mechanism of genomic imprinting is not fully understood. However, an epigenetic mechanism, DNA methylation, is felt to play at least some role in the process. The current article provides evidence that genotype may also play a role. The authors exploit a specific transgene locus background onto different genetic backgrounds. They show that the transgene locus is regulated by genotype-specific modifiers, and in some cases parental origin effects. In addition, they observe that unlike other imprinted loci, this locus fails to reprogram and restore totipotency in the germ line as is the case with other loci. Thus, this specific locus can be used as a model to investigate the epigenetic control of penetrance and expressivity in addition to the genetic basis of imprinting. *Joel M. Schindler*

Parental Origin of Mutations of the Retinoblastoma Gene
T. P. Dryja, S. Mukai, R. Petersen, J. M. Rapaport, D. Walton, and
D. W. Yandell
Nature, 333, 556—558, 1989   4-7

Retinoblastoma and osteosarcoma arise from cells that have lost both functional copies of the retinoblastoma gene. The use of cloned retinoblastoma genes and other linked polymorphic loci makes it possible to reconstruct the sequential loss of the two homologous gene copies which precedes tumor development. In hereditary tumors, the initial mutation is in the germ line, while in non-hereditary tumors, loss of each of the homologues occurs somatically.

Toguchida et al. recently found that the paternally derived copy is the first to become mutant during the development of non-hereditary osteosarcomas. The authors have found no such predilection for initial somatic mutations in analyzing retinoblastoma patients. Some 15 consecutive patients lacking a family history of retinoblastoma or osteosarcoma were studied; 8 informative patients had new germ line mutations. In each instance when an initial mutation was a new germ line mutation, it was derived from the father.

These findings cast serious doubt on the view that initial somatic mutations occur predominantly on paternal chromosome 13. In addition, they eliminate the theoretical need for genomic imprinting as a factor in somatic mutagenesis at the retinoblastoma locus.

Paternal Origin of New Mutations in Von Recklinghausen Neurofibromatosis
D. Jadayel, P. Fain, M. Upadhyaya, M. A. Ponder, S. M. Huson,
J. Carey, A. Fryer, C. G. P. Mathew, D. F. Barker, and B. A. J. Ponder
Nature, 343, 558—559, 1990   4-8

The estimated rate of new mutations producing Von Recklinghausen neurofibromatosis (NF-1), 1×10^{-4}, is one of the highest for any human disorder. The authors found the new mutation to be of paternal origin in 12 of 14 families. The maximal likelihood estimate of the fraction of mutations of paternal origin was 0.84. The average age of fathers from whom the new mutation arose, at the time the affected child was born, did not differ significantly from the average for live births in England and Wales for 1961 to 1965.

This finding is similar to that recently reported for retinoblastoma. In other genetic disorders exhibiting a bias toward a paternal origin of new mutations, the rate of mutation rises markedly with paternal age, suggesting that the mutations arise from replication errors in mitosis of spermatogonial germ cells. In retinoblastoma and NF-1, however, there is no marked paternal age effect.

Most NF-1 mutations likely arise either at mitosis in a cell that is not a self-renewing stem cell, or independently of mitosis (as in mature sperm). Paternal chromosomes probably are more susceptible to the mutation than are maternal chromosomes, for reasons that are not clear.

Genetic Imprinting Suggested by Maternal Heterodisomy in Non-Deletion Prader-Willi Syndrome

R. D. Nicholls, J. M. H. Knoll, M. G. Butler, S. Karam, and M. Lalande
Nature, 342, 281—285, 1989   4-9

Prader-Willi syndrome (PWS) is the most common form of dysmorphic genetic obesity associated with mental retardation. In about 60% of cases, there is a cytologic deletion of chromosome 15q11q13, occurring *de novo* only on the paternal chromosome. Angelman syndrome is a clinically distinct disorder associated with deletions of the same region which, however, occur *de novo* on the maternal chromosome. The parental origin of the affected chromosome in these disorders could contribute to the clinical phenotype.

Cloned DNA markers specific for the 15q11q13 subregion were employed to identify the parental origin of chromosome 15 in patients with PWS who lacked cytogenetic deletions. Probands in two of the families studied exhibited uniparental disomy for 15q11q13.

This is the first report that maternal heterodisomy can be associated with a human genetic disorder. Absence of a paternal contribution of genes in the involved region may be responsible for expression of the clinical phenotype. The authors conclude further that a gene or genes in region 15q11q13q must be inherited from each parent for normal development to take place. Isolation of the genes responsible for PWS and other disorders of genetic imprinting might be possible by identifying a gene that displays differential modification of its parental alleles.

Preferential Germline Mutation of the Paternal Allele in Retinoblastoma

X. Zhu, J. M. Dunn, R. A. Phillips, A. D Goddard, K. E. Paton, A. Becker, and B. L. Gallie

Nature, 340, 312—313, 1989   4-10

   In about two thirds of retinoblastoma (RB) cases, loss of heterozygosity for chromosome 13q14 triggers malignant proliferation. The normal RB gene (RB1) allele is lost and and a mutated allele remains in the tumor. Most patients with bilateral involvement lack a family history of RB and presumably have new germ line mutations arising in the egg, sperm, or early embryo.

   The authors examined 27 informative RB tumors for loss of heterozygosity. The parental origin of the retained chromosome was determinable in 12 of the 15 patients exhibiting loss of heterozygosity, and 9 tumors from 8 non-familial patients were finally analyzable (Figure 4-10). Six of these tumors retained the paternal allele, and three the maternal allele. Only one of three unilateral tumors retained the paternal RB1 allele, as did tumors from four of the five patients with bilateral tumors.

   There is no evidence that the paternal RB1 allele is preferentially retained in retinoblastoma, as has been proposed for osteosarcoma. The present findings suggest either that new germ line RB1 mutations arise more often during spermatogenesis than during oogenesis, or that imprinting in the early embryo influences chromosomal susceptibility to mutation. The paternal chromosome in the early embryo might be at greater risk of mutation or deficient in DNA repair.

♦ Genomic imprinting was identified first in the mouse through nuclear transplantation and genetic experiments. The latter experiments led to the identification of specific chromosomal regions showing an imprinted phenotype and more recently to the identification of the *IGF*-I locus which appears to be an imprinted gene (DeChiara et al., *Nature,* 345, 75, 1990). Subsequent to this work has been the identification of human syndromes and disease states that demonstrate an imprinted phenotype. For example, *NF*-I (Von Recklinghausen neurofibromatosis) and *RB*I (retinoblastoma) both show a predominance of new germ-line mutations which are paternally derived. The frequency appears not to be related to paternal age. Furthermore, equal transmission of the mutant gene through the male or female germ line excludes selection against maternally derived mutations as an explanation for high paternal origin of new mutations. This suggests that the mutations probably arise in a cell type (perhaps mature sperm) that is not a self-renewing stem cell. The germ-line results reported for *NF*-I and for *RB*I are unlike other disorders such as hemophilia A and Lesch-Nyhan syndrome which also show a bias for paternal origin of new mutations.

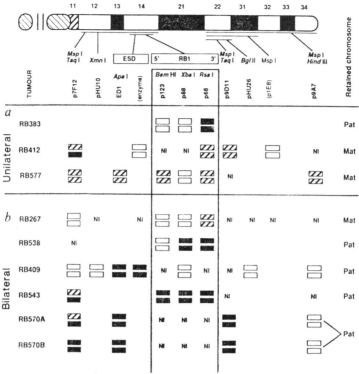

FIGURE 4-10. Parental origin of the retained chromosome 13 in RB. The approximate loci of the RFLP markers on chromosome 13 is indicated at the top. (a) Unilateral, (b) bilateral. The two markers for which only previously published data are included, are bracketed, NI, noninformative for LOH. Boxes: diagonal stripes, maternal allele; black, paternal allele; white, LOH but not NI for parental origin. Identified RB1 mutations are as follows: RB543 germline, 10-bp deletion exon 18; RB570 germline, 9-bp deletion exon 19; RB538 germline, 55-bp duplication exon 10. Methods; RB tumor DNA was obtained from surgical specimens, supplemented by xenografts in immune-deficient mice and/or in tissue culture. Constitutional DNA of the patients and parents was obtained from lymphoblasts of lymphoblastoid cell lines. DNA was prepared by a standard method. The probes used in this study were generously provided by T. Dryja, W. K. Cavenne, and J. Squire. All probes were labeled with [α-^{32}P] dCTP by random primer labeling to an activity of 4 to 6 × 10^8 c.p.m. μg^{-1} DNA. DNA samples were digested with appropriate restriction enzymes, electrophoresed in 0.7 to 1.0% agarose gel and transferred to Zeta-probe or Hybond membranes. Hybridization was performed as described previously. After hybridization, filters were routinely washed twice at room temperature in 2 × SSC, 0.1% SDA and once under high stringency conditions in 0.1 × SSC, 0.1% SDS at 65°C. Autoradiographs were developed after 24 to 48 h at –70°C. (From Zhu, X., Dunn, J. M., Phillips, R. A., Goddard, A. D., Paton, K. E., Becker, A., and Gallie, B. L., *Nature*, 340, 312—313, 1989. With permission.)

However, in these cases, a strong paternal age effect has been reported which is most easily explained if most new mutations occur at mitosis in spermatogonial stem cells. Oogonia would be spared because they do not undergo mitotic divisions once arrested in prophase early in fetal life. Other tumors (nonheritable type of Wilms tumor and osteosarcoma) also show a strong bias towards retention of the paternal allele. These results imply that paternal and maternal alleles for specific genes differ in susceptibility to mutagenic events. The differences are thought to result from mechanisms leading to genomic imprinting.

Prader-Willi and Angelman syndromes also show unusual parental origins. Both are very different clinical disorders often associated with deletions of 15q11q13. However, for Prader-Willi the deletions are exclusively of paternal origin, whereas for Angelman's the deletions occur on the maternal chromosome. Results described in the most recent report show that for 15 nondeletion Prader-Willi cases, the clinical phenotype arises from absence of paternal contribution to region 15q11q13 rather than from a specific mutation. This implies functional differences in alleles of a gene or genes from this region of the genome that depend on the sex of the transmitting parent (genetic imprinting). Thus, although genetic imprinting has been identified as a fundamental biological concept for a number of years, it is only now that individual genes are being identified as imprinted candidates. *Terry Magnuson*

Cell Interactions 5

INTRODUCTIONS

Cell interactions imply both physical and functional relationships. During development, both are important, and both can result in different developmental outcomes. When cells interact physically, some type of structural integrity is altered. This structural relationship could in itself be developmentally important, or could serve as a signal for subsequent developmental events. Thus, the act of physical contact, through an array of identified adhesion molecules, could itself initiate a cascade of biochemical signals and second messengers that result in initiating or suppressing some developmental event. Cell interactions can also occur at a distance as diffusible compounds convey information necessary for development to proceed.

Both types of interactions require a cell-surface architecture that can respond to the appropriate signal, be it direct or indirect. As a result, much effort has focused on describing the molecular nature of cell surface receptor molecules involved in the transduction of developmental signals. Such molecules could be cell-type specific, belong to large super-families of molecules (e.g., EGF-like molecules), or contain a well-characterized core that can be modified depending on cell type or developmental event.

Cell interactions are particularly critical for histogenesis. For appropriate tissues to be formed, the correct intercellular signals are essential. The development of both the immune system (cellular immunity) and the nervous system (axon outgrowth) employ sophisticated cell-interaction mechanisms to ensure normal development. Both are extensively studied to dissect the underlying molecular mechanisms.

Identification of an Octapeptide Involved in Homophilic Interaction of the Cell Adhesion Molecule gp^{80} of
Dictyostelium discoideum
R. K. Kamboj, J. Gariepy, and C. Siu
Cell, 59, 615—625, 1989   5-1

In the early stages of its development, the cellular slime mold *D. dis-*

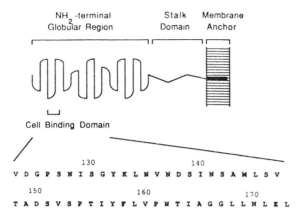

FIGURE 5-1. Schematic model of gp80 and its cell binding domain. The cell binding domain has been mapped to a segment between Val-123 and Leu-173 (Kamboj et al., 1988). The shaded amino acids represent the sequence of peptide 3. (From Kamboj, R. K., Gariepy, J., and Siu, C., *Cell*, 59, 615—625, 1989. With permission.)

coideum achieves a multicellular state through chemotactic migration and the expression of cell-cell adhesion sites on its surface. The surface glycoprotein gp80 mediates EDTA-resistant cell-cell adhesion through homophilic interaction. The authors raised an anti-gp80 monoclonal antibody directed against a 13-amino acid sequence (13-mer) near the NH$_2$ terminus of the protein in order to directly assess the cell adhesion properties of gp80.

Antibody inhibited cell reassociation, and also inhibited gp80-cell and gp80-gp80 interactions. The cell binding site was mapped to the octapeptide sequence YKLNVNDS (Figure 5-1). High salt concentrations inhibited homophilic interactions of both 13-mer and gp80, indicating that ionic interactions have a role in the forward binding process. Disruption of homophilic interactions between bound molecules required the presence of Triton X-100.

Homophilic interaction at the defined site probably involves pairing of the positively charged Lys-133 on one gp80 molecule with the negatively charged Asp-138 on the other and vice versa. Whether this mechanism of cell-cell adhesion is conserved among similar adhesion molecules remains to be determined.

Selective Elimination of the Contact Site A Protein of
Discyostelium discoideum **by Gene Disruption**
C. Harloff, G. Gerisch , and A. A. Noegel
Genes Dev., 3Z, 2011—2019, 1989   5-2

The contact site A glycoprotein is a developmentally regulated, cell-surface protein expressed during the aggregation stage of *D. discoideum* development. It has been implicated in EDTA-stable, calcium-independent adhesion of aggregating cells. The gene coding for contact site A protein was disrupted by homologous recombination using a transformation vector containing a 1-kb cDNA fragment as an insert. Transformants not expressing the protein were identified by colony immunoblotting.

The transformations not expressing contact site A glycoprotein did produce three truncated contact site A transcripts. One of these was regulated by the original contact site A promoter, while the other two were transcribed from the actin 6 promoter of the vector. EDTA-stable adhesion — examined by agitating suspended cells in an agglutinometer — was markedly reduced compared with the wild type. Nevertheless, aggregation of the transformed cells on an agar surface was not much changed.

The contact site A glycoprotein of *D. discoideum* seems to be responsible for a "rapid" type of cell adhesion that comes into play when aggregating cells are subject to shear. When cells are not mechanically disturbed, a "slow" type of adhesion, mediated by other molecules, suffices for aggregation.

♦ Nearly 30 years ago, Gerisch and colleagues introduced the use of antibodies to the cell surface of *Dictyostelium discoideum* as probes to study cell-cell adhesion and identify the molecules responsible for this important morphogenetic process. This strategy was quickly adopted by others and has resulted in much progress in the identification of the cell-adhesion molecules of the vertebrate embryo. With regard to *Dictyostelium*, Gerisch was able to show that the antibodies that blocked the so called contact site A, which is responsible for EDTA resistant cell adhesion during aggregation, recognize an 80-kDa glycoprotein (gp80). Considerable effort has gone into the characterization of the gp80 molecule and its biosynthesis, but definitive evidence for its role *in vivo* has remained elusive. Both papers reviewed here address this problem although the conclusion remains somewhat uncertain.

Kamboj et al. follow up on previous work from the Siu laboratory which identified a specific region of the N-terminus of the gp80 molecule as the homophillic binding domain. They raised antibodies to synthetic peptides representing this region and show that they can block EDTA resistant cell adhesion. In addition the peptides block the binding of gp80 to cells or immobilized gp80. Critical amino acids for binding are identified and a molecular model is presented for the homotypic binding. Most significantly, the peptides can block the formation of streams of cells during the aggregation of cells on a plastic substrate under buffer. Overall, these studies indicate that the cell binding peptide has been identified and that this region is essential for the proper cohesion of the cells during aggregation.

Taking a different approach, Harloff et al. have used homologous recombination to knock out the endogenous gp80 gene. The transformants transcribe the disrupted gene but do not make any gp80 protein or peptide fragments. These mutants are also defective in EDTA-resistant adhesion, but aggregate normally — making large streams — when allowed to develop on an agar surface. Clearly, these results are not what the peptide studies by Kamboj et al. would have predicted. One serious discrepancy are the assays used by these two groups. It is important to determine whether the gp80 mutants are defective in the liquid aggregation system. However, even if the mutants are found to be defective under these conditions, aggregation is normal on agar, which can be argued resembles more closely the normal developmental condition for this organism. Therefore, the next step seems to be to identify the molecule that is actually responsible for holding these cells together during aggregation. These studies point out the danger of equating an *in vitro* assay like EDTA resistance to *in vivo* cell adhesion. They also point out the importance of a system like *Dictyostelium* where the function of putative molecules can actually be tested in the context of the organism's normal development. *Stephen Alexander*

A Pair of Tandemly Repeated Genes Codes for gp24, a Putative Adhesion Protein of *Dictyostelium discoideum*
W. F. Loomis and D. L. Fuller
Proc. Natl. Acad. Sci. U.S.A., 87, 886—890, 1990   5-3

Cells of *D. discoideum* become mutually adhesive a few hours after the start of development. The glycoprotein gp24 is implicated in cell-to-cell adhesion in this organism. In this study, a cDNA clone coding for gp24 was used to screen cloned genomic fragments from cells that had developed for 8 h.

The study revealed two closely linked genes, GP24A and GP24B, which generates mRNAs of about 650 base pairs after excision of a small intron and the addition of poly(A). These appeared to have arisen through tandem duplication of about 800 base pairs followed by divergence. The genes were expressed within a few hours of the start of development. Their mRNAs peaked at 12 h and persisted at these levels until culmination. Both genes contained short guanine-rich sequences, or G boxes, upstream. These have been implicated in the transcriptional regulation of other genes expressed during the development of *Dictyostelium*. Their mRNAs code for proteins exhibiting 85% identity, and which have a hydrophobic domain followed by a highly charged carboxyl-terminal domain.

♦ Intercellular cohesion is a pivotal process in the development of *Dictyostelium discoideum* and has proven to be complex. At least three sequentially expressed cell adhesion molecules are involved in the acquisition and maintenance of multicellularity. The protein expressed earliest in development is gp24. Loomis and Fuller have shown now that this protein is encoded by two very similar genes (GP24A and GP24B) which arose by tandem duplication. The proteins encoded by these two genes are only 102 and 103 amino acids in length, respectively. Both genes are expressed during development commensurate with gp24 expression. The genes also contain a 5′G-box sequence similar to those found in other *Dictyostelium* genes differentially expressed during development. Interestingly, the gene products appear to be O-glycosylated and this post-translational modification results in the larger molecular weight observed on SDS-gels. The duplicity of the GP24 genes account for attempts to isolate null mutants. Perhaps antisense mutagenesis can overcome this limitation. It will be interesting to see how this small molecule functions to cause intercellular cohesion and whether the O-linked glycans are involved. *Stephen ALexander*

Characterization of a Glycosyl-Phosphatidylinositol Degrading Activity in *Dictyostelium discoideum* Membranes

A. M. Da Silva and C. Klein

Exp. Cell Res., 185, 464—472, 1989   5-4

There is increasing evidence that a number of proteins are membrane-bound through a covalent linkage with a glycolipid moiety, a glycosyl-phosphatidylinositol (GPI) anchor. Antigen 117 is a glycolipid-anchored cell-surface protein implicated in cell-to-cell cohesion in *D. discoideum* amoebae. Previous studies have shown that some of the protein is released from the cell surface during cell aggregation.

This study was designed to characterize the enzymatic activity involved in release of antigen 117. The releasing enzyme appeared to be a phosphatidylinositol phospholipase C. It is an integral membrane protein which is inhibited by thiol-reactive agents; is insensitive to divalent cation chelators; and exposes the CRD epitopes of the antigen. About 20 to 25% of the antigen can be released from membranes of aggregation-competent cells.

Antigen 117 disappears rapidly from the surface of amoebae when they complete the aggregation phase of development. This may reflect increased activity of the releasing enzyme or, alternately, conversion of the substrate to a more readily hydrolyzed form.

◆ Recent work has shown that some cell surface molecules in *Dictyostelium discoideum,* like other organisms, are anchored via a glycosylphosphatidylinositol (GIP) tail. This is true for the well-studied cell-cohesion molecule gp80 which is expressed and active during the aggregation phase of development. Approximately 20% of the cell surface gp80 protein is released from the cell surface during aggregation in intact form, but the mechanism of this release was unknown. Da Silva and Klein have now shown that there is an endogenous releasing activity associated with the plasma membranes. The activity appears to be quite specific for gp80, and although it has some of the characteristics of phospholipase C, it is clearly a different enzyme. This is not unexpected, as it was previously shown that the gp80 molecule has an unusual inositol-phosphoceramide linkage which is resistant to phospholipase C. Perhaps this enzyme is responsible for the developmental loss of gp80, and possibly other GIP-anchored molecules, from the cell surface. This is an attractive hypothesis which could account for rapid substitution of cell-surface molecules during development. It will be interesting to see whether the expression of the enzyme is developmentally regulated and why only 20% of the gp80 molecules are sensitive to cleavage. *Stephen Alexander*

C-Factor: A Cell-Cell Signaling Protein Required for Fruiting Body Morphogenesis of *M. xanthus*

S. K. Kim and D. Kaiser
Cell, 61, 19—26, 1990   5-5

During the development of fruiting bodies in *M. xanthus,* the product of the *csgA* gene is required for cell aggregation, spore differentiation, and gene expression which begins following 6 h of starvation. The question arose of whether *csgA* specifies the signal molecule itself or some other essential component of the signaling system. In order to identify the *csg* signal molecule or molecules, a bioassay based on a completion of development by a *cgsA* mutant was constructed.

A 17-kDa polypeptide, C-factor, was purified from nascent wild-type fruiting bodies. At concentrations of 1 to 2 n*M*, the polypeptide restored normal development to *csgA* mutant cells. C-factor activity was not recovered from extracts of unstarved, growing cells or *csgA* mutant cells. Amino acid sequencing of purified C-factor confirmed that it is encoded by the *csgA* gene.

C-factor is a polypeptide active over a narrow concentration range, which has the properties of a morphogenetic paracrine signal. It seems likely that the factor is transferred from cell to cell and functions at the cell surface.

The apparent tight association of C-factor with membrane suggests that intercellular signaling between wild-type cells may take place over a very short range, perhaps only between immediately contiguous cells.

♦ Cell interactions are clearly an important mechanism that guides morphogenesis and the underlying program of gene expression. However, it is often difficult to identify the molecules and mechanism responsible for specific signaling.The paper by Kim and Kaiser identifies such a pivotal molecule which is responsible for inducing specific gene expression and subsequent fruiting body construction in *M. xanthus.* This study takes advantage of earlier work that defined groups of mutants blocked at different stages of development that could be rescued by other mutants or wild-type cells. The molecule in question is made only in wild-type cells during a restricted period of development and appears to be attached to the cell surface. It has a molecular mass of 17 kDa. It rescues all complementation group C mutants and turns on genes which are not expressed in these mutants. These experiments open the way to the isolation of the signalling molecules and their receptors which are active at other stages in development and determining how the signals are transduced. *Stephan Alexander*

Cell Interactions in the Sea Urchin Embryo Studied by Fluorescence Photoablation
C. A. Ettensohn
Science, 248, 1115—1118, 1990   5-6

In the sea urchin embryo, primary mesenchyme cells (PMCs) regulate the development of secondary mesenchyme cells (SMCs) which appear late in gastrulation. The SMCs give rise to muscle cells, pigment cells, and parts of the coelomic sacs; they do not normally contribute to the larval skeleton. This study used a fluorescence photoablation technique to specifically ablate PMCs at different stages of development in order to analyze the time of cell interactions. In addition, PMCs were microinjected into PMC-depleted recipient embryos at various developmental stages, and the effects on SMC fate were recorded.

The critical interaction between PMCs and SMCs occurred briefly late in gastrulation. Before that stage, SMCs were insensitive to the suppresive signals transmitted by the PMCs. Photoablation at early stages led to a complete conversion response (Figure 5-6). Reintroduction of PMCs as late as the mid-gastrula stage, 3 h before the start of conversion, led to a reduction in the number of converted cells (Figure 5-6A).

These findings indicate that PMCs need not transmit a signal continuously during gastrulation. It may be that the SMCs become developmentally committed when they enter the blastocoel and convert to a skeletongenic phenotype, unless they interact with viable PMCs at that time. Whatever the specific mechanisms involved in the inhibitory interaction, the result is suppression of a complex array of phenotypic on the part of the converting SMCs.

♦ The secondary mesenchyme cells (SMC) of the sea urchin arise from the tip of the invaginating endoderm late in gastrulation. This lineage differentiates into cells with diverse fates including muscle, pigment, coelomic epithelium, and also has the potential for a skeletogenic phenotype. Cells of the primary mesenchyme lineage (PMC), which arise from the micromeres formed and determined at the fourth cleavage, normally construct

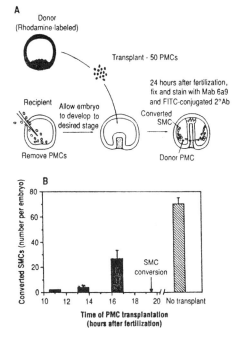

FIGURE 5-6. SMC commitment. (A) PMCs were removed from recipient embryos at the mesenchyme blastula stage (10 h after fertilization). The embryos were allowed to continue development or various times, then were loaded again into microinjection chambers and 50 to 60 RITC-labeled PMCs from mesenchyme blastula stage donor embryos were microinjected into the blastocoel. At 24 h after fertilization (14 h after PMC ingression) embryos were fixed and stained by indirect immunofluorescence with MAb 6a9. (B) The time (in hours after fertilization) at which labeled PMCs were introduced into the blastocoel and the numbers of converted SMCs. The number of SMCs that convert in embryos in which no PMCs are present is approximately 70. The start of SMC conversion as determined by immunofluorescent staining of PMC-depleted embryos with MAb 6a9 is shown. Error bars indicate 95% confidence limits on the mean. For each developmental stage, 5 to 11 embryos were scored. (From Ettensohn, C. A., *Science*, 1115—1118, 1990. With permission.)

the skeleton and suppress the potential of SMC to do so. Conversion of SMC to a skeletogenic fate was previously shown to be regulated by the number of PMC in the blastocoel (see 1989 *Year Book of Developmental Biology*, p. 133). Removal of all PMC by microsurgical techniques resulted in the SMC conversion of a full compliment of 50 to 60 cells to a skeletogenic fate, whereas removal of a portion of the PMC population resulted in conversion of only a similar number of SMC.

The nature of the signal used by the PMC for suppressing SMC conversion is unknown, but Ettensohn's paper demonstrates that this negative regulation occurs within a narrow window of development: progenitor SMC are insensitive to suppression, and their potential for conversion is quickly lost upon entering the blastocoel. Thus, the SMC lineage does not retain stem cells with a skeletogenic potential. The brief period of competence of SMC for conversion, coupled with the fact that different SMC

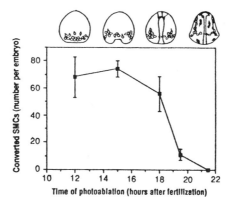

FIGURE 5-6A. Fluorescence photoablation. Numbers of converted SMCs were determined by immunofluorescent staining with MAb 6a9. Embryos were photoablated at various developmental stages and fixed 22 h after PMC ingression, after the completion of SMC conversion. PMC ingression in *L. variegatus* takes place 10 h after fertilization at 24°C. The developmental stages of the embryos at the time of photoablation are illustrated. Bars indicate standard errors (95% confidence limits on the mean). For each developmental stage, 5 to 11 embryos were scored. (From Ettensohn, C. A., *Science*, 1115—1118, 1990. With permission.)

continue to enter the blastocoel over a relatively long period to time, suggests that conversion potential is lost consecutively by different SMC in the population. This could be an important factor in regulating the number of SMC that convert. Converting SMC rapidly begins to express PMC-specific genes (C. Ettensohn, personal communication) that may contribute to the suppression of other members of their lineage who have not yet become competent. *Gary M. Wessel*

Selective Silencing of Cell Communication Influences Anteroposterior Pattern Formation in *C. elegans*
D. A. Waring and C. Kenyon
Cell, 60, 123—131, 1990

Local cell-cell interactions have a central role in pattern formation in many developing organisms. In *C. elegans* males, laterally located V cells generate a pattern of anterior alae, or cuticular ridges, and posterior rays, or mating sensilla. Two homeotic genes, *mab-5* and *lin-22,* are required for V cell patterning. The authors have found that this pattern is produced, at least in part, through the selective interruption of cell-cell interactions.

In anterior V cells, lineages leading to the production of alae are induced by cell interactions which, in specific posterior V cells, are inhibited by activity of the gene *pal-1*. This gene is required to establish a difference between the posterior V6 and the anterior V cells. The action of *pal-1* allows some posterior V cells to generate rays rather than alae.

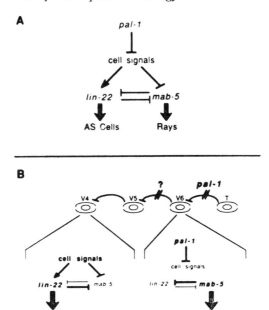

FIGURE 5-7. A model of the regulatory relationships between cell signals, *pal-1*, *lin-22*, and *mab-5* activity in the generation of the wild-type pattern of rays and AS cells. (A) Regulatory relationships between *mab-5*, *lin-22*, *pal-1*, and intercellular signals. In formal terms, *mab-5* and *lin-22* mutually inhibit one another's activities: *mab-5* specifies lineages that produce rays and *lin-22* specifies lineages that produce AS cells. This mutual antagonism is suggested by the phenotype of single *mab-5* and *lin-22* mutants, the *mab-5*, *lin-22* double mutant, and the response of the pattern to changes in *mab-5* or *lin-22* gene dosage. Signals from neighboring V cells somehow tip the *lin-22-mab-5* balance in favor of *lin-22*. In principle, cell signals could lead to the modification of *mab-5*, *lin-22*, or both activities. The signals from T to V6 are somehow silenced by the activity of *pal-1*. In formal terms, *pal-1* activity is equivalent to the removal of T. *pal-1* could inhibit the production of the T signal, as implied by the drawing. However, we emphasize that *pal-1* might just as easily affect the signal transduction machinery within V6 or act more directly to enhance *mab-5* activity and/or inhibit *lin-22* activity. Finally, as with any genetic model, we emphasize that none of these interactions need be direct ones. (B) Selective interruption of cell signaling leads to precise V cell patterning. The data suggest that V5, V6, and T each have the potential to signal their anterior neighbors to favor *lin-22* over *mab-5* activity, and thus to produce AS cells. Signaling could occur between the V and T cells themselves, or alternatively, between their descendants. In the wild type, intercellular signals cause V4 to generate AS lineages. In V6, *pal-1*+ blocks or overrides the signal from T, and as a consequence V6 generates ray and not AS cell lineages. Apparently, *mab-5* activity is naturally favored over *lin-22* activity in the absence of cell interactions (at least in this body region). The circuitry that allows V5 to produce both a ray and an AS cell is not understood. Because V5 generates its AS cell in a V6-dependent fashion, V6 must signal V5 in the wild type. In addition, it seems likely that an activity analogous to *pal-1* acts on the signal between V6 and V5 (or certain of their descendants) to allow the formation of the V5 ray. Waring, D. A. and Kenyon, C., Cell, 60, 123—131, 1990. With permission.)

Activity of this gene and of the cell signals that influence V cell fates appears to determine the state of a developmental switch involving the homeotic genes *lin-22* and *mab-5*. Mutations in *pal-1* and cell ablations elicit the same types of V cell transformations as mutations in *mab-5* and *lin-22*.

Intercellular signals cause V cells to generate adult seam (AS) lineages (which produce alae) instead of ray lineages (Figure 5-7). In anterior V cells, where *pal-1* does not function, V cells respond to signals from neighboring cells and generate AS rather than ray lineages In contrast, the interaction between T and V6 posteriorly is inhibited or bypassed by *pal-1* activity and, as a result, V6 generates rays instead of AS lineages.

Cell-Cell Interactions Prevent a Potential Inductive Interaction between Soma and Germline in *C. elegans*
G. Seydoux, T. Schedl, and I. Greenwald
Cell, 61, 939—951, 1990   5-8

Similar receptors and/or signals occasionally may serve to mediate several distinct cell-cell interactions in the same organism. In such circumstances, it may be necessary to "isolate" different sets of interacting cells from one another to ensure that only the correct interactions actually take place. Such a situation occurs in the gonad of the *C. elegans* hermaphrodite, where discrete soma-soma and soma-germ line interactions occur during development. In each gonadal arm of wild-type hermaphrodites, the somatic distal tip cell (DTC) maintains distal germ line nuclei in mitosis, while proximal nuclei enter meiosis.

Two conditions were found under which a proximal somatic cell, the anchor cell (AC), inappropriately maintains proximal germ line nuclei in mitosis: when defined somatic gonadal cells are ablated in wild-type, and in *lin-12*-null mutants. Laser ablation studies and mosaic analysis showed that somatic gonadal cells near the AC ordinarily require *lin-12* activity to prevent the inappropriate AC germ line interaction (Figure 5-8).

This interaction, like that between the DTC and germ line, requires *glp-1* activity. One model for how somatic cells neighboring the AC normally prevent it from interacting with the germ line involves an intercellular signal from the AC which interacts with the *lin-12* product in somatic gonadal cells. When *lin-12* activity is absent, the signal interacts with the related *glp-1* product in germ line (Figure 5-8).

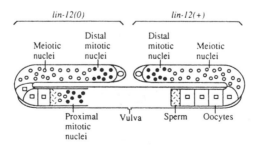

FIGURE 5-8. Germline in a Z1/Z4 mosaic hermaphrodite. The diagram depicts the germline phenotype of 13 of 14 mosaic hermaphrodites in which *lin-12* activity was absent either from the Z1 lineage or the Z4 lineage but present in the rest of the animal, including the germline (type II and III mosaics from Figure 4). In these mosaics, one somatic gonadal arm is genotypically *lin-12*(+) and the other is *lin-12*(0). The *lin-12*(+) arm is phenotypically wild-type and the *lin-12*(0) arm exhibits the proximal mitosis defect. (From Seydoux, G., Schedl, T., and Greenwald, I., *Cell,* 61, 939—951, 1990. With permission.)

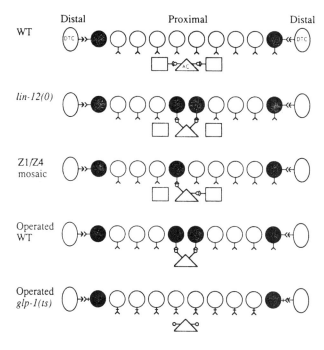

FIGURE 5-8A. A model to explain the observed phenotypes. Each diagram represents a schematic view of a late L2 gonad. At this stage all germ cells (circles) are mitotic, but some are induced to remain mitotic (solid circles), while others are not (open circles; these will enter meiosis in the L3 stage). As described in the text, this model is based on the proposal that *lin-12* is the receptor for the AC-to-VU signal and that *glp-1* is the receptor for the DTC-to-germline signal. (WT) At each distal tip, the DTC-to-germline signal (arrow) binds to the *glp-1* product (inverted arrow), thus maintaining distal germline nuclei in the mitotic cycle (solid circles). Proximally, germline nuclei are not maintained in mitosis (open circles), since *glp-1* is not bound. (It is not known where *glp-1* is located in the germline. *glp-1* may reside in the membranes that partially enclose each nuclei or in the membrane that surrounds the germline; these membranes are not shown.) The AC-to-VU signal (small circle) binds to *lin-12* (half circle) on the surface of certain somatic gonadal cells (squares). (These cells need not be only VU cells but could also be SH/SP/DIs.) The AC-to-VU signal must be present at least in the last L2 stage since the AC/VU decision is made at that time (Kimble, 1981; Greenwald et al., 1983; Seydoux and Greenwald, 1989). The AC-to-VU Signal may be locally diffusible. Oval, distal tip cell; triangle, anchor cell. *(lin-12(0))* In the absence of *lin-12*, the AC-to-VU signal binds to *glp-1*, thus stimulating proximal germline nuclei to remain in mitosis. Although *lin-12(0)* hermaphrodites have at least one extra AC (not indicated in this diagram), only one is necessary for proximal mitosis. (Z1/Z4 mosaic) One somatic gonadal arm is wild type while the other lacks *lin-12(+)* activity. Thus, in one gonadal arm the AC-to-VU signal interacts with *lin-12*, while in the other arm it is free to interact with *glp-1*. Only the proximal germline nuclei in the *lin-12(0)* arm are stimulated to stay in mitosis. (Operated WT) Since the somatic cells expressing *lin-12* have been ablated, the AC-to-VU signal is free to interact with *glp-1* and stimulate mitosis in proximal germline nuclei. (Operated *glp-1(ts)*) Although the *glp-1(ts)* product can bind to the DTC-to-germline signal and stimulate mitosis distally, it is not able to bind to or be activated by the AC-to-VU signal even in the absence of *lin-12* activity resulting from the ablation of neighboring somatic cells. (From Seydoux, G., Schedl, T., and Greenwald, I., *Cell,* 61, 939—951, 1990. With permission.)

♦ Cell-cell interactions play essential roles in pattern formation in many organisms. These two papers describe an interesting twist on the cell-cell communication theme: how specific cell-cell interactions must be prevented or silenced for development to proceed normally. The lateral hypodermal V6 cell in *C. elegans* males normally produces rays, which are part of the mating structure. Waring and Kenyon have shown that in the absence of wild-type *pal-1* gene product, V6 is induced by its posterior neighbor to make cuticular ridges instead of rays. The ability of V6 to produce rays can be restored either by ablating the posterior neighbor cell or by expressing *pal-1+* product to inhibit the signaling between V6 and its neighbor. Seydoux et al. analyzed a different cell-cell interaction, that between the anchor cell and germ cells in the proximal region of the gonad. In wild-type adult hermaphrodites, the distal tip cell of the somatic gonad maintains distal germ cells in mitosis; proximal germ cells are in meiosis. In the absence of *lin-12* gene product, the anchor cell inappropriately maintains proximal germ cells in mitosis. (This is in addition to the effect of the distal tip cell on distal germ cells.) The anchor cell can have the same effect on proximal germ cells in wild-type hermaphrodites after ablation of some of the somatic gonad cells that surround the anchor wall. Therefore, inappropriate interactions between the anchor cell and proximal germ cells are inhibited by intervening somatic gonad cells and by expression of *lin-12+* product. *Susan Strome*

Cell Interactions Coordinate the Development of the *C. elegans* Egg-Laying System

J. H. Thomas, M. J. Stern, and H. R. Horvitz
Cell, 62, 1041—1052, 1990   5-9

In *C. elegans,* many cell fates are determined by cell interactions. Egg laying requires functioning of the vulva, the gonad, the egg-laying muscles, and two HSN neurons that innervate these muscles. The authors analyzed a new mutant, *dig-1,* which displaces the gonad, and found that cell interactions coordinate spatial relationships among the various components of the egg-laying system.

Mutants homozygous for the recessive mutation *dig-1* have a fertile, morphologically normal gonad which is displaced anteriorly (Figure 5-9). The gonad induces formation of the vulva, and vulval induction by dorsal gonads strongly suggested that the inductive signal is able to act at a distance (Figure 5-9A). The gonad was found to act at a distance to regulate migration of the sex myoblasts which generate the egg-laying muscles. Initial migration is regulated by a gonad-independent mechanism, while a gonad-dependant one directs the sex myoblasts precisely to their final positions (Figure 5-9B). The positions of the axonal branch and

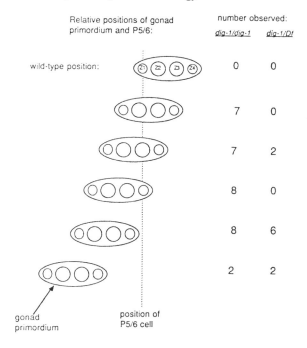

FIGURE 5-9. The somatic gonadal cells that affect the SM migration. (A) Ventral view of the cells in the somatic gonad at the time of the termination of the SM migration. The cells Z1 and Z4 give rise to all 12 somatic gonad cells (see text) by the invariant cell lineages that are shown. Their lineages are related by rotational symmetry around a point in the center of the gonad. These cells do not divide further until the end of the SM migration. (B) The positions of the SMs are shown with respect to the Pn.p cells as described for Figure 7. The distance between the left and right SMs (SML and SMR, respectively) is given in units of the distance between adjacent Pn.p cells. The anchor cell (a descendant of either Z1.pp or Z4.aa) organizes the three nearest Pn.p cells before their divisions such that the middle Pn.p cell of these three lies directly ventral of the center of the gonad. In the absence of the anchor cell (after killing either Z1.p/Z4.a or Z1.pp/Z4.aa), the Pn.p cells are not aligned as precisely at the center of the gonad. We assume that this accounts for the small degree of scatter observed after killing of Z1.pp/Z4.aa. (From Thomas, J. H., Stern, M. J., and Horvitz, H. R., *Cell,* 62, 1041—1052, 1990. With permissio.n)

synapses of each HSN neuron were displaced with the rest of the egg-laying system in *dig-1* animals.

In *C. elegans*, cell interactions control the relative positioning of the essential components of the egg-laying system. These interactions allow a hermaphrodite with a displaced gonad to maintain a functionally intact egg-laying system. Such a capacity for spatial regulation may have made possible evolutionary flexibility in the location of the nematode reproductive system.

♦ Cell interactions can be both direct (cell contact) and indirect (diffusible

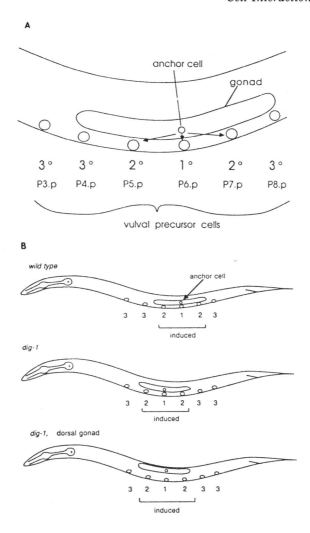

FIGURE 5-9A. Vulval induction in L3 wild-type and *dig-1* animals. (A) The model of wild-type vulval induction (modified from Sternberg and Horvitz, 1986). The gonadal size indicated is that of the late second larval stage, at about the time of vulval induction. The anchor cell induces the nearest P(3-8).p cells, as signified by the arrows, to under 1° and 2° lineages. Other cells remain uninduced. (B) Induction of P(3-8).p by anteriorly displaced or dorsal gonads in L3 *dig-1* animals. The second diagram represents the induction of a more anterior set of three of the P(3-8).p cells in a *dig-1* animal where the gonad is displaced one P(3-8).p cell anteriorly. The third diagram represents one type of induction pattern seen when the gonad is located dorsally and one P(3-8).p cell further anterior. The three closest P(3-8).p cells on the ventral side of the animal are induced to form the normal 2° 1° 2° pattern, despite the misplaced anchor cell. (From Thomas, J. H., Stern, M. J., and Horvitz, H. R., *Cell*, 62, 1041—1052, 1990. With permissio.n)

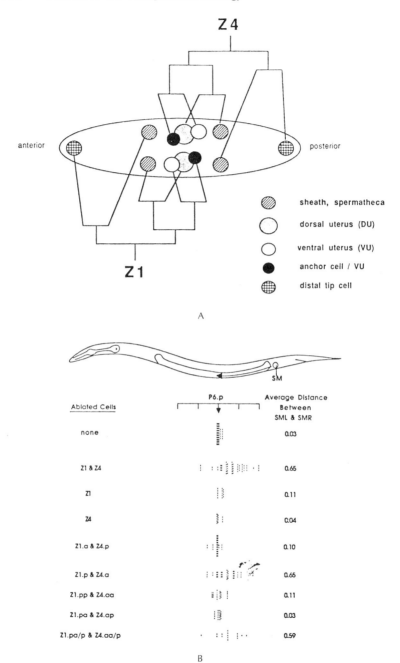

FIGURE 5-9B. Position of the gonad primordium in *dig-1* and *dig-1/Df* L1 larvae. The degree of anterior shift of the four-cell gonad primordium in L1 animals is shown for *dig-1(n1321)* homozygotes and *dig-1(n1321)/nDf16* heterozygotes. *nDf16* deletes *dig-1*. The gonad position is shown relative to one representative nongonadal cell, P5/6. In *dig-1* animals, the spatial organization of nongonadal cells is normal, and only the gonad was observed to be displaced. (From Thomas, J. H., Stern, M. J., and Horvitz, H. R., *Cell*, 62, 1041—1052, 1990. With permissio.n)

substances). The egg-laying system of the nematode involves four differ-ent organs that must interact in order to successfully extrude eggs. These four organs include the gonad, the vulva, two sets of musculature, and two neurons. All are in relative contact with each other in the wild-type worm. In the current article, the authors isolate a mutant phenotype that exhibits a displaced gonad. Despite this defect, the egg-laying system of the worm is functionally normal. This observation suggests that the inter-actions necessary for the egg-laying machinery to function are indirect — diffusible substances direct the communication between the participating organs. The isolation of additional mutants with defects in the other organs of the egg-laying machinery will help define the nature of the cell interactions necessary for this biological process to proceed. *Joel M. Schindler*

Drosophila Chaoptin, a Member of the Leucine-Rich Repeat Family, Is a Photoreceptor Cell-Specific Adhesion Molecule
D. E. Krantz and S. L. Zipursky
EMBO J., 9, 1969—1977, 1990   5-10

Cell adhesion has an important role in many stages of neurogenesis, including axonal guidance and fasciculation, cell migration, and myelination. Adhesion proteins mediate call-cell interactions, as well as interactions between cells and the extracellular matrix. In *Drosophila,* chaoptin — a member of the leucine-rich repeat family of proteins — is required for the morphogenesis of photoreceptor cells. Biochemical and genetic findings have suggested that chaoptin may function as a cell adhesion molecule. This hypothesis was tested by examining the effects of chaoptin on the adhesive properties of *Drosophila* cells.

Chaoptin cDNA was transfected into non-self-adherent *Drosophila* Schneider line 2 (S2) cells. After heat shock induction of chaoptin expres-sion, the transfected S2 cells formed multicellular aggregates. Mixing studies suggested that chaoptin-expressing cells adhered homotypically. Biochemical analyses indicated that chaoptin is linked with the plasma membrane surface through covalent attachment to glycosylphosphatidyl-inositol.

Chaoptin may, with other members of the leucine-rich repeat family of proteins, constitute a new class of cell adhesion molecules. It is likely that chaoptin mediates adhesive interactions between cell surfaces via a direct homophilic mechanism.

♦ Chaoptin was first identified as the epitope of a monoclonal antibody that exclusively stains *Drosophila* photoreceptor cells. The epitope ap-pears to be distributed along the entire surface of the cell, including the cell body, the axon, and the tightly packed microvillar portion of the cell (the rhabdomere) that functions in phototransduction. Chaoptin is expressed relatively early during morphogenesis of the eye, shortly after the onset of axon extension and some 4 to 5 d before the expression of genes involved in phototransduction such as rhodopsin. The most striking phenotype of loss-of-function mutants in the chaoptin gene is the disorganization of the rhabdomeric microvillar network. The mutants also affect the organization of other photosensitive organs, such as the ocelli and the "larval photosensitive organ". Chaoptin has been cloned and sequenced. A large proportion of the protein is made up by some 41 tandemly arranged 24 amino acid-long leucine-rich repeats. In other proteins similar leucine-rich repeats have implicated in protein-protein interactions. The leucine repeats, as well as the previously demonstrated extracellular location of the protein, suggested that chaoptin has a role in cell-cell interactions.

Krantz and Zipursky demonstrate that indeed, chaoptin can mediate cell aggregation. This probably occurs by homotypic interactions, since chaoptin-expressing cells do not aggregate with nonexpressing cells. Chaoptin is unique among neural cell adhesion molecules in that it is expressed only in one cell type. This is in contrast to other neuronal cell-adhesion molecules, such as N-CAM, which is expressed in a broad spectrum of cell types. The phenotype of chaoptin mutants and the results reported in this paper suggest that one important role of chaoptin is the organization of the microvillar membranes in the rhabdomeric portion of the cell. Defects in cell-cell interactions are observed only in the most severe mutants. Although chaoptin is expressed in axons just as they start to differentiate, no defects in axon projections were observed in mutant flies. If chaoptin plays any role in axon guidance, its function can probably be replaced by other gene products (functional redundancy has also been suggested to exist in case of other neural cell proteins, such as fasciclin I, fasciclin III, and neuroglian). Based on their demonstrated or hypothesized adhesive role, the authors propose that chaoptin and three other repeat-containing proteins (human glycoprotein 1b, human oligodendrocyte myelin glycoprotein, and the *Drosophila Toll* protein) belong to a new class of leucine-rich cell-adhesion proteins. With use of site-directed mutagenesis it might be possible to define the exact role that the leucine-rich repeats play in membrane adhesion. *Marcelo Jacobs-Lorena*

The *Notch* locus and the Genetic Circuitry Involved in Early *Drosophila* Neurogenesis

T. Xu, I. Rebay, R. J. Fleming, T. N. Scottgale, and
S. Artavanis-Tsakonas
Gene Dev., 4, 464—475, 1990   5-11

The neurogenic loci of *Drosophila* have been studied in an attempt to identify the molecular mechanisms underlying the decision of an embryonic ectodermal cell to follow a neural or an epidermal pathway. The *Notch* locus codes for a transmembrane protein which shares homology with mammalian epidermal growth factor. The existence of negative complementation between two *Notch* mutations affecting the extracellular domain of the protein was utilized to analyze the genetic circuit into which *Notch* is integrated.

Genetic screening identified a restricted set of interacting loci, including *Delta* and *mastermind*. Both these genes, like *Notch,* belong to the group of neurogenic loci previously identified as influencing early neurogenesis. Analysis of the molecular lesions of two *Notch* alleles that interact with *mastermind* mutations (*nd* and *nd*^{2}) showed that they involve changes in the intracellular domain of the protein. These lesions also interact with a mutation affecting the transducin homologous product of the neurogenic locus *Enhancer of split.*

Molecular Interactions Between the Protein Products of the Neurogenic Loci *Notch* and *Delta,* Two EGF-Homologous Genes in *Drosophila*

R. G. Fehon, P. J. Kooh, I. Rebay, C. L. Regan, T. Xu,
M. A. T. Muskavitch, and S. Artavanis-Tsakonas
Cell, 61, 523—534, 1990   5-12

Previous genetic studies in *Drosophila* have suggested that interactions occur between the gene products of the neurogenic loci *Notch* and *Delta,* each of which encodes a transmembrane protein possessing homology with epidermal growth factor. Direct evidence has now been obtained for intermolecular interactions between *Notch* and *Delta* by studying the effects of their expression on aggregation in *Drosophila* S2 cells (Figure 5-12).

Cells expressing *Notch* were found to form mixed aggregates specifically with cells expressing *Delta.* This process was calcium dependent. *Notch* and *Delta* were associated within the membrane of single cells, and they were observed to form detergent-soluble intermolecular complexes. Interactions with *Delta* did not require the intracellular domain of *Notch.*

These findings indicate that *Notch* and *Delta* proteins interact at the cell surface through their extracellular domains. Studies of tissue-specific and subcellular sites of these gene products and those of other neurogenic loci will help clarify the relative importance of such functions as intercellular adhesion and signal transduction in neurogenic gene actions.

◆ Loss-of-function mutations in any one of six zygotic neurogenic loci disrupt *Drosophila* embryogenesis in a very similar way: they cause

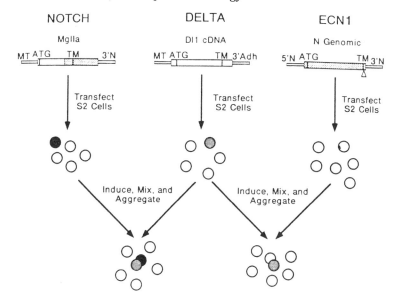

FIGURE 5-12. Expression constructs and experimental design for examining Notch-Delta interactions. S2 cells at log phase growth were transiently transfected with one of the three constructs shown. Notch encoded by the MG11a minigene (a cDNA/genomic chimeric construct: cDNA-derived sequences are represented by stippling, genomically derived sequences by diagonal-hatching; Ramos et al., 1989) was expressed following insertion into the metallothionein promoter vector pRmHa-3 (Bunch et al., 1988). Delta encoded by the Dl1 cDNA (Kopczynski et al., 1988) was expressed after insertion into the same vector. The extracellular Notch (ECN1) variant was derived from a genomic cosmid containing the complete *Notch* locus (Ramos et al., 1989) by deleting the coding sequence for amino acids 1790-2625 from the intracellular domain (denoted by Δ; Wharton et al., 1985), leaving 25 membrane-proximal residues from the wild-type sequence fused to a novel 59 amino acid tail (see Experimental Procedures). This construct was expressed under control of the *Notch* promoter region. For constructs involving the metallothionein vector, expression was induced with $CuSO_4$ following transfection. Cells were then mixed, incubated under aggregation conditions, and scored for their ability to aggregate using specific antisera and immunofluorescence microscopy to visualize expressing cells. MT, metallothionein promoter; ATG, translation start site; TM, transmembrane domain; 3′ N, *Notch* gene polyadenylation signal; 3′ Adh, polyadenylation signal from Adh gene; 5′ N, *Notch* gene promoter region. (From Fehon, R. G., Kooh, P. J., Rebay, I., Regan, C. L., Xu, T., Muskavitch, M. A. T., and Artavanis-Tsakonas, S., *Cell*, 61, 523—534, 1990. With permission.)

extensive hypertrophy of the embryonic nervous system at the expense of epidermal structures. This phenotype suggests that the "ground state" of ventro-lateral embryonic cells is the differentiation into the neural pathway and that the wild type function of this group of genes is required to divert the fate of cells toward formation of epidermal structures. The six zygotic neurogenic loci identified so far are: *Notch (N)*, *Delta (Dl)*, *Enhancer of split [E(spl)]*, *mastermind (mam)*, *big brain (bib)*, and *neuralized (neu)*. Many of these genes are also expressed at other developmental stages and accordingly, mutants can affect a number of structures including the adult wing and eye. *N, Dl,* and *E(spl)* have been cloned. *N* encodes a ~300-kDa protein with a large N-terminal extracellular domain that includes 36 epidermal growth factor EGF-like tandem repeats in addition to a number of other conserved domains. *Dl* encodes a ~100-kDa protein that has 9 (EGF)-like repeats within its extracellular domain. Although the exact function of these EGF-like repeats is unknown, they occur in a variety of proteins involved in interactions with other proteins, including those of the blood-clotting cascade. *E(spl)* encodes a transducin-like protein.

Genetic analysis of certain developmental processes in *Drosophila* (e.g., formation of the dorso-ventral embryonic pattern, somatic sex determination) suggests control by genes that can be ordered in a hierarchical regulatory cascade. Disruption of any gene alters the whole regulatory network. The zygotic neurogenic loci, all of which affect the development of the nervous system in a similar way, may form such a network. The two papers selected for this report analyze interactions among neurogenic loci.

Consistent with its large size and apparent complexity, mutations in the *N* locus can be grouped into different classes according to their phenotype and to the region of the gene affected by the mutation. One group of *N* dominant mutations that exhibit a phenotype of gapped wing veins and bristle abnormalities, are called *Abruptex (Ax)*. The experiments of Xu et al. take advantage of two *Ax* alleles, Ax^{E2} and Ax^{9B2}, that exhibit unusual behavior: although each allele is homozygous viable, the heterozygous combination is lethal (this is called negative complementation). These *Ax* alleles are each associated with the substitution of a single amino acid in the EGF-like domain. *N* is located on the X chromosome. When females homozygous for one of the *Ax* alleles are crossed to males carrying the other allele, males survive but females do not, due to negative complementation. To identify suppressors of the lethality caused by negative complementation, mutagenized Ax^{9B2} males were crossed to virgin Ax^{E2} females. In this screen, 36 fertile Ax^{E2}/Ax^{9B2} females were recovered among 12,946 sibling Ax^{E2} males. The suppressors fell into 4 classes: 21 lethal *N* loss-of-function alleles (these were expected), 9 lethal *Dl* alleles, 4 lethal *mam* alleles, and 2 unknown viable mutations on the X chromosome. Further work determined that the suppression phenotype was not restricted to the alleles of the original screen, but could be extended to other previously isolated *Ax, Dl* and *mam* alleles. Thus, *a simple reduction of Dl and mam gene dosage is capable to suppress the lethality associated with the negative complementation.* Because the suppression was not allele specific, it can be hypothesized that rescue is not due to suppression of a specific *Ax* allele but rather, that the suppression is achieved by diminishing the interaction between the two *Ax* proteins. One possible (and probably oversimplified) scenario is that the *Dl* product promotes the interaction between or among *N* protein molecules, and that certain mutant *N* protein combinations (i.e., the products of the negatively complementing Ax alleles) are deleterious to the cell. According to this model, reduction in *Dl* gene expression would decrease the efficiency of this association and thus increase viability. This is just one of many possible models.

In a set of elegant experiments, Fehon et al. extended the analysis of interactions among neurogenic mutants to the molecular level. Using cultured *Drosophila* cells transformed with the *N* and *Dl* genes, the authors show that both proteins are expressed on the cell surface and that these proteins can promote specific cell-cell interactions (aggregation) in a calcium-dependent manner. *N*-expressing cells aggregate with *Dl*-expressing cells but not with themselves. In contrast to *N*, *Dl*-expressing cells can self-aggregate. As a whole, these studies suggest very strongly that *N* and *Dl* proteins physically interact. However, the possibility that this interaction is mediated by other protein species was not ruled out. With a sensitive assay in hand, a number of interesting questions can now be asked concerning these interactions. One would like to know if the EGF repeats mediate these interactions and if the same *Dl* protein domains are involved in the *Dl-Dl* and *Dl-N* interactions. In addition to promoting cell-cell adhesions, these proteins could be involved in signal transduction. For instance, a likely candidate for performing such function is the large 1000 amino acid-long intracellular domain of *N*. *Marcelo Jacobs-Lorena*

Fasciclin III: A Novel Homophilic Adhesion Molecule in *Drosophila*

P. M. Snow, A. J. Bieber, and C. S. Goodman
Cell, 59, 313—323, 1989   5-13

Cell adhesion molecules are critical factors regulating the outgrowth and guidance of neuronal growth cones. An example is *Drosophila* fasciclin III, an integral membrane glycoprotein which is expressed on a subpopulation of neurons and fasciculating axons in the developing central nervous system — as well as in several other developing tissues. A full-length cDNA encoding the 80-kDa form of fasciclin III was isolated and used to transfect *Drosophila* S2 cells under heat shock control (Figure 5-13). These cells then were used to show that fasciclin III can mediate adhesion.

Study of transfected cells confirmed that fasciclin III mediates adhesion in a homophilic, calcium-independent manner. Expression of the molecule correlated with cell adhesion. Sequencing studies showed that fasciclin III encodes a transmembrane protein lacking significant homology with any known protein, including known families of vertebrate cell adhesion molecules. The distribution of the adhesion molecule on

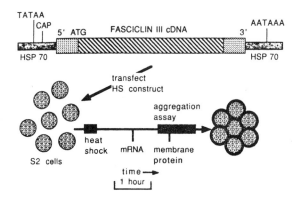

FIGURE 5-13. Strategy for examining the function of fasciclin III by transfection. The complete 2.5 kb cDNA encoding fasciclin III (Figure 1) was inserted into the KpnI site of the pHT4 vector (Schneuwly et al., 1987) as indicated. This construct was transfected into S2 cells as described in Experimental Procedures. The transfected cells were heat shocked for 15 to 30 min and allowed to recover. Cells were subsequently tested for transcripts encoding fasciclin III at 1 h of recovery or cell surface expression at 1.5 h of recovery. Subsequently, cells were tested for their ability to adhere to one another. Identical experiments were performed with the cDNA inserted in the opposite orientation relative to the *hsp70* promoter. (From Snow, P. M., Bieber, A. J., and Goodman, C. S., *Cell*, 59, 313—323, 1989. With permission.)

fasciculating axons and growth cones supports a role for fasciclin III in guidine growth cones.

Genetic analyses of cell-cell adhesion molecules should make it possible to determine whether eliminating any one adhesion system produces severe abnormalities, or whether major defects in neural development require the simultaneous deletion of two or more adhesion systems. It is possible that, apart from cell adhesion, homophilic interaction between fasciclin III molecules mediate signals involved in cell interactions, induction, and determination.

♦ Fasciclins are membrane-associated glycoproteins, identified by several monoclonal antibodies that bind to the surface of fasciculating axons in grasshopper and Drosophila embryos. This paper describes the sequence and binding properties of one of these molecules, fasciclin III. The cDNA sequence bears no sequence homology to known adhesion molecules, suggesting that this is a novel molecule.

Snow and colleagues were able to test the function of fasciclin III by transfecting the cDNA molecule into a Drosophila cell line, the Schneider 2 (S2) line. This technology has been used previously to test function of vertebrate adhesion molecules, but has not been applied successfully to invertebrates. The S2 line is particularly appropriate for testing the properties of transfected molecules because these cells have inherently low adhesivity and grow as single, unclumped cells. S2 cells transfected with

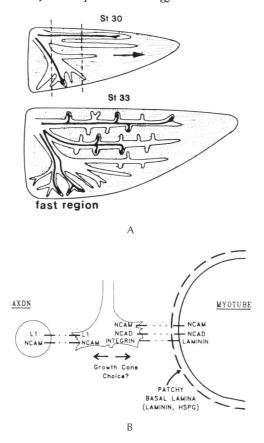

FIGURE 5-14. The development of the nerve branching pattern within the iliofibularis muscle and possible molecular interactions involved in this process. (a) Two stages in the development of the intramuscular nerve branching pattern. By stage 30, the main intramuscular nerve trunks have formed. In the slow region formation is by growth of relatively tightly fasciculated large bundles of axons parallel to the long axis of the myotubes (indicated by arrow); in the fast region, the nerve trunk is less tightly fasciculated and grows transversely across the myotubes. At this stage, individual axons (several are indicated schematically) are unbranched in either the slow or the fast region. By stage 33, a highly regulated process of branching has occurred to produce in the slow region a series of transversely oriented side branches. This is predominantly due to sprouts forming along the proximal shaft of axons whose primary growth cone is located distally in the nerve trunk (several axons are indicated). In contrast, branching in the fast region is also due to divergence of individual axons at branch points. (b) The major classes of known cell adhesion molecules present and potentially capable of influencing the fasciculative choice of growing axons. A single growth cone with its complement of cell adhesion molecules is indicated. Fasciculation with another axon could be mediated by homophilic interaction between L1 molecules or between N-CAM molecules. Homophilic interaction of N-cadherin (NCAD) could potentially contribute to axon-axon fasciculation. However, intramuscular nerve trunks have very low levels of N-cadherin at these stages, and antibodies to N-cadherin did not affect nerve trunk fasciculation (L. Landmesser, unpublished data). In contrast, growth cone preference for myotubes could be mediated by homophilic interactions between N-CAM or N-cadherin or by interactions involving integrin on the growth cone and laminin in the basal lamina surrounding the myotube. (At these stages, both laminin and heparin sulfate proteoglycan are immunocytologically detectable in basal laminae surrounding myotubes. (From Landmesser, L., Dahm, L., Tang, J., and Rutishauser, U., *Neuron,* 4, 655—667, 1990. With permission.)

fasciclin III bind to each other in a homophilic and Ca^{++} independent manner. This is the first Drosophila adhesion molecule whose function has been directly demonstrated using transfection and *in vitro* adhesion assays. Its adhesive properties together with its localization at points of contact between specific subsets of fasciculating axons suggests that this molecule may play a role in growth cone guidance in the developing insect nervous system. *Marianne Bronner-Fraser*

Polysialic Acid as a Regulator of Intramuscular Nerve Branching during Embryonic Development
L. Landmesser, L. Dahm, J. Tang, and U. Rutishauser
Neuron, 4, 655—667, 1990   5-14

N-CAM is a cell adhesion molecule found on the surfaces of various cell types — including neurons, muscle cells, and glial cells. The degree of polysialylation of this molecule is developmentally regulated. The authors studied the role of polysialic acid (PSA) during the initial innervation of chick muscle. Both N-CAM and the adhesion molecule L1 are important in balancing axon-axon and axon-muscle adhesion during this process.

Developmental changes in the pattern of innervation (Figure 5-14) did not correlate with the level of expression of L1 or N-CAM, but rather with the amount of PSA at the axonal surface. Removal of PSA using a specific endoneurominidase increased axonal fasciculation and reduced nerve branching. In contrast, the defasciculation of nerve trunks and increased branching produced by blockade of neuromuscular activity correlated with increased axonal PSA levels. The endoneuraminidase prevented these effects on nerve branching.

These findings demonstrate the potential role of PSA as a molecule regulating cell-cell interactions. A direct link exists between the morphogenetic effects of adhesion-mediated and synaptic activity-dependent processes in the chick embryo. A general conclusion is that one class of cellular properties can be markedly influenced by another.

♦ The neural cell adhesion molecule (N-CAM) is abundant on the surface of a variety of cell types including neurons and muscle. N-CAM is a glycoprotein which exhibits developmentally regulated changes in its content of polysialic acid (PSA). Large amounts of PSA are known to decrease the adhesivity of N-CAM; thus, PSA may serve as a regulator of contact-dependent cell-cell interactions.

Landmesser and colleagues have examined the role of PSA in modulating cell-cell interactions required for motorneurons to achieve their appropriate intramuscular branching patterns in the developing limb. By *in ovo* injection of function-blocking antibodies, they previously established a role for N-CAM and L1 in axon-muscle and axon-axon interactions, respectively, in the limb. During normal development, they also observed striking changes in the tightness of axonal bundling which were not reflected in the levels of N-CAM or L1. In the present study, they demonstrate that changes in axon bundling during motor axon branching in the limb are correlated with regulation of PSA levels on the nerve. They were able to experimentally alter levels of PSA in two ways: (1) blockade of neuromuscular activity, which increases PSA and (2) injection of a specific endosialidase, which decreases PSA. The former treatment increased branching of the nerves, whereas the latter treatment increased axon fasciculation and decreased nerve branching. Furthermore, treatment with endosialidase after blockage of activity prevented the observed increase in branching.

These results demonstrate that the distribution of PSA on the axon surface correlates with nerve branching and inversely correlates with tighter fasciculation. Experimental alterations of PSA leads to either increased branching (more PSA) or increased fasciculation (less PSA). Thus, PSA may act as a cell surface regulator of cell-cell interactions. This is an important concept since it points to the fact that some cell interactions are not correlated with changes in gene expression, but can be directly modulated through posttranslational changes such as alterations, in glycosylation. *Marianne Bronner-Fraser*

Localization of Specificity Determining Sites in Cadherin Cell Adhesion Molecules
A. Nose, K. Tsuji, and M. Takeichi
Cell, 61, 147—155, 1990   5-15

The cadherins are a group of homophilic intercellular adhesion molecules — integral membrane glycoproteins — which exhibit binding specificity. The authors attempted to map sites for the specificities of cadherins by constructing chimeric E- and P-cadherins; establishing L cell lines expressing the hybrid molecules, and analyzing the adhesive selectivity of cells expressing chimeric and point-mutated E-cadherin and P-cadherin.

The amino-terminal 113 amino acid region was found to be essential in determining specificities. Within this region are particularly important sites, around residues 78 and 83, where amino acid substitution alter the binding specificity of cadherins. The epitopes for antibodies capable of blocking the effects of cadherins were located in this amino-terminal region.

Probably non-conserved amino acid residues in the amino-terminal region generate a type-specific conformation for each member of the cadherin family. Cadherin molecules — each possessing a unique conformation — then would bind most strongly to those with the most similar conformation.

◆ Cadherins are a family of Ca^{++}-dependent transmembrane glycoproteins involved in cell-cell adhesion. They have been subdivided into various subtypes such as E-, P-, and N-cadherin by their unique tissue distributions. Different types of cadherins have 45 to 58% sequence identity, representing a gene family. Despite their similarity, subtypes of cadherins only recognize members of the same subtype; for example, cells that bear E-cadherin preferentially adhere to other E-cadherin bearing cells, etc.

In this study, Nose et al. constructed various chimeric E- and P-cadherins and, using transfection, established cell lines expressing these hybrid molecules. Using these chimeric molecules, they found that the amino-terminal 113 amino acids are essential for determining binding specificity of the cadherin subtypes. To determine specific sequences that were important in this process, they made point mutations in the amino-terminal region. They found that alterations in amino acid residues 78 and/or 83 changed the binding specificity, suggesting that these residues play a central role in determining the binding properties of cadherins. Other sites between residues 67 and 113 also appear to participate in determination of specificity, perhaps by conferring proper conformation. The epitopes recognized by several blocking antibodies against cadherins also were found in the amino-terminal region.

This is a significant study because it molecularly dissects cadherin molecules to establish which amino acids are important for binding specificity. From this approach, a picture emerges regarding the mechanism of homophilic cadherin binding. In addition, one gets an idea of how this gene family arose from a single precursor which was altered to form numerous subtypes. *Marianne Bronner-Fraser*

Only Dull CD3⁺ Thymocytes Bind to Thymic Epithelial Cells; The Binding Is Elicited by Both CD2/LFA-3 and LFA-1/ICAM-1 Interactions

S. Nonoyama, M. Nakayama, T. Shiohara, and J. Yata

Eur. J. Immunol., 19, 1631—1635, 1989   5-16

Thymic epithelial cells are an important part of the process by which immature lymphoid cells differentiate into immunocompetent T cells in the thymus. It has been proposed that immature T cells interact with self antigen presented by thymic epithelial cells during their maturation, and thereby are "educated" to distinguish self antigen from non-self antigen. This study was designed to analyze binding between human thymocytes, cultured thymic epithelial cells (CTEC), and adhesion molecules.

TEC expressed ICAM-1 (a ligand of LFA-1 cell adhesion molecules) immediately after separation. Expression of ICAM-1 was, however, gradually lost on culture of TEC. Interferon-gamma reinduced ICAM-1 on the CTEC and enhanced their ability to bind to thymocytes. The increased binding was inhibited by anti-ICAM-1 monoclonal antibody and by anti-LAF-1 monoclonal antibody. Only dull CD3$^+$ thymocytes bound to CTEC. CD3$^-$ cells, bright CD3$^+$ cells, and peripheral blood T lymphocytes were induced to bind by treatment with neuramidase. Binding of all these cell types was inhibited by anti-LFA-1 and anti-CD2 monoclonal antibodies.

These findings demonstrate the involvement of ICAM-1 in binding of thymocytes to TEC. This binding is an important part of the process by which thymocytes differentiate into functionally mature T cells.

♦ This study employed cultures of human thymic epithelial cells. The authors initially observed that the expression of ICAM-1 by the thymic epithelial cells decreased rapidly following their cultivation. This could be returned to preculture levels by treating the cells with gamma interferon. This experimental system provided the opportunity to study the role of ICAM-1 expression by thymic epithelial cells during the interaction of the cells with thymocytes. The levels of ICAM-1 correlated positively with the levels of thymocyte binding and appeared to be mediated by LFA-1 expressed by the thymocytes. Finally, Thymocytes were divided by cell sorting, into immature (CD3$^-$), intermediate (CD3^{dull+}), and CD3$^{bright+}$) mature subsets and tested for binding to ICAM-1-positive thymic epithelial cells. The thymocytes characterized as dull CD3$^+$ cells were found to preferentially bind onto the cultured thymic epithelial. These results are consistent with the LFA-1/ICAM-1 molecules participating during the retention of CD3^{dull+} pre T cells within the thymus, and thus allowing the appropriate cellular interactions to occur necessary for the pre-T cells to differentiate. *E. Charles Snow*

A Role for the Thymic Epithelium in the Selection of Pre-T Cells from Murine Bone Marrow

B. Bauvois, S. Ezine, B. Imhof, M. Denoyelle, and J. Thiery
J. Immunol., 143, 1077—1086, 1989   5-17

The rat thymic epithelial cell line IT45-R1 secretes soluble molecules which chemoattract rat hemopoietic precursor cells *in vitro.* Development of this *in vitro* migration assay was based on the ability of cells to migrate across polycarbonate filters in Boyden chambers. The authors have used this technique to analyze murine bone marrow cells capable of migrating toward IT45-R1-conditioned medium.

The responding cells were found to be a minor marrow subpopulation which exhibited a low ability to incorporate tritiated thymidine. The cell subset was relatively impoverished in granulocyte-macrophage CFU and pluripotent hemopoietic stem cells. Intravenous transfer studies showed that the cells have thymus-homing and colonization capacity. Following intrathymic transfer, the migrated cells generated Thy1.2$^+$-donor-type thymocytes, including all cortical and medullary-cell subsets in a single wave of repopulation, peaking after 23 to 25 d. The degree of repopulation was close to that seen with unfractionated marrow cells. Thy-1.2$^+$ cells were present in the lymph nodes and spleen of reconstituted mice.

Supernatant from the thymic epithelial cell line IT45-R1 can induce the migration of a murine bone marrow subset containing hemopoietic cells that already are committed to the lymphoid lineage. It appears that T progenitors committed to the lymphoid lineage respond to rat thymic epithelial chemoattractants.

♦ This paper characterized soluble mediators secreted by a rat thymic epithelial cell line which had previously been shown to attract hemopoietic precursor cells in an *in vitro* assay system. The presence of chemoattractant factors responsible for the migration of precursor T cells from the bone marrow to the thymus has been shown in a number of biological systems. This study sought to delineate the bone marrow-derived cell which responds to these chemotactic factors released from the thymus.

Previous work had shown that the cell whose migration was affected by these soluble mediators was Thy-1+CD4−CD8−. A very small subset of bone marrow cells (less than 0.5%) were found to migrate in the direction of the soluble mediators. The migrating cells displayed surface markers characteristic of granulocytes and T cells, and not of B cells. Also, these cells were found not to be proliferating. The migrating cells demonstrated a reduced number of detectable GM-CFU, so precursors of mononuclear cells are not attracted by these factors. The migrating cells were formally shown to possess the ability to home into the thymus following their injection into lethally irradiated mice. The animals repopulated with the migrating cells went on to develop the appropriate subsets of T cells within the thymus, and to have mature T cells within their peripheral lymphoid tissues. These results are consistent with thymic epithelial cells secreting a factor chemotactic to precursor T cells found in the bone marrow. *E. Charles Snow*

Intrathymic Signalling in Immature CD4$^+$CD8$^+$ Thymocytes Results in Tyrosine Phosphorylation of the T-Cell Receptor Zeta Chain

T. Nakayama, A. Singer, E. D. Hsi, and L. E. Samelson
Nature, 341, 651—654, 1989   5-18

Thymic selection of the developing T-cell repertoire in immature CD4$^+$CD8$^+$ thymocytes is thought to be mediated by signals transduced by T-cell antigen receptor (TCR) molecules, and possibly also by CD4 and CD8 accessory molecules. In mature T cells, CD4 and CD8 are associated with the *src*-like protein tyrosine kinase p56 *lck*. Signals transduced by TCR and CD4 activate tyrosine kinases which phosphorylate TCR-ζ chains and other intracellular substrates.

This study was designed to show whether tyrosine kinases are similarly activated in immature thymocytes. It was unexpectedly found that TCR-ζ chains from CD4+CD8+ thymocytes were already phosphorylated *in vivo*. The TCR subunit was dephosphorylated when CD4+CD8+ cells were removed from the intrathymic environment. The subunit was rapidly rephosphorylated in cultured thymocytes *in vitro*. Both cross-linking of TCR, CD4, or CD8 by specific monoclonal antibodies or cell-cell contact was effective in promoting rephosphorylation of TCR-ζ.

Tyrosine kinases are activated *in vivo* in immature CD4+CD8+ thymocytes undergoing differentiation and selection. The TCR, CD4, and CD8 molecules can act as signaling agents activating tyrosine kinases. Phosphorylated TCR-z can serve as a marker of signaling in this setting.

♦ This study follows the tyrosine phosphorylation of the zeta chain of the CD3 complex. This protein is phosphorylated on a tyrosine residue following the occupancy of the TCR. The analysis of nonstimulated thymocytes recovered from young adults revealed that the zeta chain was already phosphorylated on tyrosine residues. When fractionated into thymocyte subpopulations, a similar result was seen in the most immature, double-positive cells. The cross-linking of TCR on these cells was not found to enhance the detectable levels of zeta chain phosphorylation. The *in vitro* cultivation of thymocytes was associated with a gradual reduction in zeta chain phosphorylation. The cross-linking of TCR on these precultured thymocytes induced the rephosphorylation of the CD3 zeta chain, indicating that cross-linking the TCR complex on immature thymocytes initiates the same or similar tyrosine kinase signaling pathway seen when the receptor is cross-linked on mature T cells. This increase in phosphorylation of the zeta chain in cultured total and immature, double-positive thymocytes was also seen following the cross-linking of the αβ TCR, CD4, and CD8 proteins. Evidence was also provided that direct cell contact, presumable involving MHC class I or II proteins, initiates the phosphorylation of the zeta chain.

These results indicate that the study of signaling pathways operative during the regulation of thymocyte growth and differentiation might require a preculture period, to allow endogenously mediated signals to dissipate. In addition, these results demonstrate that immature thymocytes (CD4+CD8+TCR+) initiate tyrosine phosphorylations following ligation of the TCR, CD4, or CD8 surface proteins. *E. Charles Snow*

Engagement of CD4 and CD8 Expressed on Immature Thymocytes Induces Activation of Intracellular Tyrosine Phosphorylation Pathways

A. Veillette, J. C. Zuniga-Pflucker, J. B. Bolen, and A. M. Kruisbeek

J. Exp. Med., 170, 1671—1680, 1989   5-19

Biochemical studies of how T cells mature have focused on paths involving changes in calcium or phosphatidylinositol metabolism. Little is known about the possible role of tyrosine phosphorylation signals in this process. Recent findings indicate that the CD4 and CD8 expressed on mature T cells are associated physically with the internal membrane tyrosine protein kinase p56^{lck}, which is enzymatically activated on engagement of CD4 in the cells. The present study found that cross-linking of surface CD4 and CD8 on immature thymocytes is associated with specific signals for tyrosine protein phosphorylation.

The engagement of CD4 and CD8 expressed on CD4$^+$CD8$^+$ thymocytes was associated with a rapid tyrosine protein phosphorylation signal. The catalytic function of the *lck* expressed in CD4$^+$CD8$^+$ thymocytes was significantly enhanced upon engagement of CD4. Antibody-mediated cross-linking of surface CD8 had a less consistent effect on the enzymatic properties of bound p56^{lck} *in vitro*.

Tyrosine protein phosphorylation paths may provide a mechanism by which CD4, CD8, and, possibly, TCR/CD3 transduce relevant intracellular signal required for selection of the mature T-cell repertoire. The functional significance of phosphorylation events in T-cell differentiation remains to be established.

♦ For these studies, thymocytes were isolated from timed embryos or adult animals and separated into double-negative (CD4$^-$CD8$^-$), double positive (CD4$^+$CD8$^+$) and single-positive CD4$^+$CD8$^-$ and CD4$^-$CD8$^+$) populations and activated *in vitro* under serum-free conditions. Fresh thymocytes display a prominent tyrosine phosphoprotein migrating at 56 kDa. The cross-linking of CD4 results in a more intense band migrating at 56 kDa, along with the appearance of new bands at 120 and 72 kDa. Cross-linking of CD8 resulted in a similar pattern, while cross-linking of Thy-1 and CD45 did not induce the appearance of new tyrosine phosphoproteins. Cross-linking CD3 resulted in the appearance of the 72- and 120-kDa bands without the phosphorylation of the 56-KDa protein. When thymocytes recovered from 17-d fetal thymus were studied, the cross-linking of CD4, CD8, or CD3 resulted in the same pattern of phosphoproteins described above for thymocytes recovered from adult animals. This indicates that immature thymocytes can be induced to express tyrosine phosphorylation, since day 17 thymocytes do not express single-positive cells.

The p56 tyrosine phosphoprotein induced by cross-linking CD4 or CD8 was formally shown to be P56^{lck} by immunoprecipitation followed by immunoblot analysis. These results indicate that immature and mature thymocytes express CD4 or CD8 molecules which are coupled to tyrosine phosphorylation pathways via p56^{lck}. The authors went on to show that p56^{lck} is detectable by day 15 of fetal development, and maintained at a constant level in thymocyte populations. The physical association between p56^{lck} and CD4 or CD8 was first evident at day 19 of fetal development. This association was shown to occur in both double- and single-positive thymocytes, demonstrating that p56^{lck} is clearly associated with CD4 or CD8 in immature, double-positive thymocytes. These results suggest that the engagement of CD4 (through interacting with its ligand, MHC class II) or CD8 (through interacting with its ligand, MHC class I) initiates a tyrosine phosphorylation signaling pathway. Therefore, the physical interaction between an immature, TCR negative, CD4+CD8+ double-positive thymocyte with MHC molecules expressed on thymic epithelial cells might represent one means by which the growth and differentiation of immature thymocytes is regulated. *E. Charles Snow*

Cell Lineage and Developmental Fate 6

INTRODUCTION

What determines developmental fate remains one of the central, unanswered questions in developmental biology. Since the developmental potential of any given cell becomes more restricted as development processes, one might suggest that developmental fate is the result of genetic and cellular restrictions. However, evidence exists which suggests that at least in some cases fate is an inductive response to some external stimulus, be it cellular or molecular. In fact, it is likely that both restriction and induction play a role in establishing developmental fate.

Lineage-specific gene products, while certainly not the casual molecules in lineage determination, can provide helpful clues regarding lineage specification. What molecular motifs do they share? What functional similarities do they have? Such clues may provide insight into the real initiating event leading to lineage determination.

Clearly, "where you are" and "where you have been" play important roles in determining "where you are going to go". The real quest is for the molecules that translate those concepts into developmentally meaningful signals. The isolation of such molecules and the exploration of how they function and how they in turn are regulated will provide enormous insight into the process of cell lineage determination.

Early Inductive Interactions are Involved in Restricting Cell Fates of Mesomeres in Sea Urchin Embryos
J. J. Henry, S. Amemiya, G. A. Wray, and R. A. Raff
Dev. Biol., 136, 140—153, 1989   6-1

Mechanisms of early cell determination have been studied by following the development of mesomeres isolated at varying developmental stages in the sea urchin *Lytechinus pictus*. Previous studies of 8-cell or 16-cell sea urchin embryos have suggested that mesomeres contribute only to larval ectoderm during normal development. The authors have, however, found that pairs of mesomeres isolated from 16-cell embryos are able to differentiate endodermal and mesenchymal cells in as many as one fourth of cases.

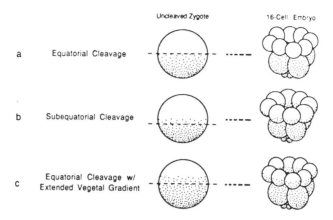

FIGURE 6-1. Hypothetical illustrations explaining the differentiation of endodermal and mesenchymal cell types from isolated mesomeres. Small dots represent the internal distribution of putative vegetal morphogenetic determinants in uncleaved eggs and 16-cell embryos. (a) Equatorial third cleavage in an embryo in which putative vegetal morphogenetic determinants are confined to the vegetal hemisphere. In this case no vegetal determinants have been segregated into the eight animal mesomeres; therefore, these blastomeres can only differentiate ectodermal germ tissues. (b) Subequatorial third cleavage in this example results in the segregation of some vegetal morphogenetic determinants into the larger animal mesomere cells. These mesomere cells gain a greater developmental potential to form such vegetal structures as endoderm and mesenchyme. In this case, vegetal morphogenetic determinants are confined to the vegetal hemisphere. (c) Equatorial third cleavage in an embryo in which there is an extended distribution of vegetal morphogenetic determinants into the animal hemisphere. The resulting mesomere cells, which have inherited some vegetal determinants, have a greater developmental potential to differentiate endodermal and mesenchymal structures. Long dashed lines indicate the relative planes of the third cleavage division in the uncleaved eggs. (From Henry, J. J., Amemiya. Wray, G. A., and Raff, R. A., *Dev. Biol.*, 136, 140—153, 1989. With permission.)

It appears that graded distributions of morphogenetic determinants exist in these embryos, since the extent of differentiation of isolated mesomeres relates to the relative position of the third cleavage plane along the animal-vegetal axis. When this plane was subequatorial and the resulting animal blastomeres inherited part of the vegetal hemisphere, nearly 40% differentiated endodermal and mesenchymal cells types. Even when mesomeres formed within the animal hemisphere, up to 14% differentiated endodermal and mesenchymal cells. Mesomeres isolated earlier in development exhibited greater differentiative ability. Aggregates of isolated pairs of mesomeres possessed less developmental potential than isolated pairs.

Animal blastomeres of *L. pictus* have a much greater early developmental potential than previously thought. The position of the third cleavage plane apparently determines how many vegetal determinative factors are included in the cells (Figure 6-1). The specification of cell fates is controlled through cell interactions which occur, not only along the animal-

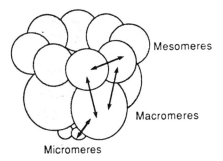

FIGURE 6-1A. Diagram illustrating various inductive interactions that take place during early sea urchin development. Illustration is of a 16-cell embryo containing 8 mesomeres, 4 macromeres, and 4 micromeres. The existence of interactions along the animal-vegetal axis, between micromeres and macromeres, and between macromeres and mesomeres has been revealed through the work of Hörstadius (1935, 1973) and Ettensohn and McClay (1988) and is shown here by the respective double-headed arrows. Lateral inductive interactions or homotypic interactions also appear to take place between the mesomere cells (this study). See text for further details. (From Henry, J. J., Amemiya. Wray, G. A., and Raff, R. A., *Dev. Biol.,* 136, 140—153, 1989. With permission.)

vegetal axis, but also laterally (Figure 6-1A). Lateral interactions may relate to the establishment of dorsoventral polarity.

♦ The amount of experimental embryology that has been carried out on sea urchin embryos is so great that one might think everything that could be discovered be testing various combinations of blastomeres would have been. Not so. The results of experiments examining the role of cell-cell interactions in determining fates of different blastomeres arrayed along the animal-vegetal axis had largely stressed the role of putative positively acting inductive forces. In recent years several examples have been described of the role of negative interactions in specifying cell fates. Ettensohn and McClay showed that primary mesenchyme cells repress the capacity of some secondary mesenchyme cells to differentiate along the primary mesenchyme pathway, and Hurley et al. showed at a molecular level that early separation of blastomeres caused secondary mesenchyme-specific cytoplasmic actin gene CyIIa to be expressed in many more cells than in the normal embryo. Negative regulation has also been implicated in the establishment of tissue-specific gene expression in differentiating cells, with the demonstration that the CyIIIa active gene, which is expressed only in aboral ectoderm, is spatially regulated by the presence of negative regulatory factors in other tissues. (See Year Book, 1989, p. 133 and 1990, p. 149 and 59, respectively).

Now Henry et al. provide evidence that interruption of lateral interactions among early blasomeres in the animal half of the embryo releases a cryptic capacity for these cells to differentiate as endoderm and mesenchyme. This work, and that of Livingston and Wilt reviewed last year

(Year Book, 1990, p. 187) also shows that the capacity to form vegetal structures extends farther toward the animal pole than had been expected. The border of the distribution of determinants required to promote differentiation of vegetal structures in animal blastomeres appears to lie, on average, slightly below the equator, because mesomere pairs derived from embryos in which the third cleavage is sub-equatorial show a much higher frequency of vegetal structures than do those from embryos in which this cleavage is equatorial. Hörstadiaus had also previously noted the effect of the third cleavage plane on the developmental potential of animal half embryos. In both these cases, variation in development of different partial embryos from the same batch of eggs with equatorial cleavage is thought to reflect corresponding differences in the distribution of determinants.

The fact that the capacity of equivalent sets of blastomeres from the animal half of the embryo show progressively restricted capacity of differentiate vegetal structures when isolated at the 8-, 16-, and 32-cell stage correlates with the fact that founder cells for oral and aboral ectoderm have completely separated by 32-cell stage (Cameron et al., *Dev. Biol.,* 137, 77—85, 1990) and that accumulation of the first ectoderm-specific mRNAs can be detected at 16-cell stage (S. Reynolds, J. Palis, L. Angerer, and R. Angerer, in preparation.) *Robert C. Angerer*

Endo16, a Lineage-Specific Protein of the Sea Urchin Embryo, Is First Expressed Just Prior to Conception

C. Noncente-McGrath, C. A. Brenner, and S. G. Ernst

Dev. Biol., 136, 264—272, 1989   6-2

Studies using cloned gene probes and monoclonal antibodies have identified an increasing number of lineage-specific molecules in the sea urchin embryo. The authors have isolated and characterized a new endoderm-specific gene, Endo16, and have used *in situ* hybridization and indirect immunofluorescence to study the spatial and temporal expressions of the Endo16 transcript and to localize the gene product.

Endo16 is first expressed in the vegetal plate — a structure consisting of endodermal and mesenchymal precursor cells which invaginates in the first part of gastrulation. Expression of Endo16 is progressively limited to a subset of endodermal cells. Indirect immunofluorescence studies in mid-gastrula-stage embryos, using a polyclonal antiserum against bacterial expressed Endo16 protein, showed that the protein was localized to the surface of endodermal and secondary mesenchymal cells. In Western blot experiments, the antiserum detected a small set of high-molecular-weight proteins. Amino acid sequencing of a partial Endo16 cDNA clone revealed a protein segment including the Arg-Gly-Asp (RGD) tripeptide which has a role in cell-binding domains of many extracellular proteins.

The Endo16 protein may be an adhesion molecule contributing to gastrulation of the sea urchin embryo.

Gastrulation in the Sea Urchin Is Accompanied by the Accumulation of an Endoderm-Specific mRNA
G. M. Wessel, L. Goldberg, W. J. Lennarz, and W. H. Klein
Dev. Biol., 136, 526—536, 1989   6-3

Invagination of endodermal cells is one of the more dramatic reorganizations occurring during gastrulation in the sea urchin embryo. This study examined spatial diversification of endoderm during gastrulation in *Lytechinus variegatus* using an endoderm-specific cDNA clone. The clone, LvN1.2, was identified using a differential cDNA screen between the ectoderm and ectoderm/mesoderm fractions of prism-stage embryos.

LvN 1.2-kilobase mRNA was first detected by Northern blot studies at the mesenchyme blastula stage just before gastrulation. The mRNA accumulated about 15-fold between gastrulation and the pluteus stage. Accumulation occurred specifically in endoderm. The mRNA was limited to the hindgut-midgut regions, and LvN1.2 protein was localized in discrete granules at the luminal aspect of the hindgut and midgut cells. The 189-amino acid open reading frame was found to represent a novel protein. The same polypeptide resulted from *in vitro* translation of synthetic LVN1.2 mRNA and Western blot analysis using antibody to the protein sequence.

Endoderm of the sea urchin embryo forms at least two distinct cell types in the course of invagination: hindgut-midgut cells containing LvN1.2 mRNA and foregut cells devoid of the transcript. The LvN1.2 protein probably has a role in the digestive function of the midgut and hindgut of *L. variegatus.*

Local Shifts in Position and Polarized Motility Drive Cell Rearrangement during Sea Urchin Gastrulation
J. Hardin
Dev. Biol., 136, 430—445, 1989   6-4

The authors studied mechanisms of epithelial cell rearrangement during elongation of the archenteron in the sea urchin embryo. This process involves both local shifts in cell position in the archenteron wall and polarized motility of the rearranging cells. Scanning electron microscopy, time-lapse videomicroscopy, cell marking (Figure 6-4), and flourescence labeling of chimeric clones served to delineate the stage-specific changes in protrusive activity, cell shape, and cell position that occurred as the cells of the archenteron were rearranged.

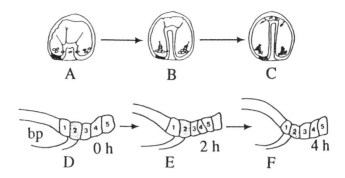

FIGURE 6-4. Measurements of involution during secondary invagination in *L. pictus*. (A to C). Application of a spot of Nile blue sulfate at the blastopore lip prior to the onset of secondary invagination. The spot does not translocate during secondary invagination. (D to F) Movements at the blastopore lip studied using time-lapse videomicroscopy. The same group of cells (labeled 1 to 5) was followed during secondary invagination (approximately 4 h in duration at 23°C); the cells do not translocate into the invaginating region. bp, blastopore. (From Hardin, J., *Dev. Biol.*, 136, 430—445, 1989. With permission.)

Labeled chimeric cells introduced into the archenteron of *L. pictus* elongated and narrowed in the course of gastrulation. The cells changed their relative positions by only one to two cell diameters during cell rearrangement. Blastopore cells rearranged at the same time as cells in the archenteron. Blastopore diameter decreased 35% as a result. Endoderm cells flattened along their apical-basal axis and, just before the start of cell rearrangement, the basal surfaces of all archenteron cells extended long lamellipodial protrusions. At the start of rearrangement, the basal cell surfaces rounded up and the cells became isodiametric. By three fourths of the way through gastrulation, the cells were stretched along the animal-vegetal axis. Finally, they returned to a less elongated shape.

Study of gastrulation in *Eucidaris tribuloides* showed that cell rearrangement in this species occurs through the progressive circumferential intercalation of cells lacking filopodia. Explosive cortical blebbing is seen at cell boundaries.

Gastrulation in the sea urchin embryo closely resembles the process in amphibians. It may be that a limited number of basic morphogenetic movements, such as invagination and convergent cell rearrangement, are shared by phylogenetically disparate species.

♦ Nocente-McGrath et al. and Wessel et al. describe the first endoderm-specific gene activities identified by cloned cDNAs in the sea urchin embryo. These papers are important because, as was shown by Horstadius and colleagues in a classic series of blastomere transplantations (reviewed by Horstadius, *Experimental Embryology of Echinoderms*, Oxford University Press, 1973), micromeres induce the specification of the endoderm lineage in macromeres. The specification of endoderm can also be induced in mesomeres, which normally give rise only to ectoderm, either by ectopic micromeres or by lithium treatment (see this *Year Book,* 1990, p. 187). Thus, the molecular probes specific for the endoderm lineage will be useful in a modern reinvestigation of this inductive phenomenon.

The third paper of this group by Hardin examines the changes in morphology of endoderm cells coincident with the activation of the Endo 16 and LvN1.2 genes. These cells change shape through a series of localized movements leading to the progressive circumferential intercalation of the epithelial cells. Such positional changes provide at least part of the motive force for gut elongation. Currently, there is no direct association between the burst of localized cell shape changes seen and the activation of Endo 16 and LvN1.2 genes. However, Endo 16 encodes a protein containing an RGD sequence, which accumulates extracellularly, around the elongation gut during gastrulation. Since endoderm differentiation at gastrulation is dependent upon molecules of the extracellular matrix (Wessel and McClay, *Dev. Biol.,* 121, 149, 1987) and the RGD sequence functions as a cell attachment domain in a number of extracellular matrix (ECM) molecules, perhaps Endo 16 is involved in endoderm-specific, cell-ECM interactions leading to the observed cell shape changes. *Gary M. Wessel*

Early Determination in the *C. elegans* Embryo: A Gene, *cib-1*, Required to Specify a Set of Stem Cell-Like Blastomeres
R. Schnabel and H. Schnabel
Development, 108, 107—119, 1990   6-5

The development of multicellular organisms requires the formation of many different cell types in a correct spatial arrangement. The early somatic blastomeres of the *C. elegans* embryo are derived from the P cells in a stem-like lineage. This study was designed to characterize seven allelic mutations defining the gene *cib-1* (changed identity of blastomeres), which is involved in specifying the stem cell-like fate of P_1-P_3 cells.

Maternal effect lethal mutations defining *cib-1* were isolated in which the P cells skipped a cell cycle and acquired the fates of only their somatic daughters. Analysis of these mutants suggested that the timing of cell cycles in the early embryo is directly coupled to cell fate. The *cib-1* mutations exhibited partial intragenic complementation. The gene product was required only in early embryogenesis.

The findings suggest that the *cib-1* gene product is required for specifying P_1-P_3 cells in a dose-dependent manner. A clock mechanism is the only feasible timing mechanism for early determination of the *C. elegans* embryo, and it is likely that two independent clocks govern events in the early embryo. The stem cell-like P destiny apparently is specified on top of the somatic decision-making machinery to supply the requisite number of cells (Figure 6-5).

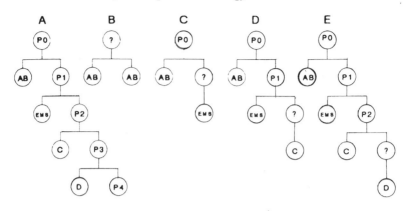

FIGURE 6-5. The logic of decision-making during early development. (A) Wild-type segregation of developmental potential. (B) The findings of Kemphues et al. (1988) suggest that the daughters of a defective P_0 cell behave like AB descendants. (C to E) A failure to specify the P_1-P_3 cells causes the cells to execute the underlying somatic fate. Thus the stem-cell-like P fate seems to be specified on top of the somatic decision-making machinery to supply the required number of cells. (From Schnabel, R. and Schnabel, H., *Development,* 108, 107—119, 1990. With permission.)

♦ The *C. elegans* zygote undergoes a series of four unequal, stem-cell-like divisions. Each division generates a large somatic founder cell and a smaller, stem-line-like P cell. It is during these early divisions that cell fates are thought to be determined. Mutations in the maternal-effect gene, *cib-1* (for changed identity of blastomeres), alter the timing and asymmetry of the last three stem-cell-like divisions and the fates of the daughters produced by those divisions. Depending on the *cib-1* allele, the stem-line-like cell P1, P2, or P3 causes for a cell cycle and appears to be transformed into its somatic daughter cell. These findings suggest that a cell-cycle-independent developmental clock participates in the process of cell-fate specification in early *C. elegans* embryos. *Susan Strome*

Novel Cysteine-Rich Motif and Homeodomain in the Product of the *Caenorhabditis elegans* Cell Lineage Gene *lin-II*
G. Freyd, S. K. Kim, and H. R. Horvitz
Nature, 344, 876—879, 1990   6-6

The gene *lin-11* of *C. elegans* is required for asymmetrical division of a vulval precursor cell type. The authors cloned *lin-11* by identifying a transposon insertion and isolating a mutane allele in TR679, a strain exhibiting a high rate of Tc*1* transposon transposition.

Putative *lin-11* cDNAs were sequenced and found to encode a protein containing both a homeodomain and two tandem copies of a novel cysteine-rich motif ($C-X_2-C-X-_{17-19}-H-X_2-C-X-_2-C-X_2-C-X_{7-11}-(C)-X_8-C$). Two tandem copies of the motif were present amino-terminal to the homeodomains in the proteins encoded by *mec-3*, required for touch neuron differentiation, and *isl-1*, which encodes a rat insulin I gene enhancer-binding protein. The motif is termed LIM (for *lin-11 isl-1 mec-3*).

LIM probably is a metal-binding domain. These proteins are unique in possessing both a potential metal-binding domain and a homeodomain. The LIM region might function by binding to a protein or nucleic acid which is differentially segregated during cell division, resulting in asymmetric segregation of a cell type-specific transcription factor.

♦ Wild-type *lin-11* product is required for the division of the 2° vulval blast cells into daughter cells that express different fates. Given its involvement in asymmetric division, it was somewhat surprising to learn that the *lin-11* gene appears to encode a transcription factor, and a novel one at that. Like the *C. elegans mec-3* product and the rat insulin I gene enhancer-binding protein, *lin-11* contains two copies of a novel cysteine-rich motif (referred to as the LIM motif after the three genes that contain it), a homeodomain, and a potential transcription activation domain. The LIM motif is probably a metal-binding domain, which may be involved in stabilization of protein-protein or protein-nucleic acid interactions. A reasonable model is that *lin-11* differentially regulates the expression of genes in the two daughters of the 2° vulval cells. Perhaps *lin-11* is expressed in the 2° vulval cells and is differentially segregated during cell division. This possibility may be tested by immunolocalization of *lin-11* and identification of the proteins and/or nucleic acids to which it binds. *Susan Strome*

Origin of Cells Giving Rise to Mesoderm in Chick Embryo
C. D. Stern and D. R. Canning
Nature, 343, 273—275, 1990   6-7

In amniotes, the esoderm and gut endoderm arise from an epiblast-derived structure, the primitive streak. The monoclonal antibody HNK-1 recognizes the cells of the primitive streak in the chick embryo. Before formation of the primitive streak, HNK-1 recognizes cells which are randomly distributed within the epiblast, one of the two cell layers of the early embryo.

The authors attempted to determine whether the HNK-1 positive cells of the epiblast are precursors of comparable cells in the primitive streak (Figure 6-7). The epiblast of the early chick embryo was found to contain two distinct populations of cells having disparate developmental fates, at a stage at which "mesodermal induction" presumably takes place. One of these populations, recognized by HNK-1, presumably forms mesoderm and endoderm. When these cells were removed, the rest of the epiblast was incapable of giving rise to mesoderm.

The epiblast of the chick embryo consists of a mixture of cells destined to become either ectoderm or primitive steak cells at a stage where the embryo can respond to rotation of the hypoblast or marginal zone by reversal of the embryonic axis. the cells of the primitive streak therefore are not induced by the hypoblast at the posterior embryonic margin at this stage of development.

♦ At early stages of development, vertebrate embryos are believed to have a great deal of plasticity with respect to cell fate. This study, however, suggests that some cells in the early vertebrate may be determined quite early to give rise to mesodermal and endodermal lineages. In avian embryos and other amniotes, the epiblast gives rise to all three germ layers. The HNK-1 antibody, which recognizes a cell surface carbohydrate epitope, identifies a population of cells randomly arranged within the chick epiblast prior to primitive streak stages. At the primitive streak stage, the HNK-1 antibody recognizes cells within the primitive streak. By immunoreactivity alone, one cannot tell whether the HNK-1 positive cells in the preprimitive streak stage embryo will give rise to the HNK-1-positive cells within the primitive streak.

Stern and Canning used two approaches to test the prospective fate of the HNK-1-positive prestreak cells. First, they labeled the cells with antibody coupled to colloidal gold. The antibody/gold complex becomes internalized and heritable marks the progeny of HNK-1-bearing cells. From this approach, they found that this population uniquely gives rise to mesoderm and endoderm, but not ectoderm. Second, they killed all HNK-1-positive cells by antibody-complement mediated lysis. Embryos treated in this way failed to form mesodermal structures. The HNK-1-complement-treated embryos could be rescued, however, by a graft of epiblast from an untreated embryo.

These results suggest that the avian epiblast contains a randomly distributed subpopulation of HNK-1 positive cells at prestreak stages that is fated to give rise to tissues derived from the primitive streak, i.e., mesoderm and endoderm. Since the prestreak stage corresponds to the time at which mesodermal induction is thought to take place, it seems likely that this population becomes determined at the time of mesodermal induction. *Marianne Bronner-Fraser*

One Synchronous Wave of B Cell Development in Mouse Fetal Liver Changes at Day 16 of Gestation from Dependence to Independence of a Stromal Cell Environment
A. Strasser, A. Rolink, and F. Melchers
J. Exp. Med., 170, 1973—1986, 1989   6-8

The early stages of B lymphocyte progenitor cell proliferation and

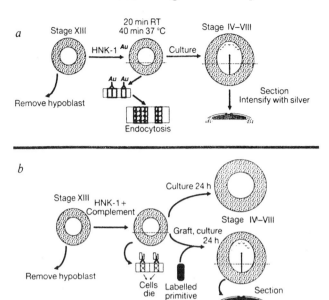

FIGURE 6-7. *a*, Experimental design for lineage mapping of HNK-1-positive cells. The HNK-1-positive cells of the epiblast were marked using the antibody coupled to colloidal gold, and the embryos allowed to develop in culture until after primitive streak formation. Histology then revealed the position of the original HNK-1-positive cells regardless of whether or not they continued to express the epitope. *b*, Ablation of HNK-1-positive cells. In this experiment, the HNK-1-positive cells of the epiblast were killed using the antibody and complement. Some of the treated embryos were grafted with HNK-1-positive cells from a quail donor before further development in culture. Histology then revealed the fate of the transplanted quail cells. Methods: *a* Stage XIII (Reference 5) chick embryos deprived of their hypoblast were incubated in HNK-1 covalently coupled to 15 nm colloidal gold (HNK-1^{Au}) for 20 min at room temperature followed by 40 min at 37°C. During this period the labeled cells endocytose the HNK-1^{Au} complex. They were then washed in Pannett-Compton saline and grown in whole embryo culture to stages 4 to 8 (Reference 16), formalin-fixed, and wax-sectioned and the sections intensified with silver reagent (IntenSE, Janssen). The sections were examined by bright-field microscopy or by confocal laser scanning microscopy in reflection mode. Method for gold-coupling to HNK-1 modified from Reference 17; the optimal pH for adsorption was 8.2. Control embryos treated identically were labeled with 15-nm AuroBeads (Janssen, diluted 1:1 in saline) and then cultured. *b*, Stage XIII (Reference 5) chick embryos deprived of their hypoblast were incubated in HNK-1 supernatant and guinea pig complement (1:1) for 1 at 37°C, washed in Pannett-Compton saline, and set up in whole embryo culture. One set of treated and washed embryos received a graft of about one third of the mesoderm from a stage 3 (Reference 16) primitive streak from a quail embryo after HNK-1-complement treatment and was then cultured. Control experiments: HNK-1 alone (1:1 in saline), complement alone (1:1 in saline). Grafted embryos were fixed in Zenker's fixative and wax-sectioned, and the sections stained with hematoxylin after acid hydrolysis to visualize the quail cells. RT, room temperature. (FromStern, C. D. and Canning, D. R., *Nature*, 343, 273—275, 1990. With permission.)

induction to immunoglobulin gene rearrangements are dependent on interactions with environmental stromal cells. These effects are, at least in part, mediated by soluble cytokines such as interleukins 3 and 7. The authors examined the development of B lineage cells in the fetal mouse liver from day 13 of gestation.

B lineage precursors were enriched at different times of gestation using a monoclonal antibody specific for pre-B cells. Enrichment of these cells was monitored by *in situ* hybridization analysis for expression of the pre-B cell-specific gene λ5, and for cells developing to LPS-reactive mature B cells. Two monoclonal antibodies against primary embryonic stromal cells from fetal liver also were employed.

Enriched purified precursor cells were not influenced by interleukins 2 through 7, alone or in combination, to develop to mitogen-reactive immunoglobulin-positive cells. Such development proved to be dependent on interactions with embryonic stromal cells from fetal liver. Precursor cell development at days 13 through 15 of gestation depended on stromal-cell interactions, but thereafter was independent of such interactions. The number of mitogen-reactive immunoglobulin-positive B lineage cells increased suddenly between days 16 and 17.

B cell development in the fetal mouse liver appears to occur in a single wave, with synchronous progress form a mitogen-insensitive, stromal cell-dependent population to a mitogen-reactive, immunoglobulin-positive, stromal cell-independent B lineage line. A pool of progenitor cells in the marrow may continuously generate cells that can participate in B cell development, at a rate which maintains the pool of primary mature B cells (Figure 6-8).

♦ This study analyzed the development *in vitro* of pre B cells recovered at varying times from murine fetal liver. In some experiments, pre B cells were isolated from total fetal liver cells by recovering cells stained with monoclonal antibody G-5-2 through the use of a cell sorter. The cells (total fetal liver or G-5-2 positive cells) were initially cultured at high cell density followed at various times with culturing at limiting dilution in the presence of the polyclonal mitogen lipopolysaccharide (LPS) and thymic filler cells and the cultures examined daily for the appearance of IgM antibody forming cells (AFC). Fetal liver cells handled in this manner always show the appearance of AFC at the equivalent of day 5 after birth. The inclusion of IL-6 and IL-7 during the preculture period was not found to alter the frequency of pre B cells which gain mitogen responsiveness. However, the longer the fetal liver cells were maintained in the preliminary, high density cultures, the higher the frequency of LPS-responsive cells seen in the limiting dilution cultures. These results suggest that some component of the fetal liver preparation contributes to pre B cell development. Day 13 and 14 fetal liver cells demonstrate such a dependence, while day 16 and 17 fetal liver B cells do not.

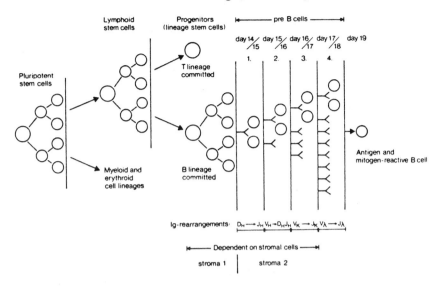

FIGURE 6-8. A proposed scheme of B lymphocyte lineage development in fetal liver. Stroma 1 and stroma 2 are two types of fetal liver stromal cells proposed to either induce proliferation of progenitors without differentiation to Ig gene rearrangements (stroma 1), or to induce Ig gene rearrangements in critical steps of divisions (stroma 2). (From Strasser, A., Rolink, A., and Melchers, F., *J. Exp. Med.*, 170, 1973—1986, 1989. With permission.)

When G-5-2-positive fetal liver cells were studied *in vitro,* it was found that neither thymus filler cells or various combinations of Il-2, -3, -4, -5, -6, and -7 could substitute for the fetal liver environment in increasing the frequency of mitogen-responsive B cells. This was directly demonstrated by showing that G-5-2-positive pre B cells cultured upon irradiated embryonic stromal cells isolated from day 13 fetal liver also displayed an increased frequency of LPS-responsive cells at the equivalent of day 5 after birth. None of the interleukins were found to alter this frequency, although IL-6 and IL-7 did increase the clone size of LPS-reactive B cells. Additional experiments demonstrated that the B-cell precursors remain dependent upon the stromal cell for their differentiation into mitogen-responsive cells until day 16 of gestation. The day 14 fetal pre B cells formed clusters upon the stromal cells until day 16, when the cells detached from the monolayers and started to divide. Two monoclonal antibodies specific for antigens expressed by fetal liver stromal cells (STR4 and STR10) interfered with both the adherance of day 14 pre B cells to the stromal monolayers, as well as their differentiation into mitogen-responsive cells. These results are consistent with the following scenario. As precursor B cells differentiate into early pre B cells, they must physically associate with stromal cells (either within the fetal liver or within the mature bone marrow). This interaction delivers important signals to these cells resulting in their proliferation and differentiation into a more mature B cell. Soluble mediators are not involved in this process. *E. Charles Snow*

αβ Lineage-Specific Expression of the a T Cell Receptor Gene by Nearby Silencers

A. Winoto and D. Baltimore

Cell, 59, 649—655, 1989   6-9

The process by which T cells differentiate into several distinct subsets having disparate functions remains unclear. T cells which express either the αβ or γδ antigen receptor (TCR) are distinct lineages. The single locus that encodes the TCRα and δ genes requires a special regulatory mechanism in order to avoid α gene expression in γδ T cells. The authors attempted to characterize the transcriptional control elements of this gene complex in order to clarify the way in which the α and δ genes are differentially regulated.

The minimal α enhancer was found to be active in both αβ and γδ T cells. It gains αβ lineage specificity through negative *cis*-acting elements 3′ of the C_α gene which silence the enhancer in γδ T cells. The negative elements at the C_α locus consist of several silencers operating independently of orientation and distance. The same silencers act on a retroviral enhancer which normally is widely expressed, restricting its activity to αβ cells. The α silencers are active in non-T cell lines including B lineage cell lines, a macrophage cell line, and a non-lymphoid-cell line.

A decision by a cell to differentiate into the αβ T cell lineage may involve specific relief from silencers. Silencers probably are important as enhancers in establishing lineage-specific gene expression. The precise specificity with which developmentally controlled genes are regulated may derive from a summation of positive and negative influences. The way in which the a silencers operate remains to be clarified.

♦ Lineage specificity is generally thought to be the result of some enhancement process. In its simplest form, lineage specificity results from the induction of a new gene or as a response to a new molecular signal. This article suggests that the opposite may also be true — lineage specificity may result if a specific gene is turned off or a specific signal is silenced. The authors investigate T-cell antigen receptors (TCRs) on specific T-cell lineage. They show that silencers which function in an orientation- and distance-independent fashion can suppress the expression of specific endogenous enhancers, resulting in a lineage-specific decision. These silencers seem to function in non-T cells, suggesting that they may play a role in T-cell differentiation. The discovery of these molecular silencers adds a new dimension to our understanding of what possible mechanisms play a role in cell lineage determination. *Joel M. Schindler*

Cytodifferentiation — Cell- and Tissue-Specific Gene Expression and Maintenance

7

INTRODUCTION

Terminal differentiation results in the expression of a phenotype that has successfully integrated all the appropriate molecular and cellular signals to ensure accurate cytodifferentiation. This terminal phenotype represents the biological compromise between instructions telling any one cell what it should be, and what it should not be. Thus, the accurate expression of the cascade of genes that ultimately defines any given phenotype reflects both inductive and repressive events.

The obvious central questions regarding cytodifferentiation are what the molecular nature of cell-specificity signals really are and when are they expressed. While it is likely that many genes expressed in a cell- or tissue-specific manner have shared DNA sequences that establish such cell specificity, the observation that some genes are expressed at different times and in different cells indicates that DNA sequence alone is not sufficient for such specificity to be established. Transcription factors, or combinations of transcription factors, may be a mechanism that can successfully confer such specificity, but what regulates their expression? What is the initial event that confers such specificity, or is there by necessity more than a single event involved? Are the mechanisms that confer cell specificity themselves unique to each individual cell phenotype?

Neurodevelopment, development of gonadal systems, and development of the immune system are examples of organ systems that include a number of well-characterized cell phenotypes. As such, they are well suited for investigating the mechanisms that regulate specific cytodifferentiation. The expression of cell-specific genes and the mainenance of their phenotypes in each of these organ systems results from the integration of many different signals. Understanding the regulation and integration of those signals is the focus of much exciting research.

A Spatial Gradient of Expression of a cAMP-Regulated Prespore Cell-Type-Specific Gene in *Dictyostelium*
L. Haberstroh and R. A. Firtel
Genes Dev., 4, 596—612, 1990   7-1

Dictyostelium discoideum grows as a single-cell vegetative amoeba until food is depleted, when multicellular development begins. A class of genes has been identified which are prespore cell-type specific in their expression in the multicellular aggragate, and are inducible by cAMP acting through cell-surface cAMP receptors. The authors have cloned and analyzed the regulatory regions controlling the expression of one of these genes, which encodes the spore coat protein SP60.

A fusion of the firefly *luciferase* gene and *E. coli lacZ* served to identify *cis*-acting regions needed for proper spatial and temporal expression in multicellular aggregates, as well as for cAMP induction. A CA-rich element and surrounding sequences, present three times within the 5′-flanking sequence, were found to be necessary for proper regulation. The *SP60* gene was expressed in mid- to late aggregation. Cells positive for the gene were localized in a doughnut-shaped ring within the forming aggregate. Sequential 5′ deletions of CA-rich elements and surrounding regions altered the level of expression of *SP60* in response to developmental signals and cAMP, and also the spatial pattern of *SP60*.

There would seem to be a gradient within the prespore zone which differentially influences the activity of promoters containing varying numbers of response elements (Figure 7-1). One possibility is a gradient of an essential transcription factor that recognizes the CA-rich element.

♦ The availability of an efficient transformation system in *Dictyostelium* has allowed the detailed study of the spacial and temporal control of gene expression in this system with a well-defined pattern of cell differentiation. Haberstroh and Firtel have studied the expression of the prespore protein SP60 and have made observations that are central to the problem of pattern formation. Using a series of 5′-deletions they show that there are three CAE (CACA-rich sequences) that are involved in both the developmental expression and the cAMP inducability of the SP60 gene. Sequential deletion of the CAE boxes (5′ to 3′) of SP60-luciferase gene fusions decreases the transcription as measured by the expression of luciferase activity. More importantly, when the SP60 deletions are coupled to the lacZ gene and the resulting multicellular slugs are assayed for the expression of β-galactosidase, it is observed that there is sequential loss of staining from the posterior to the anterior prespore region. Thus, each CAE box is apparently responsible for the expression of the gene in a restricted region of the prespore zone. The paper also presents evidence that the expression of this gene occurs early in aggregate formation and appears to depend on positional cues. The results indicate that the

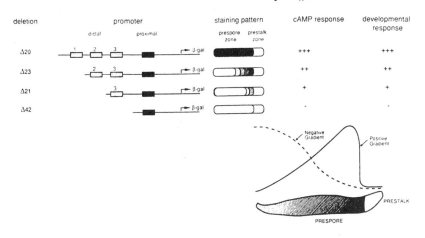

FIGURE 7-1. Model for prespore zone gradient. (Top) Cartoon of the *SP60* promoter-*lacZ* staining response (intensity and localization of expression) and cAMP and developmental response for the various promoter constructs. (Bottom) Model showing two possible gradients within the prespore zone to explain *SP60* promoter-*lacZ* responses. Positive gradient could be transcription factors or parts of the cAMP receptor/intracellular signaling pathway. Negative gradient could be DIF, adenosine, or some other metabolite/morphogen. (From Haberstroh, L. and Firtel, R. A., *Genes Dev.*, 4, 596—612, 1990. With permission.)

expression of the SP60 gene involves positional information that is present early in development and that acts on cells in different regions of the aggregate dependent on its concentration in that area. Determining the nature of these morphogen gradients, the mechanism by which their signals are transduced, and the role of cellular communications which result in this unique pattern of gene expression and eventual establishment of cellular pattern are the next challenges. *Stephen Alexander*

Localization of the Sea Urchin Spec3 Protein to Cilia and Golgi Complexes of Embryonic Ectoderm Cells

E. D. Eldon, I. C. Montpetit, T. Nguyen, G. Decker, M. C. Valdizan, W. H. Klein, and B. P. Brandhorst
Genes Dev., 4, 111—122, 1990   7-2

Ciliogenesis is a distinct and early morphologic event in sea urchin development, coming at the end of the initial stages of rapid cleavage. It coincides with the breakdown of synchronous cell division. There is evidence that some ciliary proteins may have evolved to perform specialized functions for cilia of certain differentiating cell types. Expression of the Spec3 gene of *S. purpuratus* is associated with ectodermal ciliogenesis.

Using antiserum against the amino terminus of the deduced Spec3 amino acid sequence for immunofluorescence staining, cilia as well as an apical structure at the base of the cilium in each ectodermal cell stained intensely in gastrula- and later-stage embryos. The apical staining was dispersed by microtubule-depolymerizing agents, suggesting that the Spec3 antigen is localized to the Golgi complex. Electron microscopy by the immunogold technique confirmed the presence of the Spec3 antigen on cilia and in the Golgi complex. Following synthesis, Spec3 appeared to be sequestered in the Golgi complex before appearing on cilia.

Spec3 antigen is not an essential structural component of cilia, and is not required for beating. It may have a role in interaction of the embryo with its environment. It also is not known whether Spec3 protein is restricted to echinoid embryos.

♦ The ontogeny of cilia (including the axoneme, ciliary membrane, basal body, and rootlet) is an important part of early sea urchin differentiation, yet only few of the structural proteins (e.g., tubulins, dynein, and tektin) have been identified. The paper by Eldon et al. is important because it demonstrates that Spec 3 (see the *1989 Year Book of Developmental Biology,* page 48) is a novel, ciliary surface protein that is unique to cilia of the sea urchin ectoderm; Spec 3 is not found in cilia of the endoderm. Perhaps then these cilia are functionally different, but how Spec 3 might contribute to this difference is unknown. Such a hydrophobic protein coating the cilia might simply "grease" the cilia surface to limit sticking to each other or to the cell surface. Alternatively, Spec 3 might convey sensory information from the external environment to control swimming direction. *Gary M. Wessel*

Genetic Analysis of Defecation in *Caenorhabditis elegans*
J. H. Thomas
Genetics, 124, 855—872, 1990   7-3

Defecation in *C. elegans* proceeds by a cyclical stereotyped motor program starting with contraction of a set of posterior body muscles. Subsequently a set anterior body muscles contracts and, finally, a set of specialized anal muscles opens the anus and expels the intestinal contents (Figure 7-3). The author has attempted to find mutants defective in defecation.

Testing of existing behavioral mutants and screening for new mutants that are constipated because of defecatory defects led to the identification of 18 genes involved in defecation. Mutations in 16 of these genes altered specific components of the motor program. Two mutations affected the function of posterior body muscles and four influenced the anterior muscle component. Mutations in four genes affected the final step in defecation. Six other mutations affected both the anterior muscles and final expulsion of the gut contents. Mutations in two other genes altered the period of the defecation cycle, but not the motor program per se.

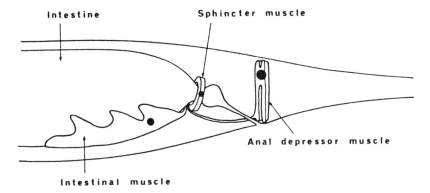

FIGURE 7-3. Diagram of the anal muscles in the hermaphrodite. The three classes of muscles that control the Exp step of defecation are shown. The anal depressor muscle is H-shaped and has muscle fibers that are oriented dorso-ventrally. It is attached at its dorsal end to the body wall and at its ventral end to the dorsal-posterior wall of the anus. It contracts to lift the dorsal surface of the anus. The intestinal muscles are sheet-like and extend anteriorly along the intestine and wrap part of the way around the intestine dorsally. They contain longitudinal muscle fibers and attach to the intestine, and along their ventral edge to the body wall. The sphincter muscle encircles the join between the intestine and the anus. It contains circumferential muscle fibers. The sphincter muscle is difficult to see using a compound microscope and its movement is very fast. It is therefore uncertain what its action is, but freeze-frame video observations suggest that it stretches open (perhaps passively) during the aBoc and contracts very fast just as expulsion occurs, simultaneously with the other Exp muscle contractions (my unpublished observations). The intestinal and anal depressor muscles send arms to the preanal region where they receive synaptic input along with the sphincter muscle. All three classes of muscles are coupled together by gap junctions on these arms. The description of the morphology of these muscles is based on White et al. (1986). (From Thomas, J. H., *Genetics,* 124, 855—872, 1990. With permission.)

A model for genetic regulation of defecation through neuronal control is illustrated in Figure 7-3A. The presence of a neuronal timer is suggested by the fact that the motor steps involved in defecation are activated periodically and in a temporally stereotyped order. Presumably, each class of mutants affects a distinct class, or classes of neurons.

◆ Identifying all of the genes that make a nervous system function may be possible in *C. elegans,* mainly because (1) the C. elegans nervous system is relatively simple and (2) most behavioral mutants are viable. This paper describes the analysis of a newly studied behavior, defecation. Defecation is achieved by a series of three sets of muscle contractions, each of which appears to be controlled by a separate set of motor neurons. Surprisingly, defecation-defective animals are viable and fertile, facilitating the isolation and analysis of defecation mutants. The hypothesis that emerges from this genetic analysis is that most of the defecation mutants affect the motor neurons and interneurons that activate the muscles involved in defecation. Laser beam microsurgery will be useful

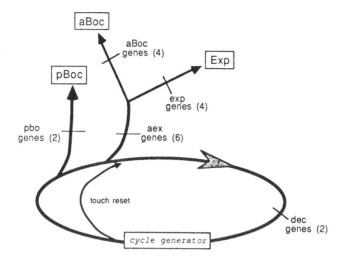

FIGURE 7-4. Model of the genetic pathway for neuronal control of defecation. This arrangement of steps in defecation is the most parsimonious explanation of the mutant classes. The point in the pathway affected by each class of genes is indicated by the cross bars. The touch reset short-circuits the pathway and returns the cycle generator to "zero," or the point just after defecation. The relative positions of the *dec* genes and the touch-reset bypass are drawn arbitrarily, no tests have been performed to distinguish their order. (From Nusbaum, C. and Meyer, B. J., *Genetics,* 122, 579—593, 1989. With permission.)

in testing this hypothesis and in identifying the "neuronal timer" that is thought to control the periodicity of the defecation cycle. *Susan Strome*

The *Caenorhabditis elegans* Gene *sdc-2* Controls Sex Determination and Dosage Compensation in *XX* Animals
C. Nusbaum and B. J. Meyer
Genetics, 122, 579—593, 1989   7-4

In *C. elegans,* both the choice of sexual fate and the level of X chromosome expression are triggered by the X/A ratio. The authors have identified and X-linked gene, *sdc-2,* which controls the hermaphrodite (*XX*) modes of both sex determination and X chromosome dosage compensation. Mutations in *sdc-2* produce extensive masculinization and death of *XX* animals, secondary to a shift of both sex determination and dosage compensation to the *XO* mode of expression.

28 independent *sdc-2* mutations had no apparent effect on *XO* animals. In *XX* animals, however, they led to masculinization and either lethality or dumpiness, the latter effects reflecting disrupted dosage compensation. Mutations in *sdc-2* produced increased levels of several X-linked transcripts in *XX,* but not in *XO* animals. Masculinization was blocked by

mutations in sex-determining genes required for male development. Lethality and the overexpression of *X*-linked genes were not, however, blocked.

These findings suggest that *sdc-2* participates in the coordinated control of both sex determination and dosage compensation in *XX* animals. It presumably acts at a step preceding divergence of the two pathways.

The Role of *sdc-1* in the Sex Determination and Dosage Compensation Decisions in *Caenorhabditis elegans*
A. M. Villeneuve and B. J. Meyer
Genetics, 124, 91—1143, 1990   7-5

In *C. elegans,* mutations in the *X*-linked gene *sdc-1* disrupt both sex determination and dosage compensation in *XX* animals, indicating that the gene acts at a step preceding the divergence of these pathways. A proposed regulatory model for the coordinated control of sex determination and dosage compensation is shown in Figure 7-5. The authors have now isolated a large number of new *sdc-1* alleles as well as a mutation which suppresses the dosage compensation defect but not the sexual transformation effects of *sdc-1*.

Analysis of 14 new *sdc-1* alleles suggested that a lack of *sdc-1* function leads to an incompletely penetrant sexual transformation of *XX* animals toward maleness, and increased levels of *X*-linked gene transcripts in *XX* animals. *XX*-specific morphologic defects were observed, but not significantly *XX*-specific lethality. All the alleles exhibited strong maternal rescue for all the phenotypes assayed. Temperature-shift studies indicated that *sdc-1* acts in the first half of embryogenesis, long before sexual differentiation occurs. On genetic mosaic analysis of *sdc-1,* genotypic mosaics failed to exhibit *sdc-1* sexual transformation phenotypes.

Chromosomal composition is the primary determinant of sexual fate in *C. elegans,* but environmental conditions such as temperature, as well as maternal influences and cellular interactions, all may have a role in the male/hermaphrodite decision.

♦ In *C. elegans* both sex determination and dosage compensation are controlled by the same primary signal, the X/A ratio. The paper by Nusbaum and Meyer shows that *sdc-2,* like *sdc-1,* is involved in controlling the hermaphrodite mode of both sex determination and dosage compensation. Based on genetic interactions with other sex determination/dosage compensation genes, both *sdc* genes are thought to act at an early step, prior to the devergence of the two pathways. Further analysis of *sdc-1,* presented in the paper by Villeneuve and Meyer, strengthens the notion that *sdc-1* acts early in the pathway of sex-determination and

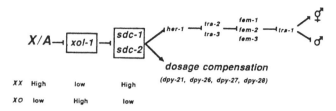

FIGURE 7-5. Proposed regulatory hierarchy for the control of sex determination and dosage compensation in *C. elegans*. In this model proposed by Miller et al. (1988) bars indicate negative regulatory interactions, while arrows indicate positive regulatory interactions; the words "high" and "low" are meant to indicate activity states of the given genes (or gene products) at high (1.0) or low (0.5) values of the X/A ratio. The *sdc* genes are involved in the hermaphrodite modes of the sex determination and dosage compensation processes; they act to promote hermaphrodite sexual development by negatively regulating the activity of *her-1,* a gene that specifies male development (Villeneuve and Meyer, 1987; Nusbaum and Meyer, 1989; Trent, Wood, and Horvitz, 1988), and to control dosage compensation presumably by activating the hermaphrodite (XX) dosage compensation functions encoded by the dosage compensation *dpy* genes (Plenefisch, Delong, and Meyer, 1989). *xol-1* controls the male (XO) modes of both sex determination and dosage compensation (Miller et al., 1988); according to the model, *xol-1* accomplishes this (at least in part) by ensuring that the *sdc* genes (or gene products) remain inactive in XO animals. The order of the genes in the branch of the pathway controlling sexual fate only (*her-1* to *tra-1*) is taken from Hodgkin (1987a); for simplicity, only the interactions pertinent to sex determination in the soma are shown. For a more complete discussion of the data and logic that form the basis of this model for the coordinate control of sex determination and dosage compensation in *C. elegans,* refer to Meyer (1988) and Villeneuve and Meyer (1990). (From Villeneuve, A. M. and Meyer, B. J., Genetics, 124, 1990. With permission.)

dosage compensation. The temperature-sensitive period of a ts *sdc-1* allele is during the first half of embryogenesis, earlier than the TSP of other downstream sex-determination genes and well ahead of the appearance of sexual dimorphisms. Mosaic analysis failed to identify a cellular focus of *sdc-1* gene function in sex determination. The possible explanations of this unusual result are that (1) the X:A ration is assessed or *sdc-1* is expressed immediately, in one- or two-cell embryos or (2) the X:A signal or *sdc-1* acts non-cell-autonomously. Discrimination between these intriguing possibilities must await future molecular analyses. *Susan Strome*

The Product of *fem-1*, a Nemtode Sex-Determining Gene, Contains a Motif Found in Cell Cycle Control Proteins and Receptors for Cell-Cell Interactions
A. M. Spence, A. Coulson, and J. Hodgkin
Cell, 60, 981—990, 1990   7-6

The *fem-1* gene is one of three genes required for adoption of the male sexual fate by *C. elegans. Fem-1* is necessary for sex determination in both germ line and somatic tissues. Clones carrying a 5.5-kilobase (kb) fragment were able to rescue the progeny of a *fem-1* mutant when injected into its oocytes. The major *fem-1* transcript in both sexes was 2.4 kb and comprised 11 exons. It encodes a soluble intracellular protein which has, near its N-terminus, six contiguous copies of a motiff found in several other genes: *cdc10* of S pombe; *SW16* of S. cerevisiae; the *Notch* gene of *Drosophila;* and the *lin-12* and *glp-1* genes of *C. elegans.*

The predicted sequence of the *fem-1* gene product resembles that of several proteins involved in cell-cycle regulation or the control of cell interactions, but it remains unclear how *fem-1* functions to determine sexual fate.

♦ An emerging story in developmental biology is the extent to which molecules with shared motifs play important, and often differing, roles in different organisms. Sex differentiation in the nematode is well character- ized genetically and requires the activity of three genes from a specific gene family, the *fem* genes. The products of these genes are essential for male development but their action in somatic tissue and germline tissue differs. One member of this gene family, *fem-1*, was cloned and in this article the authors report the predicted sequence of its protein product. Analysis of this predicted product indicates that the protein shares homol- ogy with several regulatory proteins from yeast, nematode, and *Dros- ophila*. Interestingly, those proteins are not involved in sex determination, but do play central roles in other critical developmental processes. This important observation suggests that certain conserved molecular motifs are critical biomolecules that serve differing functions in evolutionarily distinct species. *Joel M. Schindler*

Sex-Specific Alternative Splicing of RNA from the *transformer*
Gene Results from Sequence-Dependent Splice Site Blockage
B. A. Sosnowski, J. M. Belote, and M. McKeown
Cell, 58, 499—459, 1989   7-7

Alternative splicing of RNA from the *transformer* (*tra*) gene regulates sexual differentiation in *Drosophila melanogaster*. The process involves competion between two 3′ splice sites. Sex-specific splicing of the *tra* RNA is controlled by the *Sex-lethal* (*Sxl*) gene. In the absence of *Sxl* activity - as in males- only one of the two 3′ splice sites functions. In the presence of *Sxl*, as in females, both sites are functional.

Information for the sex-specific splice site choice resides within the intron itself. Deletions of the splice site in males lead to *Sxl*-independent use of the otherwise female-specific site. The relative amounts of spiced and unspliced RNA derived from the mutant genes are independent of changes in *Sxl* activity. Specific nucleotide changes in the non-sex-specific splice site found not to influence splicing activity, but they did eliminate

Sxl-induced regulation. Deletion of material between the two splice sites did not eliminate sex-specific regulation. Deletion of the female splice site led to a female-specific increase in unspliced RNA.

These findings are in accord with a model in which female–specific factors block the function of the non-sex-specific 3′ splice site. The ability to transfer sex-specific alternate splicing suggests the possibility of using the *tra* intron as part of a sex-specific conditional expression system for otherwise deleterious genes.

Binding of the *Drosophila Sex-lethal* Gene Product to the Alternative Splice Site of *transformer* Primary Transcript
K. Inoue, K. Hoshijima, H. Sakamoto, and Y. Shimura
Nature, 344, 461—463, 1990   7-8

Somatic sex differentiation in *Drosophila melanogaster* is achieved by a hierarchy of genes. One of them, *Sek-lethal (Sxl),* is necessary for functional female-specific splicing transcripts of the immediately downstream regulatory gene, *transformer (tra).* The first exon of the primary *tra* transcript is spliced to one of two acceptor sites. Splicing to the upstream site yields a mRNA which is neither sex-specific nor functional. The mRNA produced by splicing to the downstream acceptor site yields a functional female-specific mRNA.

One possible means by which the *Sxl* gene product determines the alternative splicing of *tra* primary transcripts is the blocking of non-sex-specific splicing to the upstream acceptor in female flies by sex-specific factors. Co-transformation studies have now been done in which *Sxl* cDNA and the *tra* gene are expressed in *Drosophila* Kc cells. It was found that female *Sxl*-encoded protein binds specifically to the *tra* transcipt at or near the non-sex-specific acceptor site. This region contains U-rich sequences.

These findings suggest that the female *Sxl* gene product is the *trans*-acting factor which regulates alternative splicing. It is likely that, as suggested by Sosnowski et al., female-specific splicing *in vivo* results from blocking of the dominant non-sex-specific splicing reaction through binding of the female-specific *Sxl* gene product to the non-sex-specifc acceptor site.

Regulation of Sex-Specific RNA Splicing at the *Drosophila doublesex* Gene: *cis*-Acting Mutations in Exon Sequences Alter Sex-Specific RNA Splicing Patterns
R. N. Nagoshi and B. S. Baker
Genes Dev., 4, 89—97, 1990   7-9

In *Drosophila,* the choice between male and female somatic develop-

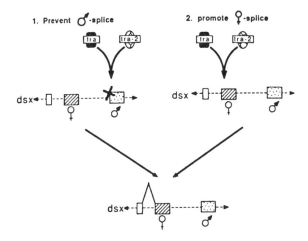

FIGURE 7-9. Models for the regulation of *dsx* RNA splicing by *tra* and *tra-2*. The portion of the *dsx* gene that contains the common splice donor, the female-specific exon, and the first male-specific exon is shown. In model 1 the *tra* and *tra-2* products act to block the male splice in chromosomally female flies. This will allow the female splice acceptor to be used. In model 2, the *tra* and *tra-2* products act positively on the female splice acceptor to promote the occurrence of female-specific splicing. Diagonal lines indicate female-specific sequences; stippled boxes designate male-specific sequences. (From Nagoshi, R. N. and Baker, B. S., *Genes Dev.*, 4, 89—97, 1990. With permission.)

ment is controlled by the number of X chromosomes relative to the number of sets of autosomes. Sex-specific alternative RNA splicing of the *doublesex (dsx)* pre-mRNA produces sex-specific polypeptides which regulate both male and female somatic sexual differentiation.

The authors have characterized a class of *dsx* mutations which act in *cis* to disrupt *dsx* RNA processing, leading *dsx* pre-mRNA to be spliced in the male-specific pattern regardless of chromosomal sex. The mutations were found to be associated with rearrangements in the female-specific exon just 3′ to the female-specific splice acceptor. They did not affect the female-specific splice sites or intron which are identical to wild-type sequences.

Sequences in the female-specific exon are important in regulating sex-specific RNA splicing in *Drosphila*, possibly by acting as sites of interaction with *trans*-acting regulators. Female-specific regulation of *dsx* RNA processing appears to occur by promoting use of the female splice acceptor site rather than by supressing use of the alternative male-specific splice acceptor (Figure 7-9).

♦ These three papers illustrate well the power and synergy of combining genetic and molecular approaches for solving major biological questions. Sex determination in somatic tissues depends on the hierarchical interaction of a small number of genes, the best characterized of which are (in hierarchical order): *Sex-lethal (Sxl), transformer (tra)* and *transformer-2 (tra-2),* and *doublesex (dsx).* Information flows down the cascade in the form of binary switches: on-of in case of *Sxl, tra* and *tra-2,* and male-female in case of dsx. An unusual feature of this regulatory network is that the binary switches do not operate at the transcriptional level, as is common for many processes. Rather, these genes are constitutively transcribed and alternate splicing regulates their expression. In females *Sxl, tra* and *tra-2* transcripts are spliced into functional mRNAs while in males the same transcripts are spliced into functionally inactive RNAs. Consequently, loss-of-function *Sxl, tra* and *tra-2* mutant males have wild-type phenotype. As for dsx, differential splicing yields functional mRNAs in both sexes: the male gene product suppresses female differentiation and, conversely, the female product suppresses male differentiation. Thus, loss-of-function *dsx* mutants develop neither as males nor as females but as intersexes. The major questions addressed by the present reports are: (1) is alternate splicing regulated by positive mechanisms (promotion of splicing at the actual site) or negative mechanism (inhibition of splicing at the alternate site)? (2) which RNA sequences mediate the proper choice of sex-specific splicing site and where do these sequences reside? and (3) is the interaction among the various genes of the sex regulatory cascade direct or indirect (does the protein from one gene bind to the RNA of the next gene or is this interaction mediated by yet-to-be-identified gene products)? As it turns out, a consensus does not always exist and the answers to some of these questions is different for different genes.

In a very careful study, Sosnowski et al. clearly established that, in case of *tra,* the information for proper choice of splice site is contained within a small RNA sequence in the regulated *intron.* By contrast, Nagoshi and Baker's molecular analysis of some *dsx* mutants indicated that sequences in the female-specific *exon* are important for sex-regulated splicing. The results obtained by Sosnowski et al. strongly suggest that the splicing of *tra* is negatively regulated, that is, splicing in the female mode requires repression of the male splicing activity and not the activation of the female site. By contrast, Nagoshi and Baker suggest that for *dsx* the male splicing pattern is the default state and that positively acting factors (perhaps the *tra* and/or *tra-2* gene products) are required to shift this balance toward the female splicing mode.

Whether the products of the different genes in this regulatory cascade directly interact with the RNAs whose splicing they control has been one of the major issues outstanding. Sequence analysis of *Sxl, tra,* and *tra-2* has shown that all of these genes code for proteins that have sequence similarity to RNA-binding proteins or to proteins involved in the control of RNA splicing. It is thus tempting to speculate that the *Sxl* protein directly binds to its own RNA *(Sxl* splicing is autoregulated) as well as to *tra* RNA, and that *tra* and *tra-2* protein bind directly to *dsx* RNA (cf. previous volume in this series). In fact, the functionally important *tra* sequence identified by Sosnowski et al. has high similarity to a *Sxl* sequence situated at the regulated splice junction. The hypothesis for direct interaction of the various gene products received strong support from a series of elegant experiments by Inoue et al. They showed that expression of the *Sxl* gene in the female mode (but not in the male mode) is sufficient to promote the correct female splicing of a *tra* RNA transcribed in the same cultured cells. Moreover, they showed that a RNA fragment containing the sequence identified as being functionally important by Sosnowski et al., is able to specifically bind to a bacterially expressed *Sxl* protein. These findings strongly suggest that *Sxl* regulates the splicing of tra *in vivo* by directly binding to its RNA.

Clearly, much progress has been made in the understanding of the involvement of regulated RNA splicing in the sex determination cascade. We may look forward to the elucidation of the other steps in this regulatory cascade. Differential RNA splicing plays important roles in regulating key biological processes in many organism, including mammals. Thus, these findings may turn out to have a broad significance. *Marcelo Jacobs-Lorena*

Isolation From Chick Somites of a Glycoprotein Fraction That Causes Collapse of Dorsal Root Ganglion Growth Cones
J. A. Davies, G. M. W. Cook, C. D. Stern, and R. J. Keynes
Neuron, 2, 11—20, 1990   7-10

Directional inhibition of axonal growth apparently is an important mechanism of neuronal development and regeneration. Segmented patterns of peripheral spinal nerves in higher vertebrates result from interactions between nerve cells and somites. Neural crest cells and motor and sensory axons all grow exclusively through anterior-half sclerotome. In the chick embryo, posterior cells bind the lectins peanut agglutinin (PNA) and Jacalin. PNA-binding glycoproteins have now been isolated from the somites of chick embryos, and implicated in inhibition of the advance of growth cones.

When liposomes containing somite extracts were applied to cultures of chick sensory neurons, growth cells collapsed abruptly and recovered within 4 h of liposome removal. Collapse was eliminated by immobilized PNA. Electrophoresis yielded two major components, 48K and 55K, which were absent from anterior-half sclerotome. Rabbit polyclonal antibodies raised against these components recognized posterior cells, and also served to eliminate collapse of growth cones.

Spinal nerve segmentation in the chick embryo seems to result from inhibitory interactions between glycoprotein components of lectin and the growth cones.

♦ In higher vertebrates, the peripheral spinal nerves (ventral roots) are segmentally arranged along the body column. As the nerves emanate from the neural tube, they preferentially move through the anterior half of each somitic sclerotome and are absent from the posterior half. This metameric arrangement has led to the idea that molecules differentially distributed in the two halves of the somites may control the pattern of axon growth. This patterning could results from attractive molecules in the anterior half of the sclerotome, inhibitory molecules in the posterior half of the sclerotome, or some combination thereof. The authors previously showed that peanut lectin (PNA) recognized a molecule(s) preferentially located in posterior sclerotome, making this a candidate for an inhibitory substance.

Here, Davies and colleagues have tested whether the PNA-binding molecules represent inhibitory molecules for axon growth. The PNA-binding molecules are present in the posterior sclerotome at the time of initial axon outgrowth. A simple bioassay was devised to detect inhibition of neurite extension. Dorsal root ganglia (DRG) were cultured on substrates containing the glycoprotein fraction eluted from a PNA column; axons grew significantly less well than on control substrates. Furthermore, incorporation of these glycoproteins into liposomes caused DRG axons grown on laminin substrates to collapse. Antibodies produced against the two major PNA-binding components (Mr 48 and 55 kDa) of the glycoprotein extract completely blocked this axon collapse.

Their results suggest that one or both of the PNA-binding glycoproteins are inhibitory for the outgrowth of peripheral axons. Thus, the segmental arrangement of axons in the peripheral nervous system may result from directional inhibition of axon outgrowth. Such inhibition represents a simple mechanism for restricting the growth of neurites or cells to prescribed regions of the embryo. *Marianne Bronner-Fraser*

Ocular Dominance Plasticity in Adult Cat Visual Cortex after Transplantation of Cultured Astrocytes

C. M. Muller and J. Best
Nature, 3432, 427—430, 1989   7-11

For a limited post-natal period, the visual cortical circuitry can be modified by experience. Changes in ocular dominance attending restricition of vision in one eye are paralleled by — and possibly mediated by — elimination of synapses and axonal sprouting. Activation of NMDA receptors and subsequent calcium influx presumably trigger synaptic changes in this setting. Immunocytochemical studies have suggested that immature astrocytes are required for plasticity of the visual cortex.

This study was designed to determine whether in fact immature astrocytes are necessary for plasticity of ocular dominance. Astrocytes cultured from newborn kittens were transplanted to the visual cortices of adult cats aged 1 year or more, and therefore past the critical period for cortical plasticity. The transplant procedure had the effect of reinducing plasticity of ocular dominance in the adult animals. It again became possible to alter ocular dominance through light deprivation. No such changes occured when previously frozen cells or cultured fibroblasts were transplanted.

These findings support the view that immature glial cells are required for a plastic visual cortex. They help explain the beneficial effects of transplanting astrocytes on behavioral recovery following cortical damage. The end of the critical period of plasticity may relate to the maturation of astrocytes.

♦ In mammals, binocular vision results from the segregation of information from the two eyes into a pattern of alternating stripes, or ocular dominance columns, in the visual cortex. The formation of these columns during development of the visual system is highly sensitive to visual experience during a critical period after birth, such that visual deprivation in one eye leads to its territory on the visual cortex being taken over by the other eye. Following this critical period, visual experience does not affect the ocular-dominance columns.

This report by Muller and Best tests the possibility that immature astrocytes may be responsible for the plasticity observed during the critical period. Maturation of the astrocytes has been proposed to correlate with the end of the critical period. To test this idea, the authors reintroduced immature astrocytes cultured from newborn kittens into the visual cortex of adult cats, 1 year or older, past the time of the critcal period. They found that ocular-dominance plasticity was regained on the side of the cortex injected with immature astrocytes, but not on the control side receiving an injection of freeze-killed cells.

Although the mechanism by which immature astrocytes support visual cortical plasticity is unclear, one possiblity is that they selectively produce neurotrophic factors which affect cortical cells. Immature astrocytes are known to secrete nerve growth factor, basic FGF, and S100 protein; one or some combination of these could be taken up by active terminals to support axonal sprouting and neurite extension. *Marianne Bronner-Fraser*

Target Control of Collateral Extension and Directional Axon Growth in the Mammalian Brain
C. D. Heffner, A. G. S. Lumsden, and D. D. M. O'Leary
Science, 247, 217—220, 1990   7-12

The means by which individual neurons in the brain send axons over considerable distances to multiple targets remains to be clarified. The mammalian corticopontine projection is a convenient system in which to study axonal outgrowth, branching, and target selection. The connection develops from parent corticospinal axons which have grown past the pons, through delayed interstitial budding of collateral branches which then grow directly into their target, the basilar pons. Explants of rat cortex and basilar pons were co-cultured at a distance within a three-dimensional collagen matrix.

When co-cultured with explants of developing cortex, the basilar pons elicited the formation and directional growth of cortical axon collaterals across the matrix. This process appeared to be target-specific and to electively influence neurons in the appropriate cortical layer.

It seemed likely that the basilar pons is innervated through control at a distance — through release of a diffusable chemotropic molecule — of the budding and directed ingrowth of cortical axonal collaterals. If so, diffusable molecules play a critical role in establishing connections in the mammalian brain.

♦ In the mammalian corticopontine projection, corticospinal axons grow past the pons into the spinal cord, but later send projections to the basilar pons by branching of existing axons. This study using an *in vitro* assay to determine whether the basilar pons target attracts cortical neurons. By growing the two tissues in a three-dimensional collagen gel, the authors found that the basilar pons selectively attracted neurons from the appropriate cortical layer. These results suggest that a diffusible, chemotropic signal stimulates cortical axons to grow toward the basilar pons. However, *in vivo*, the axons initially grow past the pons. Thus, there may be a two-stage process of axon outgrowth. In the first step, the axons bipass the pons, perhaps by responding to local cues associated with the axon tract. In the second step, collateral branches from the axons are attracted to the pons by a chemotropic mechansim. The reason that the axons do not initially respond to the pons may be due to the relatively late maturation of the basilar pons, to a preference for the axonal tract, or because they initially pass by the pons too rapidly to respond to the diffusible cues.

The idea that a diffusible, chemotropic substance selectively can attract axons is an old concept, originally suggested by Ramon and Cajal. However, supportive evidence has been lacking until recently. The present study provides further evidence that target-derived factors acting at a distance can influence the directionality of axonal projections. This is made all the more interesting because the diffusible substance made by the pons does not attract the primary axon outgrowth, but rather causes branching of existing axonal projections. Thus, the timing of release/response to a diffusible signal may be an important factor in shaping connections in the developing nervous system. *Marianne Bronner-Fraser*

Analysis of Neurogenesis in a Mammalian Neuroepithelium: Proliferation and Differentiation of an Olfactory Neuron Precursor *In Vitro*

A. L. Calof and D. M. Chikaraishi

Neuron, 3, 115—127, 1989   7-13

The problems involved in studying neurogenesis *in vivo* may be circumvented using a culture system for murine olfactory epithelium. Antibody markers are used to definitively identify putative neuroepithelial stem cells (as keratin-positive basal cells) and differentiate neurons (as N-CAM-positive olfactory receptor neurons). Along with tritiated thymidine uptake analysis, the antibodies serve to characteize the proliferation and differentiation of neuronal precursors.

The immediate neuronal precursor was distinct from basal cells and rapidly separated from them, dividing as it migrate. The precursor appeared to follow a simple lineage program. Division of an N-CAM$^-$ cell gave rise to two N-CAM$^+$ daughter neurons. Neurons were efficiently generated in defined medium, but neurogenesis eventually ceased when new precursors failed to be produced.

Neuronal differentiation in the mouse is a multi-stage process which may proceed as illustrated in Figure 7-13. In contrast to the *in vitro* situation, neurogenesis *in vivo* continues throughout life, suggesting that epigenetic factors may regulate ongoing neurogenesis.

♦ Neurogenesis is a developmental process by which neuronal precursors become post-mitotic and differentiate into the appropriate neuronal type. This process generally occurs at early stages, and most parts of the nervous system are not self-renewing. One exception is the mammalian olfactory epithelium, which generates one type of neuron, the olfactory receptor, throughtout life.

In this study, Calof and Chikaraishi describe a culture system for studying the process of neurogenesis in the olfactory epithelium under defined conditions. Using antibody markers, they identified two distinct cell types. One is a population of epithelial cells which are keratin-positive basal cells; these are likely to be the neuroepithelial stem cells. The second population are N-CAM positive, post-mitotic differentiated neurons, which represent the olfactory receptor neurons. In addition, there is an intermediate population of round cells that migrate away from the epithelial cells in medium with low calcium. These cells are both kerative and N-CAM negative and take up ^{3}H-thymidine, suggesting that they are dividing. The round, keratin/N-CAM-negative cells appear to neuronal precursor cells which are unipotent, dividing once and give rise to two undifferentiated, post-mitotic neurons that express N-CAM. Subsequently, this population expresses other neuronal markers such as GAP-

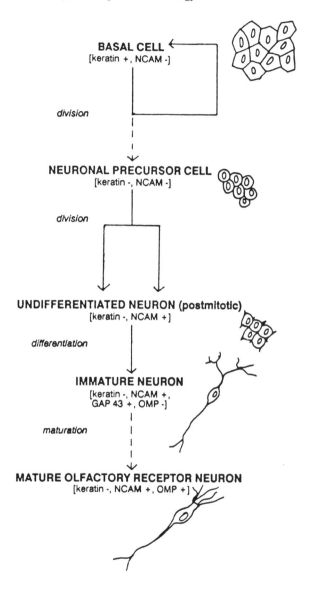

FIGURE 7-13. Neurogenesis in the olfactory epithelium: a proposed sequence of events. A proposed series of steps in proliferation and differentiation leading to the production of olfactory receptor neurons is diagrammed. Solid arrows indicate steps whose existence is directly supported by the observations in this manuscript. Dashed arrows indicate steps whose occurrence has been previously suggested by *in vivo* studies. For further discussion, see text. (From Calof, A. L. and Chikaraishi, D. M., *Neuron,* 3, 115—127, 1989. With permission.)

43. Under their culture conditions, neurogenesis only occurs for a limited time, suggesting that the basal stem cells may need additional factors to survive or proliferate.

This study represents a significant step forward in understanding neurogensesis. It suggests that neurogenesis, at least in this system, is a multistep process in which a stem cell gives rise to a dividing neuronal precursor cell which in turn becomes a post-mitotic neuron. The olfactory epithelium seems like an excellent model system to define the substances and cellular interactions that are required to induce, maintain, or inhibit the generation of neurons. *Marianne Bronner-Fraser*

Abnormal Sexual Development in Transgenic Mice Chronically Expressing Müllerian Inhibiting Substance
R. R. Behringer, R. L. Cate, G. J. Froelick, R. D. Palmiter, and R. L. Brinster
Nature, 345, 167—170, 1990   7-14

Müllerian inhibiting substance (MIS) is a glycoprotein normally secreted by testicular Sertoli cells and by granulosa cells of the post-natal ovary. Production of MIS in the male fetus leads to regression of the Müllerian ducts. *In vitro,* purified MIS induces the formation of seminiferous cord-like structures in the fetal rat ovary.

The authors have produced nine founder transgenic mice carrying a metallothionein-1-MIS fusion gene, which chronically express human MIS, in order to clarify the role of MIS in sexual development. In female animals, chronic expression of MIS inhibited Müllerian duct.differentiation and resulted in a blind vagina without uterus or oviducts. The ovaries had a subnormal number of germ cells at birth and germ cells were lost in the first 2 weeks of post-natal life. At the same time, the somatic cells were organized into structures resembling seminiferous tubules. These were not, however, detected in adult females. A majority of transgenic males developed normally but, in two lines with the highest levels of MIS expression, some of the males had feminized external genetalia, impaired Wolffian duct development, undescended testes.

MIS appears to interact with both germ cells and somatic cells of the gonad. Hypothetically, the testis-detemining gene product triggers Sertoli-cell differentiation and activates MIS gene expression. MIS then acts in an autocrine manner to induce further Sertoli-cell differentiation and testis formation. Since Sertoli and granulosa cells presumably derive from a common progenitor and express MIS receptors, improperly regulated expression of MIS in the developing female leads to transdifferentiation of granulosa cells into Sertoli cells and the formation of seminiferous tubules.

♦ The decision of the gonads to develop into testes is determined by the testis-determining gene present on the Y chromosome. In the male embryo, Müllerian inhibiting substance (MIS), a glycoprotein hormone, is synthesized early in the development of the testes and its secretion is responsible for regression of the female Müllerian duct system. Androgens are then produced by the fetal testes which stimulate the male Wolffian ducts to develop into epididymis, vas deferens, etc. The gene which encodes human MIS has been cloned, and in the work reported by the paper cited above, transgenic mice expressing human MIS under the control of the metalothionein gene promoter were produced. It turns out that female transgenic mice expressing MIS lacked oviducts and uterus. Although not shown, the assumption is that MIS was expressed in transgenic fetuses at the time of sexual differentiation. The interesting result is the absence of ovaries in most of the adult transgenic females with high levels of circulating MIS. As suggested by Anne McLaren (*Nature* 345, 111, 1990), MIS may normally be involved in the removal of germ cells that have entered into meiosis despite the inhibiting effect of the testis cords. Consequently, if oocytes are depleted in transgenic females, the follicles would degenerate since the continued presence of oocytes appears to be needed for their mainenance. Another interesting result from these experiments is that some transgenic males show feminization of the external genitalia, abnormal development of the Wolffian ducts, and undescended testes. These results suggest that MIS may have more than one role in sexual development, and the authors speculate regarding the possible effect of MIS in the transgenic males. This report is another example of dominant-like mutations being produced by over- and/or misexpressing a gene in transgenic animals. Some interesting phenotypes were produced which have led to new hypotheses concerning the role of MIS. The next obvious experiment is to "knock-out" the gene by targeted mutagenesis and to determine what happens to the developing fetus when no funcitonal MIS is produced. *Terry Magnuson*

Genetic Evidence That *ZFY* Is Not the Testis-Determining Factor
M. S. Palmer, A. H. Sinclair, P. Berta, N. A. Ellis, P. N. Goodfellow, N. E. Abbas, and M. Fellous
Nature, 342, 937—939, 1989   7-15

The testis-determining gene *TDF*, present on the mammalian Y chromosome, induces the undifferentiated gonads to form testes. The gene has been localized close to the pseudoautosomal region shared by the sex chromosome in the distal Y-specific region. A recently cloned human gene, *ZFY*, has many features suggesting identity with *TDF*, including the encoding of a protein that resembles a transcription factor. However a very similar gene, *ZFX,* is present on the X chromosome, and a *ZFY-*related sequence is present on autosomes of marsupials.

In XX males lacking *ZFY,* the male phenotype could be explained by a mutation in a gene "downstream" of *ZFY* in the sex-determining hierarchy but, in this case, no exchange of material between the X and Y chromosomes should be observed. The authors observed exchange of Y-specific sequences next to the pseudoautosomal boundary in four XX males lacking *ZFY.*

This finding argues strongly against a role for *ZFY* in sex determination. It remains possible that *ZFY* has a male-specific fuction in spermatogenesis.

Zfy Gene Expression Patterns Are Not Compatible with a Primary Role in Mouse Sex Determination
P. Koopman, J. Gubbay, J. Collignon, and R. Lovell-Badge
Nature, 342, 940—942, 1989   7-16

In mammals, a Y chromosome-linked gene influences the embryonic gonad to differentiate into a testis, initiating male development. *ZFY* is a candidate for the testis-determining gene (*TDF* in humans, *Tdy* in mice). It potentially encodes a zinc-finger protein, and has two Y-linked homologues in mice, *Zfy-1* and *Zfy-2.* It appears that *ZFY, Zfy-1,* and *Zfy-2* map to the sex-determining regions of the human and mouse Y chromosomes, but direct evidence is lacking that these genes are involved in testis determination.

The authors have found that *Zfy-1,* but not *Zfy-2,* is expressed in the differentiating embryonic mouse testis. Neither of these genes is expressed in mutant embryonic testes lacking germ cells. It is concluded that neither *Zfy-1* nor *Zfy-2* is a candidate for the mouse testis-determining gene. Instead, these genes may have a role in male germ-cell development.

A Gene Mapping to the Sex-Determining Region of the Mouse Y Chromosome is a Member of a Novel Family of Embryonically Expressed Genes
J. Gubbay, J. Collignon, P. Koopman, B. Capel, A. Economou,
A. Munsterberg, N. Vivian, P. Goodfellow, and R. Lovell-Badge
Nature, 346, 245—250, 1990   7-17

Sex determination in eutherian mammals appears to be equivalent to testis determination, with the ovarian pathway serving as the normal or "default" pathway. Expression of the gene responsible for testis development (*TDF* in humans, *Tdy* in mice) switches the fate of supporting-cell precursors in the indifferent gonad from that of follicle cells to that of Sertoli cells.

The authors have found that gene mapping to the sex-determining region of the mouse Y chromosome is deleted in a line of XY female mice mutant for *Tdy*. The gene is expressed at a stage of a male gonadel development consistent with its having a role in testis determination. The gene is a member of a family containing at least five mouse genes which share an amino acid motif homologous to known or putative DNA-binding domains.

If this gene, *Sry,* is in fact involved in testis determination, it may encode a protein capable of acting as a regulatory molecule in cells. It could be a nuclear protein which binds DNA and acts as a transcriptional switch. The authors believe that *Sry* is a good candidate for *Tdy*.

Additional Deletion in Sex-Determining Region of Human Y Chromosome Resolves Paradox of X,t(Y;22) Female
D. C. Page, E. M. C. Fisher, B. McGillivray, and
L. G. Brown
Nature, 346, 279—281, 1990   7-18

The sex-determining function of the Y chromosome lies entirely in interval 1A, since most XX persons with descended testes and normal male external genitals carry this region of the chromosome. Deletion of part of 1A at the translocation breakpoint has been described in a female having a reciprocal Y;22 translocation. The finding of four partly masculinized XX persons carrying only part of interval 1A — a part not overlapping the deletion in the X,t(Y;22) female — implies that the sex-determining function resides in the part of 1A present in these XX intersexes. The X,t(Y;22) female, then, might be explained by a chromosomal position effect or by delayed development of the gonadal soma.

The authors found that the X,t(Y;22) female has a deletion of a second portion of interval 1A, a part corresponding closely to that present in the XX intersexes. Nevertheless, phenotype-genotype correlations indicate that two or more genetic elements in interval 1A may contribute to the sex-determining function of the Y chromosome.

The X,t(Y;22) female lacks the *ZFY* gene, but does not exhibit Turner's syndrome. This argues against *ZFY* being the Turner's syndrome gene on the Y chromosome.

A Gene From the Human Sex-Determining Region Encodes a Protein with Homology to a Conserved DNA-Binding Motif

A. H. Sinclair, P. Berta, M. S. Palmer, J. R. Hawkins, B. L. Griffiths,
M. J. Smith, J. W. Foster, A. Frischauf, R. Lovell-Badge, and
P. N. Goodfellow
Nature, 346, 240—244, 1990   7-19

There is some evidence suggesting that *ZFY* is identical to *TDF,* the gene determining testis development, but recent reports have quetioned the role of *ZFY* in male sex determination. The authors searched a 35-kilobase region of the human Y chromosome, the region necessary for male sex determination, and found a new Y-specific gene which is conserved in a wide range of mammals and encodes a testis-specific transcript. The gene shares homology with the mating-type protein Mc from the fission yeast *S. pombe* and a conserved DNA-binding motif present in the nuclear high-mobility-group proteins HMG1 and HMG2. The gene is termed sex-determining region Y *(SRY).*

SRY would seem to be a candidate for *TDF.* Sequences homologous to *SRY* have been found on the Y chromsome in all eutherian mammals studied to date, and *SRY* encodes a protein possessing a potential DNA-binding domain. Proof that *SRY* actually is *TDF* will require a mutational analysis of XY females or studies of sex-reversed transgenic mice.

♦ A landmark paper published in 1987 by Page and co-workers (*Cell,* 51, 1091—1104) reported what seemed to be the long-awaited localization of the gene which encodes the testis-determining factor (termed *TDF* in humans and *Tdy* in mice) to a 140-kb region of the 1A2 region of the human Y chromosome.The gene identified (*ZFY*) encoded a zinc finger protein. Subsequently, two copies of an equivalent sequence *(Zfy-1* and *Zfy-2)* were reported to exist on the mouse Y chromosome (*Cell,* 56, 765—770, 1989; *Science,* 243, 78-80, 1989; *Science,* 243, 80—83, 1989). Some surprising findings included the existence of an X-linked homologue in many species as well as an autosomal homologue in mice. Equally confusing was the mapping of *ZFY*-related sequences to autosomes rather thant the Y chromosome in marsupials (*Nature,* 336, 780—783, 1988). Finally, several XX male patients lacking *ZFY* have been identified and these are difficult to explain if ZFY is the testis-determining factor.

These issues seem to have been resolved during the last year and another candidate gene for *TDF* has been identified. First, reports by Palmer et al. (1989) and Koopman et al. (1989) present data questioning the role of *ZFY* as the testis-determining gene. Fourteen *ZFY*-negative XX males were examined of which four were found to have Y-specific sequences localized to within 60 kb of the pseudoautosomal boundary

(Palmer et al.). It was concluded that, since these are XX males with Y-specific sequences, *TDF* must be localized within this region (1A1). The remaining XX males that do not contain the pseudoautosomal region were argued to have mutations elsewhere in the sex-determinination pathway. The second report (Koopman et al.) presented data on *Zfy* expression in the developing mouse fetus. It was presumed that the gene for testis determination would be expressed at or around day 11.5 to 12.5 of gestation when testis differentiation is first apparent. In fact, *Zfy-1* expression was detected in differentiating gonads between day 10.5 and 14.5 of gestation, with decreasing levels observed between days 14.5 and 17.5. However, *Zfy-1* transcripts were not detected in embryonic day 14.5 testes devoid of germ cells due to the *W* mutation. These results were interpreted to suggest that *Zfy-1* expression is due to the presence of germ cells and is not involved in testis determination. However, because only 14.5-d testes were examined, the question of whether *Zfy-1* could be expressed in the developing testes at earlier stages (10.5-13.5) and then turned off at day 14.5 is still open. This is an important point particularly since these investigators report decreasing levels in wild-type male fetuses beginning at day 14.5. It is possible that this process is accelerated in the mutant mice.

Since the publication of these two reports which questioned Page's assignment of *TDF* to 1A2, progress has been rapid, and three subsequent papers were published in July, 1990. The first paper (Sinclair et al.) reduced the area in which *TDF* was thought to be located to a 35-kb region of 1A1. A single-copy gene was then found which was called SRY for sex-determining region of the Y. The predicted amino acid sequence shows homologies with the fission yeast mating-type protein Mc and with the nonhistone HMG proteins. The second paper (Gubbay et al.) reports the cloning of the mouse homologue which shows a high degree of homology (80% over a 237-bp region examined) with the human sequence. Four closely related genes were mapped to autosomes, all of which were shown to be expressed at day 8.5 in development. The Y gene (*Sry*) was found to be expressed in the adult testis and urogenital ridge of the day 11.5 fetus. Furthermore, a strain of mice carrying a mutation that produces XY females shows a deletion of *Sry*. The third paper (Page et al.) clears up the paradox of the X,t(y;22) female. This patient was key to Page's assignment of *TDF* to 1A2. She was shown to be deleted for 1A2 and thought to possess the whole of 1A1 which contradicts the finding that *TDF* lies in 1A1. With the availability of additional Y-specific markers, it turns out that this patient is not only deleted for 1A2 but also possesses a deletion of the portion of 1A1 to which *SRY* has been mapped.

TDF appears to be localized to a 35-bp region of the Y chromosome. Within this region lies a gene, *SRY*, which has all of the characteristics expected for *TDF*. The experiments that are likely to take place in the coming year include the production of transgenic mice to determine whether *Sry* will transform an XX embryo into a male. Site-directed mutagenesis via homologous recombination and embryonic stem cells will also be done to determine whether inactivating *Sry* will result in XY females. Finally, expression studies in mice will likely take place to determine whether *Sry* expression is associated with somatic and/or germ cells of the testis. If *Sry* remains a strong candidate, future work will no doubt concentrate on whether *TDF* is a DNA-binding protein and whether a gene-regulatory cascade exists for mammalian sex determination. *Terry Magnuson*

Interleukin 1 Inhibits T Cell Receptor-Mediated Apoptosis in Immature Thymocytes

D. J. McConkey, P. Hartzell, S. C. Chow, S. Orrenius, and M. Jondal
J. Biol. Chem., 265, 3009—3011, 1990   7-20

T cell receptor signaling is implicated in an intrathymic cell selection process through which potentially harmful autoreactive precursors are deleted before emigrating to the periphery. Immature thymocytes undergo apoptosis, or programmed cell death, when stimulated via the T cell antigen receptor. It is possible that this mechanism mediates the deletion of self-reactive clones during T-cell development.

Interleukin 1 (IL-1), a potent co-mitogen for thymocytes, mediates signal transduction in conjunction with activation of protein kinase C. Protein kinase C activation inhibits thymocyte apoptosis. In this study, IL-1 prevented T cell receptor-mediated thymocyte apoptosis by a mechanism involving activation of protein kinase C. Il-1 blocked programmed cell death in thymocytes exposed to glucocorticoid, calcium ionophores, and antibodies to the antigen-receptor complex.

These finding indicate that IL-1 may have a role in the positive signaling which spares appropriate precursors during the generation of functional T lymphocytes. Whether lymphokine availability is a factor in thymocyte maturation remains to be determined.

♦ Thymus glands were recovered from patients under 2 years of age undergoing corrective surgery, and the thymocytes isolated and cultured by standard protocols. The addition of either a calcium ionophore or glucocorticoid to these thymocytes resulted in the elicitation of DNA fragmentation. This could be blocked by the simultaneous addition of either IL-1 or a phorbol ester. This same protective effect of IL-1 was observed when the thymocytes were stimulated with anti-CD3, and this protective effect was not mediated by IL-1 reducing the level of calcium influx induced by anti-CD3. The subpopulation of thymocytes protected from receptor-mediated apoptosis by IL-1 were $CD4^+CD8^+$ cells. The inhibition of protein kinase C (PKC) by adding either H-7 or sphingosine along with IL-1, reversed the protective effect of IL-1.

The initiation of biochemical signals to immature thymocytes via occupancy of the TCR elicits cellular events characteristic of programmed cell death. The results of this study support the developing concept that this does not occur in immature thymocytes also receiving a second growth-related signal, such as IL-1. *E. Charles Snow*

Monoclonal Antibodies to Pgp-1/CD44 Block Lympho-Hemopoiesis in Long-Term Bone Marrow Cultures

K. Miyake, K. L. Medina, S. Hayashi, S. Ono, T. Hamaoka, and P. W. Kincade

J. Exp. Med., 171, 477—488, 1990   7-21

Stromal cells in the bone marrow are able to produce a number of regulatory cytokines, but little is known about how close physical proximity is maintained between these cells and the precursor cells they influence in the marrow microenvironment. A group of cell-surface glycoproteins, classified as CD44 in humans, have been implicated in lymphocyte homing and in cell-matrix interactions. The authors prepared monoclonal antibodies recognizing a cloned murine stromal-cell line and found that they recognize epitopes on Pgp-1/CD44.

Antibodies were screened in a cell adhesion assay, and four which inhibited the binding of B lineage cells to stromal-cell monolayers were selected. These antibodies detected epitopes of the Pgp-1/CD44 antigen complex. Addition of Pgp-1/CD44 antibodies to long-term marrow cultures prevented the emergence of myeloid cells, and also blocked lymphocyte growth in culture. The antibodies did not influence the factor-dependent replication of myeloid or lymphoid progenitor cells.

Adhesive interactions likely are important for lympho-hemopoiesis in long-term marrow cultures. The present findings suggest that Pgp-1/CD44-related glycoproteins may be critical for the formation of lymphoid and myeloid cells in the bone marrow.

♦ It has become increasingly clear that families of adhesion molecules play crucial roles during cellular interactions occurring between various components of the immune system. Each of these molecules interacts preferentially with another cell-surface-expressed molecule to promote stronger adherence between cells displaying the two proteins. The current thought is that such interactions alone are not responsible for the selective coupling of different cells, but the sum total of several such interactions work together to provide the necessary strength of contact between cells to affect some differentiative function. It is also entirely possible that some or all of these molecules exert signaling capacities.

Surface adhesion molecules which have previously been shown to be important within the immune system can be divided into different families of proteins, each with its own set of ligands. The lymphocyte function-associated molecule LFA-1 (CD11a/CD18) is a member of the integrin family of cell-adhesion molecules and binds selectively to intercellular adhesion molecule-1 (ICAM-1, also referred to as CD54) and ICAM-2. CD2 (E-rosette receptor) binds to its counter receptor LFA-3. ICAM-1, ICAM-2, CD2, and LFA-3 are all members of the immunoglobulin super-family. Finally, classic homing receptors, such as MEL-14, found on some murine lymphoid cells, have been implicated in lymphocyte precursor-stromal cell interactions. One such surface protein, Pgp-1 (CD44), has been shown to be related to the Hermes families of adhesion proteins (human homing molecules). It is already clear that these families of adhesion protein will contribute to lymphocyte developmental processes.

For these experiments, a murine bone marrow-derived stromal cell clone (BMS2) was employed for the study of cell surface proteins which participate during the adhesion of lymphocytes to stromal cells found within the bone marrow. A B cell hybridoma, which adheres to BMS2, was used as the second component in a typical *in vitro* adherence assay. This assay was used to identify monoclonal antibodies possessing the ability to interfere with cells binding directly to bone marrow stromal cells.

Out of several hundred monoclonal antibodies recognizing surface components expressed by the stromal cells, the authors identified four which interfered with the ability of B cells to adhere. One of the blocking monoclonals was found to be specific for Pgp-1. The inclusion of the anti-Pgp-1 antibodies within long-term bone marrow cultures inhibited the propagation of B lineage cells. In addition, propagation of cells from these cultures was delayed following the inclusion of antibody specific for LFA-1. These antibodies did not interfere with replication of myeloid and lymphoid progenitor cells. These studies support an important role for direct contact between lymphoid precursor cells and bone marrow stromal cells during B cell development, and identify the Pgp-1 protein as one molecule which participates in this process. *E. Charles Snow*

Thymic Epithelial Cells Induce *In Vitro* Differentiation of PRO-T Lymphocyte Clones Into TCRa,b/T3+ and TCRg,d/T3+ Cells
R. Palacios, S. Studer, J. Samaridis, and J. Pelkonen
EMBO J., 8, 4053—4063, 1989   7-22

Knowledge of the initial major phase of intrathymic T-cell development is complete. It is now possible to develop prethymic and intathymic T-cell progenitor clones from the marrow of young adult mice and the thymus of 2-week-old mouse embryos. These clones represent the earliest identified stages of T-cell development. Both the "marrow-type" and "thymic-type' PRO-T lymphocyte clones contain the TCRα, β, γ, δ genes in the germ line configuration.

The authors found that PRO-T clones differentiate *in vitro* into TCR/Tc+ cells with considerable efficiency when co-cultured with monolayers of either heterogenous thymic epithelial cells or the thymic epithelial clone ET. The same PRO-T clone was able to give rise to both TCRα,β+ and TCRγ,δ+ cells. The microenvironment permitting T-cell progenitors to generate L3T4+ cells differed from that yielding L3T4+LyT2+ and LyT2+ cells.

There probably are two paths by which a given PRO-T cell generates different thymocyte subsets and peripheral T lymphocytes. One pathway involves interaction with thymic epithelial cells like ET cells, while the other involves an interaction with a different thymic epithelial subset.

♦ This study examines the possiblity that the thymic epithelial cells function during the induction of early thymocyte differentiative events. The precursor T cell arriving from the bone marrow is CD4−CD8−TCR− and must be induced to differentiate into TCR-expressing, CD4 or CD8 single expressing cells. The authors have previously prepared precursor T-cell clones from adult bone marrow and 14-d thymus. These clones have not rearranged their TCR genes, nor do they transcribe any of the CD3 genes. These clones are, therefore, antigen-receptor negative. The bone marrow-derived clones grow in the presence of IL-3 and IL-4. The addition of these clones into cultures of primary thymic epithelial cells or onto cultures of a thymic epithelial cell clone allows the study of conditions necessary for necessaary for inducing these clones to differentiate into TCR receptor expressing thymocytes.

The addition of both PRO-T-cell clones to primary thymic epithelial cultures resulted in the differentiation of cells into either CD4+CD8− or CD4−CD8+ cells, indicating that the same double negative cell can, depending upon the signals delivered following interaction with the epithelial cells, differentiate into either a CD4+ or CD8+ T cell. The primary thymic eppithelial cell cultures are heterogenous in nature. To more carefully address the role of the epithelial cell in this process, a epithelial cell clone was used to induce the differentiation of the PRO-T cell clones. The striking result was that the epithelial clone only induced the precursor cells to differentiate into CD4+CD8− cells. These results addressed two interesting issues. First, they are consistent with the differentiation of double-negative cells directly into CD4+ or CD8+ cells without going through a double-positive stage. Second, there may be microenvironmental differences resulting in CD4+ and CD8+ cell differentiation. In other words, different subsets of thymic epithelials cels may be able to induce the same precursor cell down two different diffrrentiative pathways. *E. Charles Snow*

Homeobox Genes 8

INTRODUCTION

Homeobox-containing genes continue to be the focus of much research energy and excitement. As the number of Homeobox-containing genes that are isolated and characterized continues to grow, it is likely that new functions and mechanisms of action will be discovered for members of this gene family.

Species distribution remains an exciting area of investigation and suggests some interesting questions regarding the evolutionary basis of homeobox-containing genes. While DNA sequences are conserved between species that are quite distant from an evolutionary perspective, the tissue distribution of their expression and ultimately their developmental roles appear to be quite different. Why, then, is this motif conserved, and what was its original function? Clearly the presence of this motif in nonsegmental animals suggests that determining segmentation in muticellular animals was not the first function of homeobox-containing gene products.

As transcription factors, homeobox-containing genes can directly interact with DNA to modulate gene expression. A growing body of evidence indicates that in addition to DNA interactions, homeobox-containing gene products can interact with each other to form protein complexes that can also interact with DNA. Thus, the potential number of transcription factors increases enormously when all the possible combinations of homeobox-containing genes is considered. Which combinations are actually used and how the various combination impact on their function are areas of great interest.

Caenorhabditis elegans **Has Scores of Homeobox-Containing Genes**
T. R. Burglin, M. Finney, A. Coulson, and G. Ruvkun
Nature, 341, 239—243, 1989   8-1

Homeobox-containing genes control cell identities in particular spatial domains, cell lineages, or cell types during the development of *C elegans,* and probably in vertebrates as well. More than 80 genes with homeoboxes possessing 25 to 100% sequence similarity have been isolated genetically, or by DNA hybridization to previously isolated genes.

In this study, oligonucleotides corresponding to well-conserved amino acid sequences from the helix-3 region of the homeodomain were synthesized and used as probes in screening *C. elegans* genomic libraries. 49 putative homeobox-containing loci were identified; eight of ten selected loci possessed sequences corresponding to the conserved helix-3 region plus additional flanking sequence similarity. One of these genes had a sequence corresponding to a complete *pou*-domain, and other had a sequence closely related to the homeobox-containing genes *caudal/cdx-1*. The putative homeobox loci were mapped to the physical contig map of *C. elegans*, permitting the identification of corresponding genes from the correlated genetic map.

C. elegans probably has at least 60 homeobox-containing genes, which would constitute about 1% of the estimated total number of genes.

◆ The extent to which homeobox-containing genes are represented in the genomes of eukaryotic organisms is not yet know. While such genes play critical roles in the regulation of important developmental processes, their full representation within a particular genome has not been established. As an initial attempt to determine how pervasive homeobox-containing genes are, degenerate oligonucleotides to a set of well-conserved, eight amino acid sequences from a specific region of the homeodomain were prepared and used to screen a C. elegans genomic library. Analysis of a subset of the identified clones included sequence analysis, physical mapping, and identification of possible corresponding genes from the correlated genetic map. The results suggest that for C. elegans, homeobox-containing genes constitute ≈1% of the total estimated number of genes. Such a large representation offers many interesting questions about shared structural motifs, developmental function, and evolutionary conservation. *Joel M. Schindler*

Expression of *engrailed* Proteins in Anthropods, Annelids, and Chordates
N. H. Patel, E. Martin-Blanco, K. G. Coleman, S. J. Poole, M. E. Ellis, and T. B. Kornberg
Cell, 58, 955—968, 1989   8-2

Drosophila segmentation is dependent in the homeobox gene *engrailed,* and genes homologous to *engrailed* have been identified in several other organisms including the chick and the human. A monoclonal antibody is described which recognizes a conserved epitope in the homeodomain of *engrailed* proteins in a number of arthopods, annelids, and chordates. The antibody was used to isolate the grasshopper *engrailed* gene.

Studies with the antibody demonstrated *engrailed* protein in the posterior part of each *Drosophila* segment during segmentation, as well as in a segmentally reiterated subset of neuronal cells during neurogenesis. Similar patterns of *engrailed* expression were found in the grasshopper and in two crustacean species, but not in annelids or chordates. The chordates studied included the zebrafish, frog, and chick.

The authors propose that *engrailed* is a gene ancestrally functioning in neurogenesis, whose function was co-opted during the evolution of segmentation in arthropods. Apparently, metameric development evolved sufficiently independently in arthropods, annelids, and chordates so that a regulatory gene having a critical role in segmentation in one of these phyla does not make a comparable contribution in the others.

♦ Homologues of genes involved in body plan formation in *Drosophila* have been identified in many other animal species. *engrailed,* a gene with characteristics of both a segmentation and a homeotic gene, is no exception, with several species containing two copies of the gene, others just one. Of interest to these authors was the extent to which the developmental function of the engrailed gene product, not simply the gene itself, was conserved among different species. Using a specific monoclonal antibody, the authors show that the only conserved feature of the gene product is in neurogenesis. They speculate that this was the original function of the *engrailed* gene while other functions (e. g. segmentation) were acquired later in evolution. These observations are important because they demonstrate that conservation of a specific gene may not necessarily mean that the function of that gene's may not necessarily mean that the function of that gene's product is also conserved. *Joel M. Schindler*

Binding Site-Dependent Direct Activation and Repression of *In Vitro* Transcription by Drosophila Homeodomain Proteins

Y. Ohkuma, M. Horikoshi, R. G. Roeder, and C. Desplan

Cell, 61, 475—484, 1990   8-3

Two of the genes required for normal segmentation of the *Drosophila* embryo are *fushi tarazu* and *engrailed*. Their protein products Ftz and En each contain a homeodomain and have been found — both in transient expression studies and in a *Drosophila* cell culture system — to act as transcriptional regulators. In this study an *in vitro* transcription system from human cells was used to determine whether *Ftz* and *En* act directly as transcription factors.

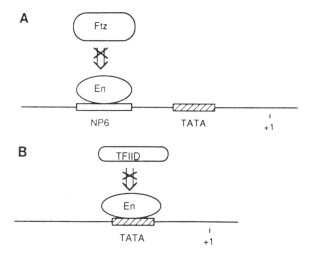

FIGURE 8-3. Model for the transcriptional repression by En. (A) Competition at the upstream sequence. At a low concentration, En binds to the homeodomain binding (NP6) sites, but not to the TATA box, and competes with Ftz for binding to the NP6 sites. Activation by Ftz is eliminated and transcription lowered to the basal level. (B) Competition at the TATA box. At a high concentration (tenfold molar excess of protein over the number of En binding sites), En binds to the TATA box and competes with TFIID. As a result, the basal level of transcription is reduced to near zero. (From Ohkuma, Y., Horikoshi, M., Roeder, R. G., and Desplan, C., *Cell*, 61, 475—484, 1990. With permission.)

Purified Ftz directly activated *in vitro* transcription by binding to homeodomain binding sites inserted upstream of the TATA box of the *Drosophila hsp70* promoter. Equimolar amounts of purified *En* suppressed activation through competing with *Ftz* for binding to these sites (Figure 8-3).

It appears that *Ftz* and *En* act as direct transcription factors. Homeodomain proteins such as these probably regulate development through combinatorial transcriptional control.

♦ Division of the embryo into clearly defined segmental units is achieved by a regulatory cascade that starts with the accumulation during oogenesis of some key maternal genes and that proceeds during embryonic development of *Drosophila* by the sequential expression of the so-called gap and segmentation genes. These genes are often expressed in sharply defined domains, sometimes in stripes of only one or a few cell diameters in width. When the first pattern-forming genes were cloned, it was not immediately obvious what regulatory mechanisms were involved in creating such precise patterns of expression. Subsequently, it was determined that the early genes in this regulatory cascade (e.g., maternal genes, "gap" genes) were expressed in broad domains and that the later-expressed gens (e.g., "segment polarity" genes) had much narrower expression domains. Moreover, and very importantly, the domains of expression of the different pattern-forming genes were found to spatially overlap. Thus, a model was proposed by several investigators that the refinement of the expression pattern of the genes in this regulatory cascade is in fact the consequence of regulatory interactions of spatially overlapping sets of genes. For instance, if genes A and B are each expressed over a band ten cells in width, and if the two domains overlapped only over a two-cell domain, it is not difficult to imagine how gene C, which requires both A and B products for its expression, would be expressed only in a narrow domain only two cells wide.

That two homeodomain gene products can interact or transcriptionally regulate a third gene is the main object of this study. This question had previously been approached by transient expression of various genes in Drosophila cultured cells (cf. report in last year's volume). The present study refines those observations by using purified *fushi tarazu (ftz)* and *engrailed (en)* proteins expressed in *E. coli.* The authors demonstrate that *en* protein inhibits the stimulatory transcriptional activity of the *ftz* proteins by effectively competing for the same DNA binding site. One important aspect of this study is the attention devoted to the quantitative aspects of the competition. Thus, it was established that *en* binds more effectively to the homeodomain binding site than does *ftz,* that binding depends on cooperative interactions, and that binding is reversible. Perhaps of equal importance, the authors show that *en* can also inhibit transcription by an entirely different mechanism; at very high concentrations *en* binds to the TATA promoter element in competition is of a different nature since once bound, TFIID cannot easily be displaced by *en* (cf. paper by the same authors in the *Proc. Natl. Acad. Sci. U.S.A., 87,* 2289—2293, 1989).

These studies lend support to a model of spatial patterning that involves the combinatorial interaction of certain homeodomain-containing proteins in regulating the transcription of genes lying further downstream in the regulatory cascade. *Marcelo-Jacobs-Lorena*

Functional Dissection of Ultrabithorax Proteins in
D. melanogaster
R. S. Mann and D. S. Hogness
Cell, 60, 597—610, 1990   8-4

Ultrabithorax (UBX) is a homeotic selector gene required for the specification of segmental identify in the posterior thorax of *Drosophila.* This study utilized an *in vivo* bioassay to study *UBX* protein function.

Expression of *UBX* proteins via a heat-inducible promoter generated homeotic transformations of segmental identities in both the embryonic

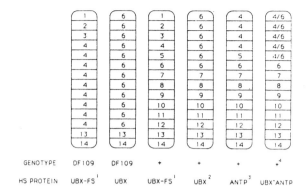

FIGURE 8-4. Summary of the various transformation phenotypes induced by UBX, ANTP, and their derivatives. Larvae are schematized by 14 boxes each representing a parasegment. The parasegmental identity of each box is indicated by the number within it. Wild-type larvae (third from the left) have 14 unique parasegments ordered 1 (anterior) through 14 (posterior). This phenotype can be altered by changing the homeotic selector protein composition of the embryos. We have altered selector protein composition by changing either the genotype with respect to the homeotic selector genes, *Ubx* or *Antp*, by heat induction of the *hsp70* promoter which drives their expression, or both. The resulting phenotypes are schematized. A wild-type genotype is designated as (+); HS PROTEIN refers to which heat-induced protein is misexpressed. 1: UBX-IVaFS expression is equivalent to wild type; 2: UBX-Ia, -IVa, -Ib, -ΔNN, or -ΔSN; 3: Gibson and Gehring, 1988; 4: wild type or *Antp*^{w10}. (From Mann, R. S. and Hogness, D. S., *Cell*, 60, 597—610, 1990. With permission.)

cuticle and peripheral nervous system of *Drosophila*. It transformed antennae into legs in the adult. *UBX* forms I and IV each induced cuticle transformations, but only form I induced the peripheral nervous system transformations. Study of the transformations produced by *UBX* deletions and by a chimeric UBX-Antennapedia (*ANTP*) protein showed that majority of the *UBX* identity information resides within the C-terminal, homeodomain-containing part of the protein.

The transformation phenotypes induced by *UBX*, *ANTP*, and their derivatives are shown in Figure 8-4. Posteriorly acting homeotic selector proteins appear to be functionally dominant over more anteriorly acting ones. The homeodomains and C-termini of *UBX* and *ANTP* contain much of the information needed to specify particular parasegmental identities.

Are Cross-Regulatory Interactions between Hemoetic Genes Functionally Significant?

A. Gonzales-Reyes, N. Urguia, W. J. Gehring, G. Struhl, and G. Morata
Nature, 344, 78—80, 1990   8-5

The first instar larva of *Drosophila* consists of a chain of segments or parasegments in which the morphologic pattern of each metamere is determined by homeotic genes. These genes are active in overlapping domains and they among themselves at the transcriptional level. It has been suggested that such cross-regulatory interactions have a role in specifying cell pattern, and therefore the identity of each metamere.

The functional significance of some of these interactions was studied by expressing the homeotic genes *Antp, Ubx,* or both under control of the heat-shock promoter. Homeotic gene products were found to escape normal regulatory controls. They were maximally expressed in regions where they normally are down-regulated. Surprisingly, however, interruption of the normal down-regulation of *Antp* and *Ubx* had no phenotypic effects in the epidermis. Similarly, the simultaneous expression of *Antp* and *Ubx* had no effect on abdominal patterning.

These findings counter the view that cross-regulatory interactions make an important contribution to determining segmental identity in *Drosophila.* The molecular basis for the functional hierarchy that arises between homeotic genes remains to be determined.

The Developmental Effect of Overexpressing a Ubx Product in Drosophila Embryos is Dependent on its Interactions With Other Homeotic Products

A. Gonzales-Reyes and G. Morata
Cell, 61, 515—522, 1990   8-6

Ultrabithorax (Ubx) is one of the homeotic genes specifying the metameric development of *Drosophila*. The goal of this study was to use an *hsp70-Ubx* construct to study the developmental effects of expressing a *Ubx* product at high levels in all body regions. Heat induction leads to high and ubiquitous expression of *Ubx* product which lasts several hours.

Interactions with resident homeotic genes determined whether or not the overexpression of Ubx had developmental effect on particular body regions of the *Drosophila* larva. In the head and thoracic region, the *Ubx* product overrode *Sex combs reduced, Antennapedia,* and probably other homeotic genes and dictated its own developmental program. In contrast, in abdominal segments A1 to A8, the overexpressed *Ubx* product, alone or combined with *Abdominal-A* and *Abdominal-B,* was consistent with a normal pattern. In segment 9, the highly expressed *Ubx* product was phenotypically suppressed by the r product of *Abdominal-B.* High levels of *Ubx* protein were irrelevant in the telson.

The molecular mechanisms responsible for phenotypic suppression remain to be delineated. A competition hypothesis based on differentiatial affinity for binding sites of downstream genes nay be a partial explanation. Alternately, different homeoproducts might compete for some intermediate molecule required for general homeotic function.

250 Year Book of Developmental Biology

♦ The organization of the embryonic body plan into an orderly succession of 14 segments (or parasegments) is mediated by the sequential action of a number of maternal and then zygotic genes. However, the identity of these segments is established by the activity of a separate class of genes, the homeotic genes. The *Deformed, proboscipedia,* and *labial* genes are required for specification of head segments, the *Sex* combs *reduced* (*Scr*) and *Antennapedia* (**Antp**) genes act to specify thoracic segments and the *Ultrabithorax* (**Ubx**), *Abdominal-A,* and *Abdominal-B* genes establish abdominal segment identity. Each gene is expressed only in a restricted domain which, as established by genetic analysis, coincides with the site of action of each gene. The above three articles use the same basic approach to investigate the role played by *Ubx* and *Antp* in the various regions of the embryo: after inducing the indiscriminate expression of the two genes in all cells of the embryo, the resulting phenotype is analyzed.

In wild-type embryos *Ubx* is maximally expressed in parasegment 6 (**ps6**), which develops into the larval **T3p** (posterior part of the third thoracic segment) and A1a (Anterior part of the first abdominal segment). *Ubx* expression is considerably lower in ps5 and in ps7 through 13 (ps7-13 differentiate into A1p-A8a). Indiscriminate expression of *Ubx* was induced throughout the embryos to address two questions: is the decreased *Ubx* expression in ps5 and ps7-13 of functional importance, and what is the importance of the absence of *Ubx* expression anterior to ps5? Surprisingly, downregulation of *Ubx* in the posterior part of develop into larvae with all structures posterior to A2 undistinguishable from wild type. The proviso is, however, that *AbdA* and *AbdB* functions are left intact. If any of these two functions is abolished by mutation, overexpression of *Ubx* causes the segments to acquire A1 character, which is the segment where *Ubx* expression is normally the highest. By contract, overexpression of *Ubx* in the anterior part of an otherwise wild-type embryo has profound effects on head and thorax development: all cephalic and thoracic segments develop as A1, even though *Antp* and *Scr* a proteins are present in normal distribution and amounts. Thus, while in the posterior parts of the embryo *Ubx* cannot override *Abda* and *AbdB*, in the anterior parts *Ubx* is dominant and overrides the action of all other anteriorly expressed genes, including *Antp*. T3 (a derivative of ps5 and that normally expresses low *Ubx*) also assumes A1 character, indicating that the increase of Ubx protein is sufficient to induce change in identity. The sensitivity of transformation by *Ubx* overexpression increased from anterior to posterior segments (ps1 was the least sensitive and ps5 the most sensitive).

In wild-type embryos *Ubx* is expressed as a family of related proteins that have a common N-terminal region of 247 amino acids and a common homeodomain-containing region of 99 amino acids. The proteins differ from each other by the presence or absence of middle elements 9,

17, and 17 amino acids in length as a result of alternative splicing of the 77-kb primary transcript. Do these proteins have separate functions? By expressing from heat shock promoters different members of the *Ubx* family, Mann and Hogness determined that every Ubx family member tested led to the same larval cuticle phenotype. However, when transformations of the peripheral nervous system (PNS) where assayed with monoclonal antibodies, *Ubx-Ia* but not *Ubx-IVa* was able to transform PNS of thoracic segments into PNS with abdominal characteristics. Thus, not all Ubx family members appear to be equivalent. Mutational analysis led to some surprising observations. Deletion of substantial N-terminal portions of the *Ubx* protein (up to 215 amino acids or 57% of the protein) had little effect on its ability to transform anterior structures into A1. Transformations were qualitatively the same, although the extent of transformation was lower, especially in more anterior segments. This shows that the major *Ubx* functional elements reside in the homeodomain-containing C-terminal portion of the protein. Remarkable, the two polyalanine stretched, the polyglycine stretch, all known phosphorylation sites, and the "transcriptional activator domain" (as defined by cotransfection experiments in *Drosophila* cultured cells; cf. last year's volume), are all missing from the truncated protein. In view of this outcome, the result from a second experiment is perhaps less surprising. When the C-terminus of the *Ubx* protein was substituted by that of the Antp protein, the chimeric gene induced denticle belt patterns that were indistinguishable from those induced by the intact *Antp* gene (see below). However, the chimeric gene, but not the intact *Antp* gene, suppresses the differentiation of sensory organs, suggesting that the N-terminal *Ubx* portion of the chimeric protein can have a determinative effect on the phenotype. It should be noted that the homeodomains of *Ubx* and *Antp* proteins differ in only 7 out of 61 residues and that the so-called third "recognition helix" of the two homeodomains are identical. One interpretation of the results is that the C-terminus confers Antp-like DNA-binding properties and that its N-terminus mediates interactions with factors normally associated with the *Ubx* protein. Other possibilities also exist and will have to be tested.

In wild-type embryos *Antp* is maximally expressed in ps4 (T1p + T2a) and in lower amounts in ps3 and 5. Ubiquitous expression of high levels of *Antp* throughout the embryo causes extensive changes toward a T2 character, of segments anterior to T2. On the other hand, *Antp* overexpression had no effect T2 or any of the segments posterior to it. Thus, *Antp* function is dominant and overrules the genes that determine head and anterior thoracic structures but is recessive to genes expressed more posteriorly, including those of the *Ubs* complex. Taking this outcome into consideration, the experiment by González-Reyes et al. that induced the concomitant overexpression of both *Ubx* and *Antp,* had the predicted result: the phenotype of these embryos was indistinguishable from that of embryos where only *Ubx* was overexpressed (*Ubx* is normally the more posteriorly expressed gene of the two).

In summary, these experiments provide further evidence that posteriorly acting homeotic genes are functionally dominant over more anteriorly expressed ones. Thus, *Ubx* can overrule *Antp* and other more anteriorly expressed genes, but *Antp* cannot overrule *Ubx* function. Moreover, although *Ubx* is required in segments posterior to ps6, its down regulation in these more posterior segments appears to have no functional significance. A major functional domain of the *Ubx* protein lies in its homeodomain-containing C-terminal half and this portion of the protein is sufficient to perform most of its functions. However, the experiments also identified a clear role for the N-terminal part of the protein. Finally, one should remember that all conclusions apply only to the parameters that were measured in these studies. In most cases the observations ere restricted to cuticular phenotypes and sometimes extended to the nervous system. Thus, the possibility remains that homeotic genes may perform other important roles that were not measured by these assays. *Marcelo Jacobs-Lorena*

Control of the Initiation of Homeotic Gene Expression by the Gap Genes *giant* and *tailless* in *Drosophila*
J. Reinitz and M. Levine
Dev. Biol., 140, 57—72, 1990   8-7

The segmentation process in *Drosophila* involves the specification of finely localized patterns of expression through the combinatorial actions of more diffusely distributed regulatory products. Both maternal and zygotic genes contribute to segmentation in this species. Members of the gap class of segmentation genes have a key role in interpreting maternal information and regulating the expression of pair-rule genes and homeotic genes. This study was intended to analyze the patterns of expression of various homeotic, pair-rule, and gap genes in *tailless* and *giant* gap mutants.

It was found that *tailless,* in its anterior domain, suppresses the expression of *fushi tarazu, hunchback,* and *Deformed.* In its posterior domain of action, *tailless* was shown to be responsible for establishing Abdominal-B expression and for demarcating the posterior boundary of the initial expressive domain of *Ultrabithorax.* The gene *giant* is an early zygotic regulator of the gap gene *hunchback;* in *giant* embryos, the anterior domain of *hunchback* expression was altered by the start of cycle 14. In addition, *giant*_regulates establishment of the expression patterns of *Antennapedia* and *Abdominal-B.* In the case of *Antennapedia,* giant regulates the anterior limit of early gene expression.

a

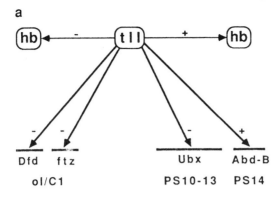

b

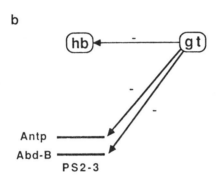

FIGURE 8-7. Summary of regulatory relationships. (a) Regulatory actions of *tll* described in this work. Each action is represented by an arrow labeled with a plus to indicate activation or a minus to indicate repression. The approximate segment primordia in which the targets are regulated are indicated at the bottom. *tll* represses *hb*, *Dfd*, and *ftz* in the primordium of the optic lobe and first gnathal segment. It activates *hb* and *Abd-B* in approximately the presumptive parasegment 14 and represses *Ubx* in a domain extending from about parasegment 10 to parasegment 13. (b) Regulatory actions of *giant* described in this work. Regulatory actions are indicated by arrows with the appropriate sign. *gt* represses *Abd-B*, *Antp*, and *hb* in the primordia of parasegments 2 and 3. Abbreviations: ol, optic lobe; C1, first gnathal segment; PS, parasegment. (From Reinitz, J. and Levine, M., *Dev. Biol.*, 140, 57—72, 1990. With permission.)

Both *giant* and *tailless* are integral parts of the gap-gene system (Figure 8-7). Both exhibit cross-regulatory interactions with other gap genes, and each also has important regulatory effects on the pair-rule and homeotic genes.

♦ The body plan of *Drosophila* results from the interaction of a series of regulatory genes expressed from both the maternal and the zygotic gene. A general characteristic of this genetic hierarchy is the combinatorial action of the regulatory products. Within this hierarchy, the homeotic genes are primarily responsible for segmental identity. In the current article, the authors use a series of mutations to further define the roles of two additional gap genes, *giant* and *tailless,* in establishing proper segmentation. The authors describe the domains of expression of these genes and identify other genes with which they interact and upon which they exert regulatory control. Of particular interest is the observation that both genes exhibit cross-regulatory interactions with hunchback, another well-characterized gap gene. This article is important because it further defines the genetic circuitry necessary to define the *Drosophila* body plan. *Joel M. Schindler*

Craniofacial Abnormalities Induced by Ectopic Expression of the Homeobox Gene *Hox-1.1* in Transgenic Mice
R. Balling, G. Mutter, P. Gruss, and M. Kessel
Cell, 58, 337—347, 1989   8-8

Hox-1.1 is a murine homeobox-containing gene which is expressed in a time- and cell-specific manner during embryogenesis. This gene was expressed ectopically in transgenic mice utilizing the transcriptional activity of the chicken beta-actin promoter.

The expression of *Hox-1.1* was nearly ubiquitous in transgenic mice. These animals usually died shortly after birth and exhibited multiple craniofacial anomalies — including cleft secondary palate, incomplete eyelid fusion, and detached pinnae. One of the animals had a poorly formed maxilla. The phenotype resembled what is seen following the systemic administration of retinoic acid during gestation.

The developmental defects caused by ectopic expression of *Hox-1.1,* a potential developmental control gene, appear to share a common pathogenetic mechanism with retinoic acid embryopathy. There is experimental evidence that retinoic acid is a potential morphogen during vertebrate embryonic development. It is possible that the *Hox* genes are induced by retinoic acid during embryogenesis.

Variations of Cervical Vertebrae After Expression of a *Hox-1.1* Transgene in Mice
M. Kessel, R. Balling, and P. Gruss
Cell, 61, 301—308,1990   8-9

Vertebrates become segmented early in the course of embryogenesis by the formation of somites — condensations of mesoderm lateral to the neural tube. Schlerotomal cells from the caudal half of one somite join with cells from the cranial half of the one following to form a prevertebra

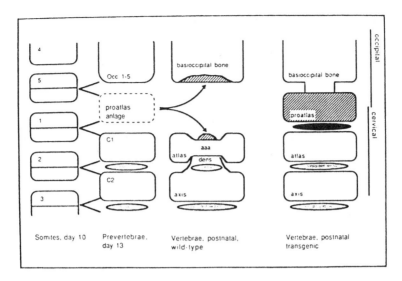

FIGURE 8-9. Development of cervical somites into vertebrae. The last occipital somites (4 and 5) and the first cervical somites (1, 2, and 3) are depicted with their anterior and posterior halves. They develop into the precursors of the occipital bones (Occ 1-5), the proatlas anlage, and cervical prevertebrae (C1 and C2) as indicated. The vertebral body of C1 fuses with C2, resulting in the typical axis with the dens axis. Therefore, the neonatal atlas does not contain a vertebral body. Unique to the atlas is the anterior arch (aaa), which is only a transient embryonic structure for the other vertebrae. Also, the proatlas anlage is transient: these cells fuse to the occipital bones or form the tip of the dens (hatched). Intervertebral discs are shown as ellipsoids (stippled), and the additional disc in the scheme for the transgenic animals is shown in black. A further explanation of the transgenic phenotype is given in the text. (From Kessel, M., Balling, R., and Gruss, P., *Cell,* 61, 301—308, 1990. With permission.)

and finally a vertabra, as illustrated in Figure 8-9. The authors produced dominant gain-of-function mutations by deregulating the spatial expression pattern of a *Hox* gene in transgenic mice. Genomic sequences of *Hox-1.1* were introduced into mice under the transcriptional control of a chicken beta-actin promoter.

The transgenic animals were nonviable after birth and exhibited craniofacial abnormalities. The basioccipital bone, atlax, and axis all were malformed, and an additional vertebra — a proatlas — was present in the craniocervical transitional region. The deformation of the basioccipital bone resembled human occipital dysplasia. In some transgenic animals, there was no ossified supraoccipital bone.

The *Hox-1.1* transgene appears to interfere with development at about day 9 of gestation, the time of neural crest migration and somite differentiation. The vertebral abnormalities clearly result from an effect on the somitic mesoderm. Malformations of anterior derivatives of both germ layers develop when the anterior border of *Hox-1.1* expression is extended (Figure 8-9A). Some of the variations seen after ectopic expression

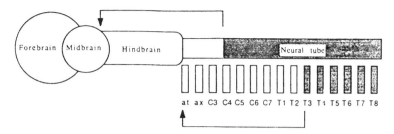

FIGURE 8-9A. Anterior expression boundaries and sites of malformations. The anterior neuroectoderm (CNS) and the prevertebrae as metameric units are schematically represented. The expression domains of *Hox-1.1* (Mahon et al., 1988) are shown as stippled areas in the CNS (up to the fourth cervical ganglion) and in the prevertebrae (up to T3). The arrowheads point out on the anteroposterior axis the sites of malformations occurring in transgenic mice: at the level of the rostral hindbrain (first visceral arch) and at the craniovertebral transition (at). For further explanation see text. (From Awgulewitsch, A. and Jacobs, D., *Development,* 108, 411—420, 1990. With permission.)

of the *Hox-1.1* transgene are consistent with a homeotic function of the gene.

♦ The genetic control of mammalian development has traditionally been approached using techniques of random mutagenesis. Generally, mutations were identified because of phenotypic effects associated with the heterozygous state of the mutant gene. The phenotype associated with the homozygous state was often more severe and quite different from that seen in heterozygotes. The main problem with this approach was the inability to identify the actual gene product or the gene itself. More recently, work has focused on predicting what types of cloned genes would likely be important in development and then determining their stage and tissue-specific pattern of expression. The types of genes that have been analyzed in this manner include homeobox, growth factor and proto-oncogenes. Based on *in situ* expression patterns, some predictions have been make regarding the role of these genes. For example, the overlapping domains of homeobox gene expression have led investigators to suggest that these genes are important for establishing positional information within the developing embryo. The main problem with this approach was the lack of mutations to see what happens to the embryo if these genes are not expressed correctly. With targeted mutagenesis in embryonic stem cells being a reality, it is now possible to come full circle to ask what happens to embryos not expressing a functional protein product. Another approach taken by the Peter Gruss's group is to generate dominant gain-of-function mutations by deregulating or misexpressing genes in transgenic mice. The specific goal was to create transgenic mice carrying the *Hox-1.1* gene under control of the chicken β-actin promoter. The gene was found to be expressed in an ubiquitous pattern, and mice carrying the transgene were found to die shortly after birth with cranio-facial abnormalities and variations of the cervical column. Variability in the severity of the phenotype was observed and this was thought to be due to influences of the integration site on the transgene. The craniofacial abnormalities were attributed to a disruptive effect of the deregulated *Hox-1.1* gene on neural crest cells. In contrast, the vertebral problems were associated with transformations of mesodermally derived structures. Thus, *Hox-1.1* may function as a developmental control gene by switching on different genetic programs or by interfering with developmental programs specifying the vertebral column. These types of variations are consistent with a homeotic-like function of *Hox-1.1*. These results are the first to validate the prediction that some of the homeobox genes may be master regulatory genes establishing positional information within the embryo. *Terry Magnuson*

Differential Expression of *Hox 3.1* Protein in Subregions of the Embryonic and Adult Spinal Cord
A. Awgulewitsch and D. Jacobs
Development, 108, 411—420, 1990   8-10

Most, if not all, of the homeobox genes examined to date are expressed at a high level in the developing murine central nervous system (CNS). An example is *Hox 3.1,* which maps to chromosome 15 and is localized in a subregion of the CNS corresponding to the cervical and anterior thoracic spinal cord of newborn mice. In this study, synthetic oligopeptides derived from the predicted Hox 3.1 protein coding sequence served to produce antibody that recognizes Hox 3.1 protein in tissue sections. The antibody was used immunohistochemically to monitor the expression of Hox 3.1 protein in the CNS of embryonic and adult mice.

Studies of 12.5-day embryos demonstrated congruency between Hox 3.1 RNA and protein expression patterns in the developing spinal cord. Spatially restricted expression of the protein was first noted at about 10.5 d of development. The distribution of expression later changed markedly, within certain anteroposterior limits, at the same time as cytoarchitectural changes took place in the developing spinal cord. The protein accumulated predominantly within the nuclei of neuronal cells. Persistent accumulation also was seen in nuclei of mature neuronal cells of the adult CNS.

The persistent expression of murine homeobox genes in the adult CNS is not consistent with their functioning as developmental control genes. It would be of interest to learn whether the mature neuronal cells that accumulate Hox 3.1 protein are direct descendants of the differentiating neuroblasts which express gene-specific antigen during embryogenesis.

A Murine *even-skipped* Homologue, *Evx 1*, is Expressed During Early Embryogenesis and Neurogenesis in a Biophasic Manner

H. Bastion and P. Gruss

EMBO J., 9, 1839—1852, 1990   8-11

Compared with *Drosophila,* little is known about the molecular basis for vertebrate morphogenesis. The *Drosophila even-skipped (eve)* homeobox was used as a probe to isolate two murine genes, *Evx 1* and *Evx 2,* from a genomic library *Evx 1, Evx 2,eve,* and the Xenopus *Xhox-3* comprise a family of related genes based on similar homeodomain sequences. In addition, *Evx1* and *Evx2* share extended amino acid sequences outside the homeobox.

Evx 1, present near the *Hox 5* locus on mouse chromosome 2, is expressed in undifferentiated F9 stem cells but not in cells differentiated by retinoic acid and cAMP. It has a biphasic expression pattern during embryogenesis, first emerging at the posterior end of the embryo within the primitive ectoderm on days 7 to 9. Subsequently, it appears in the mesoderm and neuroectoderm. On days 10 to 12.5, *Evx 1* transcripts are limited to specific cells within the neural tube and hindbrain, coinciding in time and space with the maturation of early-form interneurons (possibly commissural interneurons).

The transcription pattern of *Evx 1* is consistent with a role for this gene in specifying posterior positional information along the embryonic axis (Figure 8-11), similar to *Xenopus Xhox-3.* Additionally, it specifies neuronal cell fates within the differentiating neural tube, in analogy with the role of eve in the embryonic central nervous system of *Drosophila.* If homeobox genes set up positional information, Hox genes may be important in defining anteposterior regions of the embryonic axis, while *Evx 1* helps establish this axis.

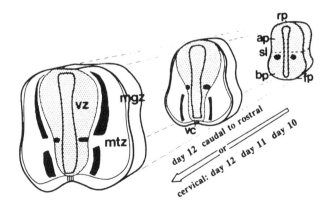

FIGURE 8-11. Schematic illustration showing *Evx 1* expression in sections of the neural tube at different stages of development. Abbreviations: ap, alar plate; bp, basal plate; fp, floor plate; rp, roof plate; sl. sulcus limitans; mtz, mantle zone; mgz. marginal zone; vz. ventricular zone; vc, ventral commissure. Black areas represent *Evx 1* expression. (From Bastion, H. and Gruss, P., *EMBO J.,* 9, 1839—1852, 1990. With permission.)

Coordinate Expression of the Murine *Hox-5* Complex Homeobox-Containing Genes during Limb Pattern Formation

P. Dolle, J. Izpiua-Belmonte, H. Falkenstein, A. Renucci, and D. Duboule
Nature, 342:767-772, 1989   8-12

There is increasing evidence that vertebrate homeobox-containing genes may have a critical role in limb development and regeneration. Genes of the murine *Hox-5* complex are strongly expressed in all stages of limb morphogenesis. They are expressed with the same cellular specificities, but there are subtle differences in the extent of their domains of expression. The more 5' the position of a given gene within the complex, the later and more distally it is expressed. Anteroposterior differences in gene expression also are observed.

The authors propose a model which accounts for the establishment of gene expression domains in relation to a morphogene released by the zone of polarizing activity. Each gene of the *Hox-5* complex has a specific domain of expression in the mesoderm of the developing limb bud. One area of the limb bud contains all five *Hox-5*-complex gene transcripts. This zone of maximum overlap is at the posterior side of the limb bud, coinciding with the region containing the zone of polarizing activity. The balance between the time needed for limb bud growth and sequential opening of the *Hox-5* genes is decisive (Figure 8-12).

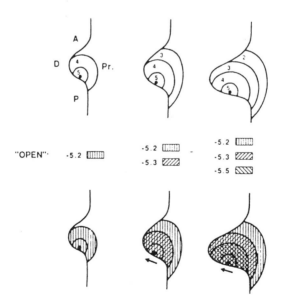

FIGURE 8-12. Discussion scheme. This hypothetical scheme illustrates the time-dependent successive "opening" of the *Hox-5*-complex genes during limb-bud outgrowth, leading to a more distal expression of the genes expressed later (or located 5'; see the text for details). (From Dolle, P., Izpiua-Belmonte, J., Falkenstein, H., Renucci, A., and Duboule, D., *Nature*, 342, 767—772, 1989. With permission.)

It is likely that similar molecular mechanisms are involved in positional signaling along the axes of both the embryonic trunk and the fetal limbs. Hensen's node could be a key zone in establishing the expression domains of genes of the *Hox* complexes along the rostrocaudal axis, analogous to the role of the zone of polarizing activity in limb development.

Pax2, a New Murine Paired-Box-Containing Gene and its Expression in the Developing Excretory System
G. R. Dressler, U. Deutsch, K. Chowdhury, H. O. Nornes, P. Gruss
Development, 109, 787—795, 1990   8-13

Like *Drosophila,* the mouse genome contains multiple paired-box-containing genes, or Pax genes. *Pax1* is a member of this gene family which is expressed in segmented structures of the developing vertebral column. There is evidence that *Pax1* is the murine developmental mutation *undulated.* The mouse genome contains multiple copies of the paired box, and the authors now report a study of a second member of this gene family, *Pax2,* which also is expressed in embryogenesis.

Two overlapping cDNA clones were isolated and sequenced, and at least two forms of *Pax2* protein were deduced from the cDNA sequence. A highly conserved paired domain and a downstream octapeptide sequence were observed. Expression of *Pax2* was chiefly restricted to the excretory and central nervous systems of the developing embryo. Expression of *Pax2* during formation of the kidneys was transient, correlating with polarization and the induction of epithelial structures.

Structure and Expression Pattern of the Murine *Hox-3.2* Gene
J. R. Erselius, M. D. Goulding, and P. Gruss
Development, 110, 629—642, 1990   8-14

The homeobox is a highly conserved sequence fist describes in genes controlling pattern formation in *Drosophila.* More than 30 murine homeobox genes (*Hox* genes) have been isolated to date. As in Drosophila, many of them are arranged in clusters, an example being *Hox*-3 on chromosome 15. The authors carried out a detailed analysis of the murine *Hox-3.2* gene and its expression during development.

Hox-3.2 is the most 5' member of the *Hox-3* complex thus far isolated. A 260-amino acid protein was identified, lacking the conserved hexapeptide found in most homeobox genes. Three transcripts were found in 9- to 15-day embryos. Transcripts first appeared in the posterior part of the embryo and subsequently were present in the ventral part of the neural tube, with a sharp anterior boundary at the level of the third thoracic prevertebra. Unlike *Hox-3.1, Hox-3.2* was not expressed in the dorsal horns containing sensory neurons at day 14.5. *Hox-3.2* transcripts were detected in the posterior prevertebrae, the hindlimb buds, and the cortex of the developing kidney.

Hox-3.2 exhibits a restricted pattern of expression in the posterior part of the developing mouse embryo. It is expressed in post-mitotic cells of the neural tube and also in spinal ganglia, prevertebrae, and the developing kidney. The findings agree with previous results indicating that genes located 5′ in homeobox gene clusters are expressed more posteriorly than those from the 3′ region.

Isolation of the Mouse *Hox-2.9* Gene; Analysis of Embryonic Expression Suggests That Positional Information Along the Anterior-Posterior Axis is Specified by Mesoderm

M. A. Frohman, M. Boyle, and G. R. Martin
Development, 110, 589—607, 1990   8-15

The vertebrate neural tube — especially the hindbrain — develops segmentally in conjunction with segment-specific gene expression, but the positional cues that instruct the neural tube to express genes in a restricted manner are unknown. The authors have cloned a murine homeobox-containing gene, *Hox-2.9,* which is expressed only in a segment of the neural tube corresponding to rhombomere 4 of the hindbrain.

Prior to the start of neurulation, *Hox-2.9* was expressed within and posterior to the embryonic mesoderm destined to participate in hindbrain formation. At the onset of neurulation, *Hox-2.9* was expressed in the part of the neural plate overlying the *Hox-2.9*-expressing mesoderm. It remained absent from the more anterior neuroectoderm destined to form the midbrain and forebrain.

It appears that, in the developing mouse, the mesoderm provides cues which instruct the overlying neuroectoderm with respect to its position along the anteposerior axis. Subsequently, mesoderm provides a more localized signal inducing *Hox-2.9* expression. The specific mechanisms of pattern formation in mammals are not fundamentally unlike those observed in amphibians and avians. Demonstration of a direct relation between the expression of *Hox* genes and specification will require experimental manipulation of the embryo.

The Mouse *Hox-1.4* Gene: Primary Structure, Evidence for Promoter Activity and Expression During Development

B. Galliot, P. Dolle, M. Vigneron, M. S. Featherstone, A. Baron, and D. Deboule
Development, 107, 343—359, 1989   8-16

Genes located in the 3′ part of murine *Hox* gene clusters have anterior expression domains, while those located near the 5' end of the clusters are expressed more posteriorly. The authors analyzed the structure and sequence of *Hox-1.4* gene, one of those having an anterior expression boundary, and studied its expression pattern during embryonic and fetal development.

The *Hox-1.4* gene has two major exons, the second of them encoding the homeodomain. The putative *Hox-1.4* protein has similarities to the products of homologous genes located at the same relative postions on other *Hox* gene clusters. A fragment extending 360 bp upstream of a transcriptional start site was able to promote transcription in transcripted cells. *Hox-1.4* was expressed in the fetal central nervous system, in sclerotomes and prevertebrae derived from somitic mesoderm, and in several mesodermal components of such structures as the lungs, the gut, and the developing kidney.

The *Hox-1.4* gene is one of a subfamily of homeogenes expressed anteriorly in the murine embryo. The origin and nature of *Hox-1.4*-positive structures reinforces the view that homeogenes may serve as positional cues during vertebrate development.

Mouse *Hox-3.4:* Homeobox Sequence and Embryonic Expression Patterns Compared With Other Members of the *Hox* Gene Network

S. J Gaunt, P. L. Coletts, D. Pravtcheva, and P. T. Sharpe
Development, 109, 329—339, 1990   8-17

A putative murine homeobox gene, *Hox-3.4,* was previously identified 4 kb downstream of the *Hox-3.3* gene. The authors have now sequenced the *Hox-3.4* homeobox region. The predicted amino acid sequence exhibited the closest homology with the murine *Hox-1.3* and *Hox-2.1* genes. *Hox-3.4* appears to be a homologue of the *Xenopus* gene *Xlhbox5* and of the human cp11 gene. Studies using a panel of mouse-hamster somatic cell hybrids mapped *Hox-3.4* to chromosome 15.

In situ hybridization studies were carried out to delineate the distribution if *Hox-3.4* transcripts also was determined for seven other members of the *Hox* gene network. The anterior limits of *Hox-3.4* were related to the position of *Hox-3.4* within the *Hox-3* locus. In the central nervous system, the anterior limit of *Hox-3.4* expression was similar to that for *Hox-2.1* and *Hox-1.3*. The tissue-specific patterns of expression, however, differed substantially. The patterns of *Hox-3.4* expression in the spinal cord and testis were very similar to those of *Hox-3.3,* but quite different from those of *Hox-1* genes.

There is increasing evidence that tissue specificity in the expression of a homeogene, known to be variable in co-members of a subfamily, may be

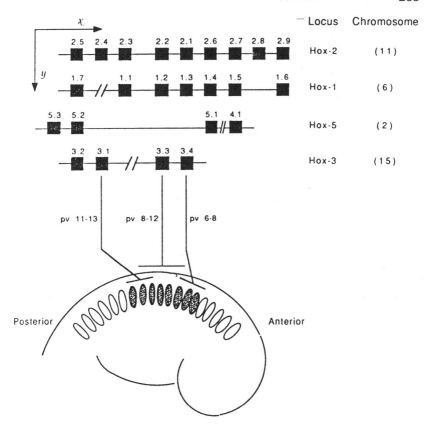

FIGURE 8-17. Diagrammatic representation of the relationship between the Hox-gene network and the expression within the prevertebral column of *Hox-3.4, Hox-3.3* (Sharpe et al. 1988; Gaunt et al. 1988), and *Hox-3.1* (Gaunt et al. 1988; Holland and Hogan, 1988a). For each gene, the figure shows the prevertebrae over which transcripts increase in abundance at the anterior boundary of the expression domain. Arrangements of the Hox-gene clusters are from Schughart et al. 1989; Graham et al. 1989 and Duboule and Dollé and Duboule, unpublished). (From Gaunt, S. J., Coletts, P. L., Pravtcheva, D., and Sharpe, P. T., *Development,* 109, 329—339, 1990. With permission.)

relatively constant for co-members of a locus — or at least for subclusters of adjacent genes within a locus (Figure 8-17).

Segment-Specific Expression of a Homeobox-Containing Gene in the Mouse Hindbrain
P. Murphy, D. R. Davidson, and R. E. Hill
Nature, 341, 156—159, 1989   8-18

Segmentation is a very prevalent form of early embryonic develop-

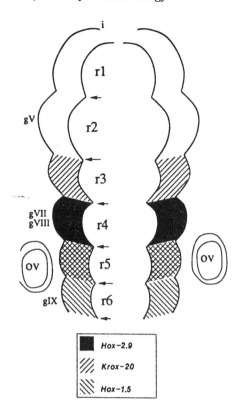

FIGURE 8-18. Diagrammatic representation of a frontal section through the hindbrain of a 9.5-d mouse embryo showing the position of the neuromeres (rhombomeres) with respect to the otic vesicle (ov), the hindbrain-midbrain junction (isthmus, i), and the cranial ganglia which are represented alongside the rhombomeres from which they extend (gV to gIX). The arrows indicate rhombomere boundaries, and the rhombomeres are numbered from anterior to posterior (r1 to r6). The shaded areas indicate the domains of expression of *Hox-2.9, Hox 1.5,* and *Krox-20* in the hindbrain (see text for references). *Hox-1.5* expression extends posteriorly into the neural tube. (From Murphy, P., Davidson, D. R., and Hill, R. E., *Nature,* 341, 156—159, 1989. With permission.)

ment. In *Drosophilia*, homoeotic genes have a role in all levels of segmental organization and also in the process of determining segment identity. In contrast to *Drosophila*, the embryonic expression of homoeobox-containing genes in vertebrates does not appear to follow a segmental pattern. In vertebrates, segmentation is evident in the mesodermal somites, and is starting to be recognized in the central nervous system. Neuromeres in the hindbrain (rhombomeres) have been identified as segmental units through their patterns of nerve formation and gene expression.

The authors found that the murine homeobox-containing gene *Hox-2.9* is expressed in a segment-specific manner in the developing mouse hindbrain. The gene is expressed in a region flanked by the regions of expression of *Krox-20,* a gene encoding a zinc-finger DNA-binding protein (Figure 8-18). Expression of *Hox-2.9* appeared to be limited to a single neuromere, rhombomere 4.

It seems possible from these findings that other homoeobox-containing genes are involved in specifying neural segments. It has been proposed that the overlapping domains of homoeobox-containing gene expression transmit positional information in the form of unique combinations of gene products distributed in blocks along the anteroposterior axis. It remains to be learned whether these blocks relate to neuromeres.

**Spatially and Temporally Restricted
Expression of *Pax2*
During Murine Neurogenesis**
H. O. Nornes, G. R. Dressler, E. W. Knapik, U. Deutsch, and
P. Gruss
Development, 109, 797—809, 1990   8-19

Many segmentation and homeotic genes of *Drosophila* share highly conserved protein domains that have permitted the isolation of vertebrate genes possessing similar domains. The *Pax2* gene, a murine paired-box-containing gene, is an example. It is expressed in the developing excretory system, a transiently segmented structure of mesodermal origin. The present study examined the pattern of expression of *Pax2* during neuronal differentiation, using the technique of *in situ* hybridization.

Expression of *Pax2* was observed along the boundaries of primary divisions of the neural tube. Initially, it appeared in the ventricular zone in two cell compartments on either side of the sulcus limitans, and along the entire rhomencephalon and spinal cord (Figure 8-19). Later in development, *Pax2* expression was limited to progeny cells that had migrated to specific regions of the intermediate zone, These included the ventral half of the optic cup and stalk, and subsequently the optic disc and nerve. Expression of *Pax2* in the ear was limited to areas of the otic vesicle that form neuronal components.

There may be as many as four different *Pax2* transcripts and at least two forms of the protein. It is not yet clear whether the expression of a particular form of mRNA and protein correlated with ectoderm- or mesoderm-derived structures. Different forms of *Pax2* protein could have multiple functions on the local cellular environment and on interactions with other factions.

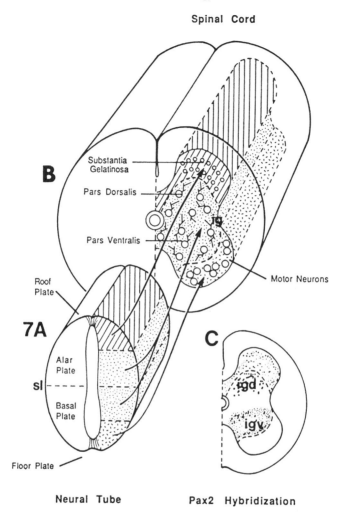

FIGURE 8-19. Schematized illustration of the longitudinal organization of the neural tube. Arrows indicate how each of these longitudinal columns gives rise to (B) the longitudinally organized functional systems in the adult spinal cord. In early development, the lateral plate is partitioned into a basal (ventral) and alar (dorsal) half at the sulcus limitans (sl). The most ventral column in the basal plate gives rise to motor systems, the two columns on either side of the sulcus limitans give rise to the intermediate grey region (ig), and the most dorsal column gives rise to the substantia gelatinosa. The motor neurons are the output system, the intermediate grey neurons form intersegmental and long ascending systems, and the substantia gelatinosa neurons form an intersegmental system within this same functional layer (Brown, 1981). (C) Schematized pattern of *Pax2* expression in the adult spinal cord. Initially, *Pax2* is expressed on either side of the sulcus limitans and later in their progeny in the pars ventralis (igv) and pars dorsalis (igd) of the intermediate grey. The latest forming *Pax2* expressing cells settle in the substantia gelatinosa and the medial intermediate grey. (From Nornes, H. O., Dressler, G. R., Knapik, E. W., Deutsch, U., and Gruss, P., *Development,* 109, 797—809, 1990. With permission.)

Segmental Expression of *Hox-2* Homoeobox-Containing Genes in the Developing Mouse Hindbrain

D. G. Wilkinson, S. Bhatt, M. N. Cook, E. Boncinelli, and
R. Krumlauf

Nature, 341, 405—409, 1989   8-20

The vertebrate hindbrain develops segmentally, with distinct groups of neurons arising from different segments..The authors found that members of the *Hox-2* cluster of murine homeobox genes are expressed in a segment-specific manner in the developing hindbrain. The expression of successive genes exhibits boundaries at two-segment intervals.

The expression of *Hox-2* genes was analyzed in coronal sections of 9.5-day mouse embryos. At this stage, the presence of the otocyst allows rhombomeres to be unambiguously identified. The boundary of *Hox-2.1* expression was caudal to the most posterior hindbrain segment. The boundaries of expression of *Hox-2.6, Hox-2.7,* and *Hox-2.8* lay at two-segment intervals. The limit of *Hox-2.6* expression lay at the most posterior segment boundary.

The fact that the 3′-most members of the *Hox-2* gene cluster are expressed in segment-restricted patterns suggest that they have a role in specifying segment phenotype. The two-segment periodicity of gene expression limits (Figure 8-20) has also been demonstrated in studies of neuronal developmental patterns and of the expression of *Krox-20*. These homoeobox genes probably have an ancient conserved role in specifying regional variations in pattern along the body axis. The coupling of homoeobox gene expression with segmentation may have evolved independently in *Drosophila* and in mice.

♦ Several classes of homeobox genes have been discovered during the past few years and numerous papers have been published regarding expression of these genes. The following represents a brief synopsis of some of the more important points regarding homeobox genes and pattern formation in the mouse. A more detailed review has been published recently by Kessel and Grus (*Science,* 249, 374—379) and should be consulted for a complete list of papers. The *antennapedia-like (Antp)* class of homeobox genes are organized into four clusters each of which spans more than 100 kb on chromosomes 6 (*Hox-1*), 11 (*Hox-2*), 15 (*Hox-3*), and 2 (*Hox-4*). More sequence conservation exists between rather than within the *Hox* clusters suggesting that duplication of an ancestral cluster occurred during evolution. In general, expression of the *Hox* genes begins during early gastrulation (day 7.5) with significant levels observed during organogenesis (days 9 to 12). Expression occurs primarily but not exclusively in the neural tube, somites, and sclerotomes. Other domains include the developing kidney, lung, testes, intestine, thymus, sternum, and

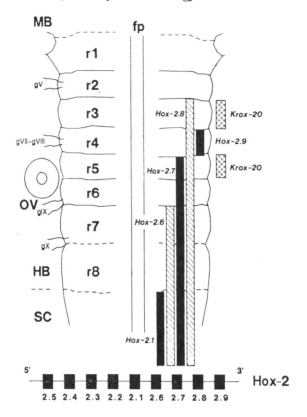

FIGURE 8-20. Summary of Hox-2-cluster gene expression in the 9.5-d-old mouse hind-brain. The lower part illustrates the physical order of the Hox-2-cluster genes on chromosome 11. In the upper part, rhombomeres (r1-r8), the otic vesicle (OV), and the cranial ganglia (gV, gVII/VIII, gIX and gX) are represented on the left and the domains of Hox-2-gene and *Krox*-20 expression are shown on the right. The anterior limits of r1 and r8 are indicated with dashed lines as they have not been shown to represent segment boundaries. MB, midbrain; HB, hindbrain; SC, spinal cord; fp, floor plate. For simplicity, the relationship between branchiomotor nerves and rhombomeres is not indicated; see Figure 4 of Lumsden and Keynes. (From Wilkinson, D. G., Bhatt, S., Cook, M. N., Boncinelli, E., and Krumlauf, R., *Nature*, 341, 405—409, 1989. With permission.)

germ cells. During midgestation, sharp anterior boundaries of *Hox* gene expression coupled to less well-defined posterior boundaries are ground in the ectoderm and mesoderm. For example, rhombomere boundaries have been shown to correlate precisely with the anterior expression boundaries of *Hox 2.1, 2.6, 2.7,* and *2.8.* Furthermore, *Hox 2.9* is expressed in rhombomere 4 and is flanked by expression of the zinc-finger *Krox-20* in rhombomeres 3 and 5. More recent evidence indicates that *Hox 2.9* is also expressed in mesoderm underlying neuroectoderm which subsequently expresses the gene. The linear order of most of the Hox genes clustered along the chromosome correlates with the spatial order of their anterior borders of expression. The more 5′ a gene is, the more posterior its anterior boundary is located. Correlation between expression domains and chromosomal locations of *Hox* genes can also be followed in the developing limb. For example, *Hox 4.2, 4.4, 4.5, 4.6,* and *4.7* (formerly known as the *Hox 5* complex) show a dynamic, temporally restricted pattern of expression. In the posterior limb area, RNA from 3′ genes appears earlier and more proximal followed by 5′ genes with more distal patterns of expression. More divergent homeoboxes include two genes similar to the *Drosophila engrailed* homeobox (*En-1* and *En-2* on chromosomes 5 and 1, respectively) and the even-skipped box (*Evx-1*) on chromosome 2. *En-1* is expressed in the developing brain and neural tube, whereas *En-2* and *Evx-1* expression is limited to the midbrain, and to the hind brain and neural tube, respectively. The mouse Cdx-1 gene (chromosome 18) was cloned using the Drosophila caudal box and is expressed in the intestine. The mouse *Pax* genes are located on chromosome 2 (*Pax-1*), chromosome 7 (*Pax-2*), and an unknown position (*Pax-3*). In addition to the *paired*-type of homeobox, these genes contain a second motif known as the paired box. *Pax* gene expression has been found in the intervertebral disc, sternum, thymus, hindbrain, neural tube, and kidney. Finally, as described elsewhere in this yearbook, the *Oct* genes contain a homeobox distinct from the Antp class as well as a second conserved region known as the POU domain. These genes encode transcription factors some of which are expressed in the developing central nervous system (*Oct-2*) as well as oocytes, inner cell mass, germ cells, and neuroectoderm (*Oct-4*). The prevailing thought is that overlapping domains of homeobox gene expression, for example in the central nervous system, underlying mesoderm, or in the developing limb bud, define positional information needed for pattern formation. Direct proof for this hypothesis is only beginning to accumulate (for example, see papers on ectopic expression of Hox-1.1 and targeted mutagenesis of *En-2* described in this yearbook), and future experiments are likely to concentrate on the effects of disrupted or misexpressed genes on expression of other genes, and subsequently on pattern information. *Terry Magnuson*

Morphogenesis and Pattern Formation 9

INTRODUCTION

The true essence of "biological elegance" occurs when all the pieces fit together in the puzzle and the final product appears perfect and as expected. The number of events that need to occur correctly is astounding. The potential for a mistake is enormous, making the high rate of success all the more impressive. Yet, with few exceptions, each product is unique, unlike any other. The real goal of developmental biology is to try and understand al the aspects of sameness, and all the aspects of difference.

While the biology of morphogenesis has been well described in many organisms for some time, the molecules that orchestrate this elegant process are just now being described. One morphogen in particular, retinoic acid, has become the focus of much investigation because it can serve as a paradigm for other possible morphogenic molecules. The presence of this molecule in several vertebrate species and the knowledge we have gained into its possible mechanism of action provide strong insight into how a single molecule can be capable of influencing the types of events that morphogenesis entails.

Cogenital malformations remain the leading cause of death for infants under the age of 1 year. Many of those malformations are manifest in the disfiguring deformities that result from abnormal pattern formation. Through continued research on the molecular basis of morphogenesis and pattern formation, it may ultimately be possible to prevent such tragic events from occurring.

Origins of the Prestalk-Prespore Pattern in *Dictyostelium* Development

J. G. Williams, K. T. Duffy, D. P. Lane, S. T. McRobbie, A. J. Harwood, D. Traynor, R. R. Kay, and K. A. Jermyn

Cell, 59, 1157—1163, 1989

9-1

Dictyostelium amoebae enter development with the potential to differentiate into either stalk or spore cells. An anteroposterior pattern of prestalk and prespore cells is well established by the migrating slug stage

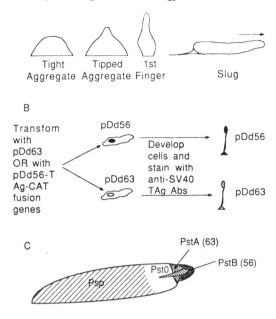

FIGURE 9-1. The dorsal-ventral pathway. All of the genes upstream of *dorsal*, including *dorsal*, are maternally active. Genes that are underlined have a ventralized loss-of-function phenotype. (From Steward, R., *Cell*, 59, 1179—1188, 1989. With permission.)

(Figure 9-1). It is not clear whether patterning depends on positional differentiation of prestalk and prespore cells, or whether the prestalk-prespore pattern arises through initial random differentiation of the two cell types according to cell-autonomous differences, after which they are sorted out to different regions of the aggregate.

The authors used specifically expressed genes as cell-autonomous markers to trace the origins of prespore cells and two types of prestalk cells during slug formation. Both cell sorting and positional information were found to contribute to morphogenesis. The initial pattern — established at the mound stage — differed topologically from that of the slug. Prespore cells occupied most of the aggregate, but were absent from a thin layer at the base and from the emerging tip. One type of prestalk cells were almost entirely localized to the basal region in the early stage of tip formation. The other prestalk cells were scattered throughout the aggregate when first detected, and then migrated to the apex at the site of tip formation.

The prespore cells of the slug and some prestalk cells differentiate in response to localized morphogenetic signals. These signals may arise from the base as well as from the tip.

♦ The multicellular assemblies of *Dictyostelium discoideum* have relatively simple structures — anterior prestalk cells and posterior prespore cells — and thus are ideal organisms for the study of pattern formation. Indeed, it is now known that a small chlorinated organic molecule termed

DIF (differentiation inducing factor) is a natural morphogen responsible for inducing the expression of prestalk specific genes. Other small molecules including cAMP, adenosine, and ammonia also participate in the determination of pattern. Nevertheless, the underlying mechanism of patterning is far from clear. Several hypotheses have been suggested ranging from morphogen gradients to predetermined cell types that later sort to form the terminal pattern. Using powerful molecular methods, Williams et al. now show that cell determination is a relatively early position dependent process and that subsequent cell movements are required to establish the final pattern. In previous work, they have shown that there are actually at least two types of prestalk regions as defined by the expression of DIF-inducible prestalk genes. They now use the promoters from these genes, as well as a prestalk gene, coupled to a B-galacto-sidase reporter genes to observe where and when the initial gene expression takes place. The results indicate an unexpectedly complex temporal and spacial regulation. These are exciting results which portend further significant advances in this field. *Stephen Alexander*

The Graded Distribution of the Dorsal Morphogen Is Initiated by Selective Nuclear Transport in *Drosophila*
C. A. Rushlow, K. Han, J. L. Manley, and M. Levine
Cell, 59, 1165—1177, 1989   9-2

About 40 zygotically active regulatory genes control the position-dependent establishment of diverse cell types in the early *Drosophila* embryo. The maternal morphogen *dorsal (dl)* has a key role in establishing dorsal-ventral polarity. The authors now report that development of the *dl* gradient involves a process of selective nuclear transport. Anti-*dl* antibodies were used to determine the distribution of *dl* protein.

In early embryos, the *dl* protein was uniformly distributed in the ooplasm, but, as soon as 90 min after fertilization, *dl* protein present in the ventral region was selectively transported to the nucleus. This occurred during cleavage cycle 10. In dorsal regions, the protein retained a cytoplasmic location. Mutations in maternally active genes which regulate *dl* disrupted the transport process and resulted in an inactive, cytoplasmically located form of *dl* protein. Truncated *dl* proteins lacking C-terminal sequences accumulated mostly in the nuclei of transfected Schneider cells, while full-length protein was largely restricted to the cytoplasm. Co-transfection assays suggested that *dl* activates expression from several promoters, apparently in a sequence-independent manner.

The means by which *dl* activates gene expression remains uncertain, but there is considerable evidence that dl protein functions in the nucleus. Persistence of the *dl* protein in advanced-stage embryos suggests that it may play a direct part in regulating target genes needed for differentiation of the ventral mesoderm.

Relocalization of the *dorsal* Protein from the Cytoplasm to the Nucleus Correlates with its Function
R. Steward
Cell, 59, 1179—1188, 1989   9-3

Initial formation of dorsal-ventral asymmetry in the *Drosophila* embryo is regulated by about 20 maternally active genes which function during oogenesis and early embryogenesis. One of these genes, *dorsal,* is homologous to the vertebrate proto-oncogene c-*rel.* Antibody studies have been carried out to delineate the distribution of *dorsal* protein in early embryos.

In wild-type embryos, *dorsal* protein is present in the cytoplasm during cleavage. Following the migration of nuclei to the embryonic periphery, in the syncytial blastoderm stage, there is a ventral-to-dorsal gradient of nuclear *dorsal* protein. Distribution studies in several maternal dorsal-ventral polarity mutants suggested that nuclear localization is required for the function of *dorsal* protein. Only cytoplasmic protein is present in dorsalized embryos, while in ventralized embryos the nuclear gradient is shifted dorsally.

The maternal dorsal-ventral polarity genes of *Drosophila* probably act upstream of *dorsal* and appear to have a role in the nuclear localization of *dorsal* protein, and/or in generating the graded distribution of protein in the nuclei (Figure 9-3). It is not clear whether the level of dorsal protein determines the differential regulation of zygotic genes or whether the time for which protein is present in the nucleus is a critical factor. It is likely, however, that, because of the way the protein is distributed in the

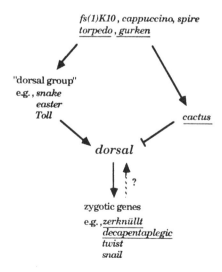

FIGURE 9-3. The dorsal-ventral pathway. All of the genes upstream of dorsal, including dorsal, are maternally active. Genes that are underlined have a ventralized loss-of-function phenotype. (From Steward, R., Cell, 59, 1179—1188, 1989. With permission.)

nuclei, zygotic genes are activated or inactivated in domains along the dorsal-ventral axis.

A Gradient of Nuclear Localization of the *dorsal* Protein Determines Dorsoventral Pattern in the *Drosophila* Embryo
S. Roth, D. Stein, and C. Nusslein-Volhard
Cell, 59, 1189—1202, 1989   9-4

The dorsoventral axis of the *Drosophila* embryo is formed through the establishment of a morphogen gradient by the action of 12 maternal-effect genes, the dorsal group genes, and *cactus.* One of the dorsal group genes, *dorsal (dl),* encodes the putative morphogen. The other dorsal group genes probably act to establish the morphogen gradient in a correct orientation. While the *dorsal* mRNA is distributed evenly in early embryos, the protein product reportedly is distributed in a nuclear gradient in the syncytial blastoderm-stage embryo, peaking at the ventral side of the egg. Dorsally, the protein remains in a cytoplasmic location.

Nuclear localization of *dl* protein and, consequently, formation of a gradient is blocked in dorsalizing alleles of the other dorsal group genes. In ventralizing mutants, nuclear localization of *dl* protein extends to the dorsal side of egg.

The correlation between *dl* protein distribution and embryonic pattern in mutant embryos suggests that the nuclear concentration of *dl* protein is what determines patterning along the dorsoventral axis. It is possible that the *dorsal* protein acts as a transcriptional regulator.

♦ Maternal gene products play a pivotal role in establishing the embryonic body plan. Twelve maternal genes are of key importance for determining dorso-ventral polarity. Loss of function in any of 11 of these genes leads to an identical phenotype, formation of dorsal structures along the entire circumference of the embryo (embryo dorsalization) while loss of function in mutants of the *cactus* gene have the opposite, ventralized phenotype. Genetic evidence indicates that the 11 "dorsal group" genes form a functional cascade (akin to the blood clotting cascade). Moreover, *dorsal,* that gives the name to this group of genes, appears to be the last component of this cascade and is therefore considered to be the "morphogen". *dorsal* was cloned and shown to share common sequences with the avian *v-rel* oncogene. *dorsal* mRNA is uniformly distributed at all developmental stages. However, the *dorsal* protein was found to form a gradient that has its highest point around the ventral midline. The basis for the formation of this gradient was unknown (a report on this subject appears in last year's volume). The three research groups above have make a startling finding concerning the asymmetric distribution of the protein": it is now clear that the *dorsal* protein already starts to accumulate during oogenesis and that it is uniformly distributed up to embryonic nuclear division 9 (about 90 min after fertilization) when the outwardly migrating syncytial nuclei reach the periphery of the embryo. At this time the *dorsal* protein starts to migrate into the nuclei, but only around the ventral regions of the embryo. Cross-sections of these embryos immunostained for the dorsal protein show a gradient of nuclearly located protein that has its highest point at the ventral midline (were the mesodermal precursors lie) and gradually decreases so that nuclear staining past the equator is hardly detectable. *dorsal* protein that fails to enter the nuclei is eventually degraded. Thus, the basis for the uneven distribution of the *dorsal* gene product is not regional activation of transcription (as is common for genes involved in the establishment of the of the anteroposterior embryonic axis), or to the localization of the mRNA (as is the case for *bicoid;* reported in last year's volume), or to translational control of the dorsal mRNA (as had been hypothesized earlier); differential nuclear transport followed by destabilization of the protein that is left in the cytoplasm seems to regulate regional accumulation. The other ten dorsal-group genes must be involved in regulating this differential nuclear uptake since mutants in these genes produce normal amounts of *dorsal* protein that does not migrate into the nucleus or form a gradient. By contrast, in ventralizing mutants (e.g., gain-of-function Toll and cactus mutants) nuclear transport of dorsal extends well into the dorsal half of the embryo.

How is regional activation of nuclear transport achieved? The answer to this question is not yet available. Some preliminary evidence from Rushlow et al. suggests that the carboxy-terminus of the protein may be involved. When wild-type *dorsal* protein is expressed in *Drosophila* cultured cells, it remains in the cytoplasm. However, a truncated protein that lacks as few as six amino acids from the carboxy terminus becomes nuclearly localized. It is possible that covalent modifications of *dorsal* (e.g., phosphorylation, glycosylation) regulate nuclear entry. It should be noted in this connection that the mobility of the native protein upon gel electrophoresis is lower than that predicted from its size. It is also possible that the *cactus* gene product serves as a cytoplasmic anchor for dorsal. The source of the nuclear localization signal is also not know. *Toll,* a transmembrane protein with similarity with platelet glycoprotein 1b, is very close to *dorsal* (or even directly precedes it) in the regulatory cascade. It is not unreasonable to hypothesize that products from earlier members of this genetic cascade (for instance, *snake* and *easter* that have similarity to serine proteases) reside outside of the oocyte (in the perivitelline space) and that regional activation of Toll transmits the critical signal to the inside of the oocyte.

What does *dorsal* do once it is in the nucleus? Even though *dorsal* does not possess recognizable sequences typical of transcriptional regulators (e.g., zinc fingers, homeodomains), it is clear that it affects the expression of downstream genes. Alterations in expression patterns of certain dorsally expressed (e.g., *zerknullt, decapentaplegic*) and ventrally expressed (e.g., *twist, snail*) genes in dorsal mutants suggest that dorsal acts to repress the former and activate the latter group of genes. What remains to be established is whether this is a direct or indirect relationship.

Although regulation of gene activity at the nuclear transport level is not without precedent (see "minireview" by T. Hunt in *Cell,* 59, 949—951, 1989), it is not a commonly occurring regulatory mechanism. Moreover, this is the first example where regulated nuclear transport has been shown to be involved in determination of embryonic cell fate. In the near future we should be able to gain a better understanding on how the components of this fascinating genetic cascade interact and fit together. *Marcelo Jacobs-Lorena*

The Drosophila *patched* Gene Encodes a Putative Membrane Protein Required for Segmental Patterning
J. E. Hooper and M. P. Scott
Cell, 59, 751—765, 1989   9-5

The *patched (ptc)* gene is one of several segment polarity genes needed for correct patterning within segments of *Drosophila.* When ptc function is absent, the fate of cells in the central part of each segment is altered so that they form pattern elements typical of cells positioned about the segment border. Analysis of the mutant phenotype suggested that both

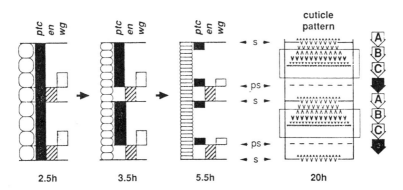

FIGURE 9-5. Summary of the ectodermal pattern of *ptc* transcript within the segment and relative to the cuticle pattern. A two-segment repeat is shown for blastoderm stage (2.5 h), early extended germband (3.5 h), mid-extended germband (5.5 h), and mature embryo cuticle pattern (ventral abdominal segments). The transcript patterns for *ptc* (solid shading), *en* (hatched shading), and *wg* (light shading) at each stage are indicated relative to the segment borders (solid lines labeled "s"), parasegment borders (dashed lines labeled "ps"), and the cells (circles and ovals). (From Hooper, J. E. and Scott, M. P., *Cell,* 59, 751—765, 1989. With permission.)

segment and parasegment borders are included in the duplicated pattern of *ptc* mutants.

The authors have cloned the *ptc* gene and deduced that the product is a 1286-amino acid protein having at least 7 putative transmembrane alpha helices. The *ptc* RNA was not detected in embryos before nuclear cleavage 13. During cellularization of the blastoderm, the RNA was present throughout most of the cortical region of the embryo and also was associated with yolk nuclei. Later in development, *ptc* RNA was expressed in broad stripes of segmental periodicity which subsequently split into two stripes per segment primordium. The expression pattern was not directly predictive of the transformation seen in *ptc* mutant embryos.

In normal embryogenesis, *ptc* organizes the pattern within the anterior three fourths of the segment (Figure 9-5). Mosaic analysis suggests that the *ptc* protein is involved in signaling between neighboring cells. Studies of segment polarity gene function at the molecular level will show how cellular interactions establish position and pattern within segments.

A Protein with Several Possible Membrane-Spanning Domains Encoded by the *Drosophila* Segment Polarity Gene *patched*
Y. Nakano, I. Guerrero, A. Hidalgo, A. Taylor, J. R. S. Whittle, and P. W. Ingham
Nature, 341, 508—513, 1989   9-6

Generation of segment patterns in the *Drosophila* embryo depends on the action of segment-polarity genes. Mutations in these genes produce site-specific deletions of various parts of each segment and duplication of the remaining structures. The segment-polarity genes control the exchange of information between cells, which is critical for their correct patterning.

The authors have molecularly characterized one of these genes, *patched (ptc),* which encodes a large protein possessing several possible membrane-spanning domains and is expressed in a complex manner during embryogenesis. The *ptc* transcripts are first expressed in the cellularizing blastoderm, when they are nearly uniformly distributed about the embryonic periphery. By stage 10, they are expressed in broad bands in both the ectoderm and mesoderm of each parasegment. The way in which *ptc* protein is oriented relative to the membrane maximizes the number of possible glycosylation sites located extracellularly.

Possibly, *ptc*-expressing cells lose their ability to respond to a *wg*-encoded signal because of an early response to another signal for which *ptc* protein acts as a receptor. Alternatively, the expression of *ptc* may cause cells to respond differently to the *wg*-encoded signal. The *ptc* protein also could interact directly with the receptor for the *wg* signal, coupling it to a different pathway.

◆ Differentiation of the early *Drosophila* embryo along the anterior-posterior axis is established by a set of maternal genes that influence the global organization of the embryo, and is refined by three classes of zygotic genes that act in the following order: gap genes, that affect defined broad regions; pair-rule genes, that affect alternate segments; and segment polarity genes, that affect every segment. Mutations in segment polarity genes delete defined pattern elements from each segment and these are often replaced by mirror-image duplications of remaining elements. About a dozen segment polarity genes have been identified, of which four had previously been cloned: two, *engrailed (en)* and *gooseberry (gsb)*, encode homeobox-containing proteins, *armadillo (arm)* encodes a putative cytoskeleton-associated protein, and *wingless (wg)* encodes an extracellular protein with similarity to growth factors. These reports add a fifth gene, *patched (ptc)*, which encodes a putative transmembrane protein. Because segment polarity genes act after cellularization of the blastoderm, they must rely more than the other groups of genes on cell-to-cell communication in order to establish segmental pattern elements. For instance, there is strong evidence that *en* expression is dependent on the expression of *wg* in neighboring cells. The possible transmembrane localization of the *ptc* protein suggests that this gene may also be involved in cell-to-cell communication.

In the early embryo *ptc* transcripts are uniformly distributed. Following the activation of *en* in the posterior cells of each segment, *ptc* transcripts disappear only from *en*-expressing cells, suggesting that *en* may be directly responsible for transcriptional repression of *ptc*. However, this is not a simple relationship. Later in embryogenesis *ptc* expression in the mesoderm is out of register with that of the ectoderm; and there is good evidence to suggest that in the mesoderm *en* and *ptc* are probably expressed in the same cells. A further complexity relates to the discrepancy between patterns of *ptc* expression and mutant phenotypes: the anterior-most cells of each segment are unaffected in ptc mutants, even though they express *ptc* at both 3.5 and 5.5 h of development; on the other hand, cells in the middle of each segment, are severely affected in *ptc* mutants even though they do not express *ptc* at 5.5 h. These properties are probably due to long-range effects of the *ptc* gene and are consistent with its hypothesized role in cell-to-cell communication. The *en* and *wg* genes are likely to participate in the same regulatory network that involves *ptc*. Obviously, much remains to be learned. However, the cloning of the *ptc* gene is major step toward the understanding on how cell-to-cell communication related to segmental patterning. *Marcelo Jacobs-Lorena*

Kruppel Requirement for *knirps* Enhancement Reflects Overlapping Gap Gene Activities in the *Drosophila* Embryo

M. J. Pankratz, M. Hoch, E. Seifert, and H. Jackle

Nature, 341, 337—340, 1989   9-7

During segmental pattern formation in *Drosophila,* the anterior system is strictly dependent on the product of the maternal gene *bicoid.* The posterior system, in contrast, appears to lack such a morphogen. The known posterior maternal determinants simply define the boundaries within which abdominal segmentation can take place. Active generation of the abdominal body pattern may occur entirely through the interactions between zygotic genes. The most likely candidates for initiating posterior pattern formation are the gap genes — the first segmentation genes to be expressed in the embryo.

The authors have examined interactions among the gap genes *Kruppel (Kr), knirps (kni),* and *tailless (tll).* Expression of *kni* was repressed by *tll* activity, but was directly enhanced by *Kr* activity. *Kr* activity was present throughout the domain of *kni* expression and formed a long-range protein gradient. Both this gradient and *kni* activity are required for abdominal segmentation to proceed normally.

Because of the extensive overlap of gap gene activities, the relative concentrations in a given combination of gap gene products could direct diverse pattern-forming events. This could account for a relatively few gap genes being able to control a wide range of pattern-forming processes.

Gradients of *Kruppel* and *knirps* Gene Products Direct Pair-Rule Gene Stripe Patterning in the Posterior Region of the Drosophila Embryo

M. J. Pankratz, E. Seifert, N. Gerwin, B. Billi, U. Nauber, and H. Jackle

Cell, 61, 309—317, 1990   9-8

Abdominal segmentation in the Drosophila embryo requires the activities of the gap genes *Kruppel (Kr), knirps (kni),* and *tailless (tll).* These genes control the expression of the pair-rule gene *hairy (h)* through the activating or repressing independent *cis*-acting units which generate individual stripes. In this study, stripe-specific *cis*-acting control elements of *h* were identified using promoter fusion constructs expressing a reporter gene. The constructs were introduced in *Kr, kni,* and *tll* mutant embryos to determine how the gap gene regulate the expression of particular *h* stripes in the posterior embryonic region.

It was found the *kr* activates stripe 5 and represses stripe 6. At the same time, *kni* activates stripe 6 and represses stripe 7, while *tll* activates stripe 7. The *kr* and *kni* proteins bound strongly to h control units which generate stripes in areas where the concentration of the respective gap

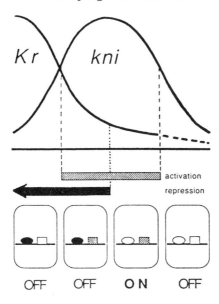

FIGURE 9-8. Generation of the sixth *h* stripe through superimposition of distinct gap gene activities. *Kr* and *kni* protein gradients overlap in the posterior region of the embryo. *Kr* activates, while *kni* represses, transcription from the control element that drives expression of the sixth *h* stripe (line with oval and square inside the four schematized nuclei). The critical factor is the different threshold levels at which the two gap proteins act. *kni* can activate transcription only at relatively high concentrations (found in a region delimited by the stippled bar). However, *Kr* can repress transcription even at very low concentrations (down to the level denoted by the black arrow). This is effected through the different binding strengths that the two gap proteins have for this particular *h* control element: *Kr* binds strongly (*Kr* binding sites represented by ovals), whereas *kni* binds very weakly (*kni* binding sites represented by squares). This means that the *kni* binding site will be occupied only in a region of high *kni* concentration (stippled squares), while *Kr* binding sites will be occupied in areas of both high and low *Kr* concentrations (black ovals). If repression can override activation, this particular promoter element will then be turned on only in a region of the embryo that has sufficient levels of *kni* to activate, but insufficient levels of *Kr* to repress, transcription. As a result of the distributions of *Kr* and *kni* proteins in the blastoderm embryo, this situation would be found only in a narrow region near the posterior end, i.e., where the sixth *h* stripe is located (*Kr* is expressed at low levels down to ~33% egg length, *kni* protein domain is visible from about 27 to 43%, and the sixth *h* stripe is located at around 27 to 33%). Given the present data, this is the minimum requirement for generating the sixth *h* stripe. Regulatory inputs from other sources, such as the gap genes *tll* and *giant*, or the pair-rule genes *eve* and *runt* may also be required. (From Pankratz, M. J., Seifert, E., Gerwin, N., Billi, B., Nauber, U., and Jackle, H., *Cell*, 61, 309—317, 1990. With permission.)

gene products are low. The proteins bound weakly to *h* control units, generating stripes in areas of high gap gene expression.

These findings indicate that the Kr and kni proteins form overlapping concentration gradients which generate the periodic pair-rule stripe patterning characteristic of the posterior region of the Drosophila embryo (Figure 9-8). Initial periodicity could serve to "template" further periodic patterns through the local activation and/or repression of downstream pair-rule genes.

♦ Organization of the anterior-posterior embryonic body plan in *Drosophila* is dependent on the sequential action of maternal genes followed by the gap, pair-rule, and segment polarity genes, in this order. Molecular analysis has shown that the earliest genes are expressed in relatively broad regions of the embryo and that the patterns of gene expression gradually evolve to cover increasingly narrower portions of the embryo. For instance, the early-expressing gap gene *Kruppel* (*Kr*) is expressed over a broad band covering at least 30% of the embryo while the relatively late-expressing segment polarity gene *engrailed* is expressed over stripes that are only one-cell wide. These patterns evolve in great part while the embryo is a syncytium and no cell membranes separate the nuclei expressing the particular genes. The central question addressed by these experiments is, how is the banded or striped pattern of gene expression achieved in a morphologically uniform syncytium? The experiments strongly suggest that the sharpening of the striped pattern is achieved by the cross-regulatory interactions among genes that are expressed in overlapping patterns.

Mutations in the gap genes *Kr, knirps* (*kni*), and *Tailless* (*tll*) cause large pattern-deletions in the abdomen: *Kr* affects the development of thoracic (T) and abdominal (A) segments T1 to A6, *kni* affects abdominal segments A1 to A7, and *tll* affects segments posterior to A7. To deduce possible cross-regulatory interactions among these various genes, Pankratz et al. (1989) determined how the expression of one gene changes by the genetic inactivation of another gene. For example, in mutant *Kr* embryos, the posterior expression of kni is considerably weakened. Since *Kr* and *kni* expression physically overlap, the simplest interpretation of this result is the the *Kr* protein acts as an enhancer of *kni* transcription. Indeed, additional experiments demonstrated that the *Kr* protein can physically bind to a *kni* DNA fragment situated upstream from its transcription site, a result that reinforces the above interpretation. By contrast, removal of *tll* activity expands the region of *kni* expression posteriorly into the region where *tll* is normally expressed. Although molecular data are not yet available, the results suggest that the *tll* protein normally acts to repress *kni* transcription. Overall, these experiments provide partial explanation for how a band of *kni* expression is generated in the embryo: it requires activation by the *Kr* gene product (and probably by other factors as well) and the posterior border is established at least in part, by the transcriptional repression of the tll protein.

The paper by Pankratz et al. (1990) examines the role of *Kr* and *kni* in the establishment of the seven stripes of *hairy* (*h*) expression. *h* is a pair-rule gene. Pair-rule genes have been subdivided into two classes, according to their position in the regulatory hierarchy. Expression of primary pair rule genes such as *h, even skipped,* and *runt* is completely independent from the activity of other pair rule genes; primary pair rule genes are probably regulated by the activity of gap genes. On the other hand, expression of secondary pair-rule genes, such as *fushi tarazu,* depends at least in part on the activity of other pair rule genes (e.g., the primary pair-rule genes). Molecular analysis has revealed a striking difference between the organization of these two classes of pair rule genes: while in primary pair-rule genes separate promoter elements appear to be responsible for directing gene expression in each region (stripe) of the embryo, in secondary pair-rule genes the metameric pattern of expression appears to driven by a single and complex promoter element. This model appears to apply to the *h* gene. Pankratz et al. (1990) demonstrate that *Kr, kni,* and *tll* each act to activate or repress specific *h* stripes along the anterior-posterior axis of the embryo. Interestingly, the same protein (e.g., *Kr* and *kni*) can act as repressors of *h* gene expression in one region (stripe) of the embryo and as an activator in another region. These opposing effects are mediated by different promoter elements and are likely to require additional factors. Moreover, by joining promoter elements to a reporter β-galactosidase gene, the authors demonstrate that these elements can act in each strip independently from other promoter elements. Overall, these experiments identify promoter elements of the h gene and suggest roles for three gap genes, *Kr, kni* and *tll,* in the generation of the posterior stripes. *Marcelo Jacobs-Lorena*

Spatial Aspects of Neural Induction in *Xenopus laevis*
E. A. Jones and H. R. Woodland
Development, 107, 785—791, 1989   9-9

As an alternative to using nucleic acid probes in studying formation of the nervous system in *Xenopus,* the authors employed a monoclonal antibody marker, 2G9, as a marker of neural differentiation. The antibody may be used to stain the entire adult central nervous system and also peripheral nerves.

Staining first appeared in embryos at stage 21 in the thoracic region. By stage 29, the entire central nervous system except the tail tip was stained. Neural induction was observed as early as stage 10, and occurred in embryos lacking gastrulation. The epitope is present in a 65-K M_r protein and includes sialic acid. The monoclonal antibody also reacted with neural tissue of mice, axolotls, and newts. Both notochord and somites appeared capable of neural induction. Attempts to demonstrate induction of neural tissue by the developing nervous system itself failed.

These findings provide no reason to rule out the possibility that the lateral extent of the neural plate is determined solely by the lateral extent of the inducing tissue. At the same time, neural induction occurs rapidly and diffusion of signals is a possible complicating factor.

Neural Expression of the *Xenopus* Homeobox Gene Xhox3: Evidence for a Patterning Neural Signal that Spreads through the Ectoderm

A. R. Altaba
Development, 108, 595—604, 1990   9-10

The need for mesoderm in neural induction has been demonstrated in exogastrulated amphibian embryos. In *Xenopus,* exogastrula ectoderm remains a small, but critical, junctional area with axial mesoderm, corresponding to the organizer region — in contrast to isolated ectoderm cultured *in vitro.* The ectoderm is neuralized despite the absence of axial mesoderm beneath the ectoderm of exogastrulae.

The homeobox gene Xhopx3 is expressed chiefly in the axial mesoderm in the gastrula and neurula stages, and later is expressed in the nervous system and tail bud (Figure 9-10). Mesodermal expression is not detectable by the late neurula-early tailbud stage. *In situ* hybridization studies showed Xhox3 to be restricted within the neural type and the cranial neural crest in the tailbud-early tadpole stages. Later, it was expressed only in the mid/hindbrain region. Studies of exogastrulated embryos showed that Xhox3 is expressed in the apical ectoderm, a region which develops in the absence of anterior axial mesoderm.

These findings suggest the existence of a neural-inducing signal that spreads through the ectoderm and contributes to anterior-posterior patterning of the neural ectoderm. Since the anterior-posterior polarity of

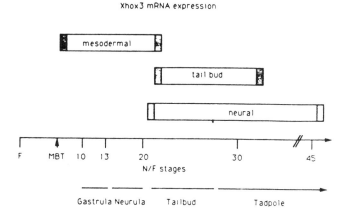

FIGURE 9-10. Diagram showing the expression profile of Xhox3 during embryonic development. Note the two distinct periods of expression (boxes), first in the axial mesoderm of gastrula and neurula stage embryos and later on in the tail bud and nervous system. The filled ends of the boxes denote the approximate periods of beginning and end of Xhox3 expression. The diagram is based on the data of Ruiz i Altaba and Melton (1989a). Numbers refer to stages according to Nieuwkoop and Faber (1967). (From Altaba, A. R., *Development,* 108, 595—604, 1990. With permission.)

the axial mesoderm may derive in part from gradients of growth factor-like molecules, it is possible that similar gradients of diffusible factors derived from the organizer region have a role in neural ectodermal patterning.

Cellular Contacts Required for Neural Induction in *Xenopus* Embryos: Evidence for Two Signals
J. E. Dixon and C. R. Kintner
Development, 106, 749—757, 1989   9-11

In amphibian embryos, neurogenesis begins at about the time of gastrulation, when part of the ectoderm receives an inducing signal from dorsal mesoderm. There are two suggestions for how ectoderm comes into contact with dorsal mesoderm to permit passage of the signal (Figure 9-11). One is that normal gastrulation movements bring dorsal mesoderm into apposition with the overlying ectoderm. Alternatively, only a small area of contact may be required for the signal to pass laterally and spread within the ectoderm, perhaps even before gastrulation.

Studies of embryonic tissue explants favoring one or the other of these types of contact, along with a quantitative RNAse protection assay to measure neural-specific RNA transcript levels, showed that neural tissue

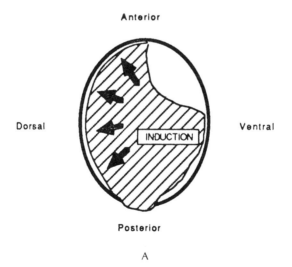

A

FIGURE 9-11. Potential interactions between ectoderm and presumptive dorsal mesoderm. A and B diagram two potential routes through which an inducer could pass between dorsal mesoderm and ectoderm. In A, the inducer would pass between the invaginating mesoderm and overlying ectoderm during gastrulation. In B, the inducer would cross laterally the IMZ-NIMZ boundary, perhaps before gastrulation, and then spread within the ectoderm. (From Dixon, J. E. and Kintner, C. R., *Development,* 106, 749—757, 1989. With permission.)

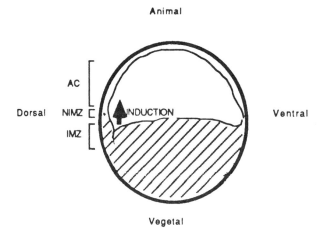

FIGURE 9-11B.

is efficiently formed when the ectoderm and dorsal mesoderm interact laterally within a tissue plane. When ectoderm was placed in apposition to involuting anterior-dorsal mesoderm, ectoderm formed very poorly. A synergistic effects was observed when both types of contact were possible.

It appears that contact between ectoderm and anterior-dorsal mesoderm during gastrulation in *Xenopus* makes only a minor quantitative contribution to the induction of neural tissue. Nevertheless, apposition may permit contacts which are important for regionalizing the neural ectoderm. Some of the events involved in neural tissue induction may take place before the start of gastrulation.

Progressive Determination during Formation of the Anteroposterior Axis in *Xenopus laevis*
H. L. Sive, K. Hattori, and H. Weintraub
Cell, 58, 171—180, 1989   9-12

The cement gland is an ectodermal organ in the head of frog embryos, which lies anterior to any neural tissue. The cement gland of *Xenopus* serves well in studies designed to elucidate the induction of an anterior-specific dorsal ectodermal structure. The authors isolated a set of DNA clones which are early markers of cement gland differentiation.

Like neural tissue, the cement gland was found to be induced by the dorsal mesoderm. The mesoderm possessing the strongest cement gland-inducing potential lay posterior to the ectoderm destined to form the organ, indicating that induction takes place at a distance from the inducer

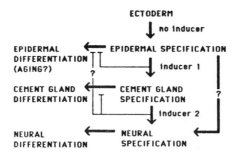

FIGURE 9-12. Temporal progression of ectodermal inductions during early Xenopus embryogenesis. Inductions are arrowed; barred lines indicate repression interactions. Induction of mesoderm is not included. During cleavage stages, presumptive ectoderm that is not induced to form mesoderm becomes specified as epidermis. Specified epidermis is transiently competent for induction to cement gland or neural tissue, upon which suppression of epidermal differentiation occurs. Loss of competence presumably accompanies irreversible commitment to epidermal differentiation. Ectoderm that becomes induced as cement gland can be altered by further stimuli to a neural fate. We do not know whether specified epidermis can directly be induced to neural tissue, and whether the repression of epidermal differentiation seen with neural induction is a result of prior cement gland induction. It is also not known whether repression of particular pathways is a direct consequence of the inducer involved, or whether it is dependent on the new cell type specified. Further, since we have been unable to reverse neural induction with an artificial inducer of cement gland, NH_4Cl (data not shown), it seems that in the hierarchy of possible superinductions, nerve to cement gland is not a permissible conversion. (From Sive, H. L., Hattori, K., and Weintraub, H., *Cell,* 58, 171—180, 1989. With permission.)

source. The first evidence of cement gland induction was in early gastrulation. Nevertheless, most of the initially induced cells did not contribute to the mature gland, but instead formed part of the neural plate. Specification of the cement gland could be altered by continued contact with mesoderm.

Part of the anteroposterior axis of *Xenopus* is determined progressively. Future neural ectoderm is first induced to a cement gland-like state. Further induction by mesoderm may override this state as gastrulation proceeds. The findings support the conversion of presumptive cement gland to neural tissue. The temporal sequence of inductive and suppressive interactions that determine the ectoderm is linked with the relative positions of inducing and responding tissues (Figure 9-12).

♦ Probably the most familiar example of embryonic induction is the respecification of dorsal ectoderm as neural plate by the dorsal mesoderm, primarily notochord, that comes to lie underneath it during gastrulation. A less famous but closely related and important concept is that the neural inducing signal can spread through the neural plate. Nieuwkoop proposed (see the article by Jones and Woodland for a discussion), that the notochord induces a narrow strip of adjacent neural tissue on the dorsal midline, and the rest of the neural plate is induced by a neuralizing signal emanating laterally from this strip, via a process known as "homeogenetic" induction.

Jones and Woodland present results which cast some doubt on the lateral neural induction model. They carried out a series of explantation and transplantation experiments to show that somite tissue as well as notochord is capable of neural induction of ectoderm. This means that most of the neural plate, including the lateral portions, is in close contact with inducing mesoderm, so there is no need for expansion of the neural plate by homeogenetic induction. Furthermore, they show that small pieces of neural plate tissue (taken from gastrula stages when the homeogenetic induction ought to be occurring) transplanted into early gastrula ventral ectoderm do not neuralize surrounding tissues.

These results suggest that neural plate tissue does not generate a neural inducing signal, or at least does not produce a neuralizing signal sufficiently potent to neuralize ventral ectoderm. This is an important qualification, since it has been shown (Sharpe et al., 1987, see the 1989 Year Book pp. 188—189) that ventral ectoderm is relatively refractory to neural induction compared to dorsal ectoderm.

The papers by Ruiz i Altaba and by Dixon and Kintner seem at first glance to contradict the conclusions of Jones and Woodland. *Xenopus* embryos can be make to undergo complete exogastrulation by culturing in hypertonic medium. This results in physical separation of ectoderm from mesoderm, except for the region where the base of the ectodermal sac is attached to the remainder of the embryo. One might predict that this would prevent neural induction, but as shown earlier by Kintner and Melton (*Development, 99, 311—325, 1987*) molecular neural markers are in fact induced in such embryos. Ruiz i Altaba show further that the homeobox gene Xhox3 is activated in exogastrula ectoderm at a site spatially separated from the mesoderm contact zone, suggesting inductive activity "at a distance". The Dixon and Kintner paper describes the results of various tissue recombination experiments indicating that the most extensive neural induction in such "sandwiches" is obtained when a piece of dorsal mesoderm is inserted between two sheets including dorsal ectoderma and some superficial marginal zone tissue. The authors interpret the quantitatively distinct behavior off the differently constituted sandwiches as evidence for two neural induction signals, one of which is propagated laterally from the dorsal lip.

Do these findings mean that a homoiogenetic neural inducing signal spreads through the ectoderm? Perhaps; the Ruiz i Altaba data seem explicable in no other way. It is possible to reconcile this with the Jones and Woodland results if one assumes that the ectoderm is acting as a passive conduit for the neuralizing signal rather than an actual secondary source. It is not clear how this might occur, but possibly the extracellular matrix on the inner ectodermal surface could facilitate lateral diffusion of inducers secreted by mesoderm.

The complexity of neural induction is further highlighted by the paper by Sive et al. in which evidence is presented suggesting that as dorsal mesoderm invaginates, the overlying ectoderm is first converted to cement gland. Subsequently, more posterior ectoderm is respecified as nervous system, which could be regarded as a "posteriorization" phenomenon. These authors make the curious observation that posterior mesoderm is the most potent inducer of cement gland. One might have predicted that anterior mesoderm would have this property. The significance of this is unclear, but one interpretation is that cement gland induction can be transmitted laterally. Alternatively, it could be that the most effective cement gland inducer tissue happens not to be the one that actually does the job in the embryo. This would be inefficient, but Occam's razor should be used as a rule, not as a law. *Thomas D. Sargent*

Retinoic Acid Causes an Anteroposterior Transformation in the Developing Central Nervous System

A. J. Durston, J. P. M. Timmermans, W. J. Hage, H. F. J. Hendriks, N. J. de Vries, M. Heideveld, and P. D. Nieuwkoop
Nature, 340, 140—144, 1989   9-13

Retinoic acid (RA), a known teratogen, appears to be an endogenous signal molecule and a morphogen which specifies the anteroposterior axis during limb development. RA and other retinoids also influence the development of other organs, including the central nervous system (CNS). The authors examined the effects of RA on development of the CNS in *Xenopus laevis*.

RA was found to transform anterior neural tissue to a posterior neural specification. A low concentration of RA produced microencephaly in early Xenopus Embryos. The minimal concentration required was similar to the endogenous RA concentration. Microencephaly occurred when RA pulses were applied up to the early neurula stage, following the completion of gastrulation. A level of RA producing extreme microencephaly failed to induce neural differentiation in competent ectoderm, and did not inhibit differentiation in competent ectoderm combined with a small piece of dorsal mesoderm — the natural neural inducer. Effective concentrations of RA inhibited the formation of anterior neural structures as well as sense organs. Treatment of excised neuroectoderm with RA after neural induction had occurred suppressed eye formation.

The direct effect of retinoic acid on neuroectoderm and the timing of RA effects suggest that it influences neural transformation directly. RA also may influence anteroposterior specification of the mesoderm.

Interaction between Peptide Growth Factors and Homeobox Genes in the Establishment of Antero-Posterior Polarity in Frog Embryos

A. R. Altaba and D. A. Melton
Nature, 341, 33—38, 1989   9-14

Expression of the homeobox gene *xhox3* of *Xenopus* is an early response to mesoderm induction (Figure 9-14) by peptide growth factors.

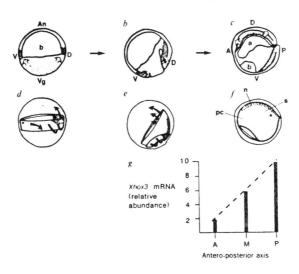

FIGURE 9-14. Drawings of *X. laevis* blastula and gastrula embryos showing inductive and morphogenetic events. *a,* Mesoderm (hatched area) is formed in the blastula by an inductive interaction (open arrows) between endoderm (dotted area) and ectoderm (white area) along the animal (An)-vegetal (Vg) axis. The induced mesoderm forms a torus around the vegetal end of the blastocoel (b) as shown in *d. b,* Differences in the dorso-ventral axis established shortly after fertilization are evident in the different gastrulation movements (solid arrows) of dorsal (D) and ventral (V) mesoderm. Dorsal mesoderm invaginates first and migrates farther. *c,* At the end of gastrulation, the dorsal axial mesoderm has finished its invagination, creating the antero-posterior axis and forming the archenteron (a), the primitive gut. Note that the blastocoel is pushed ventrally during gastrulation. By this time, the axial mesoderm has induced neural structures in the overlying dorsal ectoderm; anterior (A) mesoderm induces brain and posterior (P) mesoderm induces spinal cord (open arrows). *d* to *f.* The three-dimensional aspects of mesodermal movements during gastrulation, adapted after Kellen. Note that prospective notochordal (heavy dots) and somitic (light dots) mesoderm is drastically rearranged by convergent extension (arrows). Consequently, the orthogonal relation between the dorsoventral and anteroposterior axes of the late gastrula *f* is not found in the early gastrula *d.* As an example, the dorsal and ventral positions of white and black circles in the early gastrula are converted into anterior/dorsal and posterior/dorsal positions in *f,* the late gastrula. pc, Prechordal plate; n, notochord; and s, somitic mesoderm. *g,* Diagram of the graded expression of *xhox3* mRNA in the axial mesoderm of late gastrula-early neurula embryos. The histogram shows the relative abundance of *xhox3* mRNA assayed by RNAse protection. The highest value corresponds to the posterior third (P) and was equated arbitrarily to 10 and the values for middle (M) and anterior (A) parts were set proportionately. *Xhox3* is expressed in the mesoderm of the early gastrula but its distribution in the prospective axes has not been determined. (From Altaba, A. R. and Melton, D. A., *Nature,* 341, 33—38, 1989. With permission.)

The level of *xhox3* expression marks the anteroposterior character of induced mesoderm. In this study, levels of *xhox3* were used as a mesodermal marker to learn how different peptide growth factors coordinately determine the level; of *xhox3* mRNA and the anteroposterior pattern in the developing *Xenopus* embryo.

Expression of *xhox3* was readily detected by the early gastrula stage. The gene was expressed very early and transiently in mesoderm. Injection of synthetic *xhox3* mRNA directed ectopic expression of the gene in animal caps, but did not induce mesoderm. At any level of peptide growth factor sufficient to induce muscle, bovine fibroblast growth factor produced higher levels of *xhox3* mRNA than did XTC-MIF, a mesoderm-inducing factor of the transforming growth factor-beta family. Anterior and posterior mesoderm induced the corresponding neural/epidermal structures after implantation. Anterior mesoderm, for instance, induced brain and eyes, while posterior mesoderm induced spinal cord and dorsal fin. Overexpression of *xhox3* mRNA in the anterior end of normal embryos led to a failure to form normal anterior structures, not the production of posterior structures at the anterior end.

The diffusible graded signal involved in establishing anteroposterior polarity in the *Xenopus* embryo is likely to consist of one or more peptide growth factors. Gradients of such factors could set positional information and polarity along the anteroposterior axis in the responding axial tissues through regulating the level of homeobox gene expression.

Interference with Function of a Homeobox Gene in *Xenopus* Embryos Produces Malformations of the Anterior Spinal Cord
C. V. E. Wright, K. W. Y. Cho, J. Hardwicke, R. H. Collins, and
E. M. De Robertis
Cell, 59, 81—93, 1989   9-15

The *XlHbox 1* gene of *Xenopus* is expressed in a narrow band across the cervical region of the central nervous system, neural crest, and mesoderm. The gene produces two related proteins, the "long" and "short" homeodomain proteins. The function of this gene was studied by injecting long *XlHbox 1*-specific antibodies into the frog embryo at the one-cell stage. This resulted in tadpoles having cervical spinal cord — located just posterior to the hindbrain — structurally transformed into a hindbrain-like structure. The fourth ventricle was greatly expanded and appeared as a huge vesicle covered by epidermis.

The phenotypic alteration consequent to injecting antibody against long *XlHbox 1* protein was limited to the region normally expressing this protein. Injection of long protein mRNA disrupted segmentation and tissue organization, but did not inhibit cell proliferation. Injection of short protein mRNA into one-cell embryos produced distinct but similar spinal cord malformations. Immunostaining for N-CAM in tadpoles bearing extended hindbrains consistently demonstrated defective organization of spinal nerves over the affected region. The nerves arising from the transformed part of the CNS did not behave as spinal nerves, but rather formed a diffuse network.

This abnormal phenotype is limited to the normal cervical region of expression of the long *XIHbox 1* protein. A lack of properly organized spinal nerves over the transformed region was a consistent finding.

♦ A fundamental problem in developmental biology is understanding the mechanisms whereby the anterior-posterior (AP) and dorsal-ventral (DV) axes are established and used by the embryo to guide differentiation. In reviews of amphibian development, one can find diagrams which imply that the anterior of the embryo is specified by the animal pole of the egg and the posterior by the vegetal. This is misleading. The central role in forming the AP axis is played by the involuting chordamesoderm. The early involuting tissue specifies anterior, and later specifies posterior mesoderm and, indirectly, neural plate. Chordamesoderm arises from the dorsal marginal zone, and the first cells to involute are located more vegetally than the later cells. Thus, the incipient anterior is actually oriented towards the vegetal pole of the egg (see Nieuwkoop, *Curr. Top. Dev. Biol.,* Vol. 11, 115—132, 1977).

How does this axis specification work at the molecular level? Important progress on this problem has been make on two fronts. The paper by Durston et al. describes the very interesting finding that retinoic acid (RA) can posteriorize anterior tissue in frog embryos (see Figure 9-15). The dose required for this effect *in vitro* is in the 10^{-7} M range, and these investigators have shown by HPLC that similar levels of RA exist in early neurula stage embryos, a time when AP specification is taking place in the neural plate. They have also found that ectoderm can respond directly by treating this tissue with a pulse of RA prior to combining it with dorsal mesoderm; posterior neural tissue is induced rather than anterior, which would otherwise result from such a recombination.

It has been known for several years that RA can act as a morphogen in vertebrate limb development. In 1982, Tickle et al. (*Nature*, 296, 564—566) showed that RA treatment posteriorized chick limb bud, a process normally carried out by a cluster of cells at the posterior margin referred to as zone of polarizing activity. By analogy, there should be such a zone in frog embryos, and its identification ought to receive a high priority.

Another area of continuing interest is the expression and role in pattern formation of homeobox-containing proteins in *Xenopus*. Ruiz i Altaba and Melton have extended their studies of Xhox3, which is distributed in a posterior-anterior gradient in early dorsal mesoderm, and has been shown to play an important role in specifying posterior tissue identity, apparently by suppressing anterior characteristics. These authors have found that Xhox3 can be rapidly and efficiently induced in ectoderm by treatment with fibroblast growth factor (FGF) or with substantially less efficiency by "XTC-MIF" (now identified as activin A, see section on mesoderm inducers). Ruiz i Altaba and Melton propose a model for axis formation in which FGF induces posterior/ventral mesoderm and activin A induces anterior/dorsal mesoderm. They suggest the critical early step in this is to adjust the expression level of region-specific homeobox proteins such as Xhox3, which in turn modulate expression of target genes in a manner appropriate for the position along the AP axis.

Wright et al. also report experiments which bear on this tissue, in particular with respect to AP patterning of the central nervous system. They manipulated in various ways the expression of another homeobox protein, X1Hbox1, resulting in conversion of the anterior end of the spinal cord into tissue resembling the hindbrain. This could be considered an "anteriorization" event brought about by reducing the level of the long form propose that posteriorization of anterior neural tissue might involve an increase in X1HBox1, analogous to the effects of Xhox3 expression. The observation that injections complicates this scenario, however.

The diagram below is a speculative (and uncritical) attempt to fit the above results into a coherent picture. In it, endoderm generates two growth factor signals, FGF and Activin A. Adjacent cells recognizes these signals and respond by elevating (FGF) or depressing (Activin A) Xhox3 levels, which in turn converts the responding tissue to posterior or anterior mesoderm, respectively. Anterior/dorsal mesoderm contains Spemann's organizer region and manufactures neural inducer, which converts adjacent ectoderm into an anterior type of central nervous system. Meanwhile, the posterior/ventral mesoderm secretes retinoic acid which results in respecification of posterior nervous system as spinal chord.

A model such as this, while somewhat fanciful, can be useful in helping to identify knowledge gaps. For instance, Xhox3 expression alone does not convert ectoderm into mesoderm. If marginal zone cells are equivalent to ectoderm, a point that is not really settled, then mesoderm, induction must depend on additional control genes, one possible candidate being Mix.1 (Rosa, *Cell*, 57, 965—974, 1989; but see the 1990 Year Book for a discussion). Another assumption in this model is that retinoic acid is manufactured and secreted by posterior mesoderm. This leads to the prediction that this morphogen is absent in eggs and very early embryos, an easily tested hypothesis. We also do not know much about the early genetic response to neural induction, nor do we know the identity of the neural inducer(s). Clearly there is a lot to be done, a favorable situation which makes for exciting science. *Thomas D. Sargent*

The Limb Deformity Gene Is Required for Apical Ectodermal Ridge Differentiation and Anteroposterior Limb Pattern Formation

R. Zeller, L. Jackson-Grusby, and P. Leder
Genes Dev., 3, 1481—1492, 1989   9-16

The mechanisms underlying limb formation are evolutionarily conserved. Experimental findings to date suggest that the two major axes in the limb bud are established by a series of ectodermal-mesenchymal interactions. The authors studied the role of the limb deformity *(ld)* gene in limb morphogenesis by a detailed examination of the morphologic aspects of early embryonic limb formation in the mutant *ld/ld* mouse.

Morphologic differences between wild-type and homozygous ld embryos were apparent on gestational day 10, a time during which anteroposterior limb morphogenesis takes place. A shortened anteroposterior axis caused the mutant limb to appear more pointed than its wild-type counterpart. The apical ectodermal ridge (AER) — which has a critical role in anteroposterior and proximodistal limb development — failed to properly differentiate in mutant *ld* embryos. The limb ectoderm contained a level of *ld* transcripts fivefold higher than the mesenchyme. Studies of the expression of *ld* transcripts in other regions of the developing embryo and in primitive streak embryos suggested a possible role for this gene in the earliest determinative events.

These findings indicate that normal expression of the *ld* gene is necessary for proper differentiation of the murine apical ectodermal ridge and for correct anteroposterior limb patterning.

Disruption of Formin-Encoding Transcripts in Two Mutants *limb deformity* Alleles

R. L. Maas, R. Zeller, R. P. Woychik, T. F. Vogt, and P. Leder
Nature, 346, 853—855, 1990   9-17

Identification of a gene at the murine limb deformity *(ld)* locus permits a test of the hypothesis that disruption of the gene produces an inherited anomaly affecting embryonic pattern formation. The gene produces alternatively processed mRNAs which can be translated as a family of related protein products, the formins.

The authors analyzed transcripts of *ld* in four independently isolated mutant alleles. In two of them, *ld^Hd* (produced by inserting a transgene) and *ld^In2* (created by a translocation-inversion involving mouse chromosomes 2 and 17), a common subset of ld transcripts was abolished while others appeared to be unchanged.

The correlation of altered transcripts in two independent ld mutants supports the view that one or more altered formins are responsible for the phenotype observed. Limitation of the defect to the limb and kidney, despite the expression of *ld* mRNA in unaffected organs, suggests that the mutant alleles represent only partial loss of *ld* function.

"Formins": Proteins Deduced from the Alternative Transcripts of the *limb deformity* Gene

R. P. Woychik, R. L. Maas, R. Zeller, T. F. Vogt, and P. Leder
Nature, 346, 850—853, 1990   9-18

Transgene insertion studies previously demonstrated a mutation at the mouse *limb deformity (ld)* locus which disrupts embryonic pattern formation and results in reduction and fusion of the distal bones and digits of all extremities, as well as renal apolasia. The authors have now characterized the *ld* locus at the molecular level.

The *ld* locus contains evolutionary conserved coding sequences which are transcribed in embryonic and adult tissues as a population of low-abundance mRNAs created by alternative splicing and differential polyadenylation. These transcripts were associated with the gene causing the mutant phenotype by demonstrating that they are disrupted in two independently arising *ld* alleles. Several novel proteins, termed formins, were characterized from the long, open reading frames encoded by various *ld* transcripts. A search of protein sequence databases failed to demonstrate significant homologies with known proteins or amino acid motifs.

The presence of different *ld* transcripts in different tissues suggests that the formins have a role in the formation of several organ systems.

♦ The anteroposterior/proximodistal axes of the vertebrate limb bud are established through a series of ectodermal-mesenchymal interactions controlled by two morphogenetically active regions. One region, known as the apical ectodermal ridge, is a thickened epithelial structure found at the limb-bud tip and establishes axis formation by interacting with the underlying mesenchyme. The mechanisms by which the apical ectodermal ridge exerts its effect is not know. The zone of polarizing activity is a second region located at the posterior margin of the limb bud that acts as a signaling region specifying anteroposterior limb pattern. Retinoic acid has been proposed to be the morphogen produced by the zone of polarizing activity. Although there are several mutations in the mouse that affect limb development, only the limb-deformity locus on chromosome 2 has been identified at the molecular level. The ability to clone the locus was due to a chance insertion of a transgene (described earlier by Woychik et al., *Nature,* 318, 36—40, 1985). There are four known alleles of the limb deformity locus, all of which are recessive. The phenotype is one of synostosis of long bones of all limbs, oligodactyly and syndactyly, and high frequency of uni- and bilateral renal aplasias. These observations suggest that the mutations interfere with anteroposterior limb pattern formation and organogenesis of kidney. The most recent results reported on the locus are a developmental analysis of the limb phenotype in mutant embryos and a description of gene expression in wild-type embryos. A comparative analysis of wild-type and mutant embryos showed that anteroposterior pattern in mutant limb buds is affected as early as day 10 in development. The apical ectodermal ridge appears nondistinct in its cellular morphology from adjacent nonridge limb ectoderm. The level of limb-deformity transcripts is higher in ectoderm relative to mesenchyme of wild-type limb buds and is reported to be abolished in two of the limb-deformity alleles. These data correlate with poor differentiation of the apical ectodermal ridge in mutant embryos. Interestingly, the gene is expressed in early primitive-streak stage embryos (day 7) but the phenotype is not observed until day 10. Furthermore, during limb bud formation, expression of the gene is not restricted to the developing limb but is seen throughout the embryo. Multiple transcripts resulting from alternative splicing and differential polyadenylation of a primary transcript are detected both in the embryo as well as several tissue types of the adult. Alterations in transcripts have been detected in two of the four limb-deformity alleles. The deduced amino acid sequence indicates that the locus encodes novel proteins named formins. Although expression is wide spread, phenotypic abnormalities are associated with limb and kidney. One explanation for this would be that these alleles and resulting proteins retain some function. This question can now be addressed with targeted mutagenesis. The ability to cone a gene affecting limb development is an important step towards understanding pattern formation in the vertebrate limb. However, the difficult work is yet to come in defining a function for the novel formins. *Terry Magnuson*

AUTHOR INDEX

A

Abbas, N. E., 234
Adler, D. A., 54
Albert, P. S., 106
Allen, N. D., 162
Altaba, A. R., 284, 290
Amador-Perez, J. F., 152
Ambros, V., 6
Amemiya, S., 201
Anderson, D. J., 122
Anderson, D. M., 143
Anderson, P., 6, 104
Andre, C., 140
Angerer, L. M., 102
Angerer, R. C., 102
Artavanis-Tsakonas, S., 186, 187
Asami-Yoshizumi, T., 101
Asashima, M., 117
Asouline, G., 63
Atkins, H. L., 149
Awgulewitsch, A., 257

B

Baird, A., 143, 148
Baker, B. S., 224
Balderelli, R. M., 75
Balling, R., 82, 254
Ballinger, D. G., 14
Baltimore, D., 214
Barde, Y.-A., 121
Barker, D. F., 163
Baron, A., 261
Bastian, H., 258
Bauskin, A. R., 61
Bauvois, B., 196
Becker, A., 165
Behringer, R. R., 233
Beier, D. R., 146
Bejsovec, A., 104
Belote, J. M., 223
Ben-Neriah, Y., 61
Benian, G. M., 103
Benzer, S., 14
Berends, F., 34
Bernstein, A., 141
Berta, P., 234, 237
Besmer, P., 141, 146, 147

Best, J., 228
Bhatt, S., 87, 267
Bieber, A. J., 190
Billi, B., 280
Birkenmeier, E. H., 35
Birkett, N. C., 147, 149
Birren, S. J., 122
Bix, M., 45
Blackwell, J., 56
Bolen, J. B., 199
Bolger, G., 63
Boncinelli, E., 267
Bondurant, M. C., 129
Bonifacino, J. S., 92
Bosselman, R. A., 149
Boswell, H. S., 143, 148
Botstein, D., 59
Boyle, M., 261
Bradbury, M. W., 29
Bradley, A., 42
Brady, C., 141
Brandhorst, B. P., 217
Brenner, C. A., 204
Brinster, R. L., 233
Brockdorff, N., 55
Broudy, V. C., 149
Brown, L. G., 236
Brown, N. H., 75
Brown, S. D. M., 49, 55
Brzeska, H., 98
Buccione, R., 137
Buck, J., 146, 147
Bullard, C. D., 60
Burgess, G. S., 148
Burglin, T. R., 243
Burmeister, M, 53
Butler, M. G., 164
Byrne, G. W., 32

C

Calof, A. L., 231
Canning, D. R., 209
Cantor, C., 59
Capecchi, M. R., 43
Capel, B., 235
Carey, J., 163
Carlson, J. R., 24, 25
Cate, R. L., 233

297

Cattanach, B. M., 149
Cavanna, J. S., 49
Ceccarelli, A., 4
Ceci, J. D., 51, 52
Chalfie, M., 10
Chandrasekhar, A., 2
Chang, A. C. M., 4
Chapman, V. M., 54, 57
Chikaraishi, D. M., 231
Cho, B. C., 143
Cho, K. W. Y., 291
Chow, S. C., 239
Chowdhury, K., 85
Chowdhury, K., 260
Chu, T., 146
Coleman, K. G., 244
Coletts, P. L., 262
Collignon, J., 235
Collins, R. H., 291
Cook, G. M. W., 227
Cook, M. N., 267
Cooke, J., 119
Cooley, L., 22
Cooper, G. M., 138
Copeland, N. D., 61
Copeland, N. G., 51, 143
Cosman, D., 143, 148
Costantini, F., 67
Coulson, A., 222, 243
Cox, D. R., 53
Crenshaw, III, B., 30
Culotti, J. G., 71
Cupples, R. L., 147

D

Dahm, L., 193
Da Silva, A. M., 173
d'Auriol, L., 140
Davidson, D. R., 263
Davies, J. A., 227
Deboule, D., 261
Dechant, G., 121
DeChiara, T. M., 37
Decker, G., 217
de Leeuw, W. J. F., 34
De Lozanne, A., 97
Denoyelle, M., 196
De Robertis, E. M., 291
Desplan, C., 245
D'Eustachio, P., 60
Deutsch, U., 260
Deutsch, U., 265
de Vries, N. J., 289
Devreotes, P. N., 100

Dexter, T. M., 132
Diaz, R. J., 75
Distech, C. M., 54
Dixon, J. E., 285
Dixon, R. A., 94
Doetschman, T., 39
Doll, R. F., 17
Dolle, P., 259, 261
Donaldson, D., 136
Donovan, P. J., 143
Dosch, H.-M., 89
Dressler, G. R., 260, 265
Dron, M., 94
Dryja, T. P., 163
Duboule, D., 61, 259
Duffy, K. T., 271
Dunn, J. M., 165
Durston, A. J., 118, 289
Dynes, J. L., 3

E

Economou, A., 235
Efstratiadis, A., 37
Egelhoff, T. T., 4
Eisenman, J., 148
Eldon, E. D., 217
Ellis, M. E., 244
Ellis, N. A., 234
Eppig, J. J., 137
Erjavec, H. O., 147
Ernst, E. G., 204
Erselius, J. R., 260
Eshel, I., 126
Ettensohn, C. A., 175
Ezine, S., 196

F

Fain, P., 163
Falkenstein, H., 259
Faure, M., 69
Featherstone, M. S., 261
Fehon, R. G., 187
Fellous, M., 234
Figueroa, F., 62
Finney, M., 243
Fire, A., 73
First, N. L., 39
Firtel, R. A., 3, 216
Fisher, E. F., 147
Fisher, E. M. C., 236
Flanagan, J. G., 145
Fleming, R. J., 186
Flores, J. C., 147

Foe, V. E., 108
Forejt, J., 62
Forrester, L, 141
Foster, J. W., 237
Franke, J., 69
Freyd, G., 208
Frischauf, A., 237
Froelick, G. J., 233
Frohman, M. A., 261
Fryer, A., 163
Fukui, Y., 97, 98
Fuller, D. L., 172

G

Galibert, F., 140
Galli, S. J., 149
Gallie, B. L., 165
Galliot, B., 261
Gandy, S., 54
Gariepy, J., 169
Gaunt, S. J., 262
Gehring, W. J., 248
Geissler, E. N., 149
Georgi, L. L., 106
Gerhart, J. C., 114
Gerisch, G., 170
Gerwin, N., 280
Giddens, E., 141
Gilbert, D. J., 143
Gimpel, S. D., 143
Gisselbrecht, S., 140
Goddard, A. D., 165
Goff, S. P., 41
Goldberg, L., 205
Golic, K.G., 11
Gonzales-Reyes, A., 248, 249
Goodfellow, P. N., 234, 237
Goodman, C. S., 190
Goodwin, S. F., 15
Gordon, J. W., 29
Gossen, J. A., 34
Goulding, M. D., 260
Graham, G. J., 136
Grant, S. G., 54, 57
Green, E. D., 65
Green, J. B. A., 119
Greenberg, M. E., 37
Greenfield, A. J., 49
Greenwald, I., 7, 179
Griffiths, B. L., 237
Gros, P., 56
Gruss, P., 82, 85, 254, 258, 260, 265
Gubbay, J., 235
Guerrero, I., 278

Gundersen, R. E., 100
Gurdon, J. B., 79
Gwynn, B., 35

H

Haberstroh, L., 216
Hagaman, J. R., 39
Hage, W. J., 289
Hagemann, L. J., 39
Hall, A. L., 69
Hall, D. H., 71
Hamada, H., 83
Hamaoka, T., 240
Hamvas, R. M. J., 55
Han, K., 273
Hand, R. E., Jr., 54
Handyside, A. H., 28
Hardin, J., 205
Hardwicke, J., 291
Hardy, K., 28
Harloff, C., 170
Hartzell, P., 152, 239
Harvey, R. P., 80
Harwood, A. J., 271
Hasbold, J., 130
Hastie, N. D., 61
Hattori, K., 286
Hatzopoulos, A. K., 82
Hawkins, J. R., 237
Hayashi, S., 240
Hayashi, S.-I., 47
Hedgecock, E. M., 71
Heffner, C. D., 229
Heideveld, M., 289
Helfand, S. L., 24, 25
Hendriks, H. F. J., 289
Henry, J. J., 201
Herrera, C., J., 147
Herrmann, B. G., 86, 87
Hewick, R., 136
Hibi, T., 89
Hidalgo, A., 278
Hill, D. P., 155
Hill, R. E., 263
Hirsh, D., 9
Hoch, M., 280
Hodgkin, J., 6, 222
Hoffmann, J. W., 35
Hogness, D. S., 247
Hood., L., 59
Hooper, J. E., 277
Hoppe, P. C., 35
Hopwood, N. D., 79
Horikoshi, M., 245

Horvitz, H. R., 181, 208
Hoshijima, K., 224
Howard, T. A., 60
Hower, G., 119
Hsi, E. D., 197
Hsu, R., 149
Huang, E., 146
Huang, S., 39
Huang, S.-Y., 9
Hui, M. F., 89
Huson, S. M., 163
Huylebroeck, D., 118

I

Imhof, B., 196
Ingham, P. W., 278
Inoue, K., 224
Ishihara, K., 101
Ishii, K., 117
Ishimura, Y., 127
Isola, L. M., 29
Izpiua-Belmonte, J., 259

J

Jackle, H., 280
Jackson-Grusby, L., 294
Jacobs, D., 257
Jacobsen, F. W., 147, 149
Jadayel, D., 163
Jaenisch, R., 44, 45
Jenkins, N. A., 51, 52, 61, 143
Jenkinson, E. J., 131
Jeong, H. D., 90
Jermyn, K. A., 271
Johnson, D. K., 54
Johnson, K. J., 49
Johnson, M. J., 147
Johnson, R. S., 37
Jonas, E. A., 81
Jondal, M., 152, 239
Jones, E. A., 283
Jordan, C. T., 134
Justice, M. J., 51, 61

K

Kafatos, F. C., 75
Kaiser, D., 174
Kaiser, K., 15
Kamboj, R. K., 169
Kappler, J. W., 40
Karam, S., 164
Karkare, S. B., 149

Karpen, G. H., 110
Kasahara, M., 62
Kaufman, P. D., 17
Kawaguchi, H., 62
Kay, R. R., 271
Keer, J. T., 55
Kenyon, C., 177
Kessel, M., 254
Kessin, R. H., 69
Keynes, R. J., 227
Kiessling, A. A., 138
Kiff, J. E., 103
Kim, S. K., 174, 208
Kim, S., 53
Kincade, P. W., 240
King, D. L., 75
King, T. R., 86
Kingsley, D. M., 52
Kingston, R., 131
Kinoshita, K., 117
Kintner, C. R., 285
Kizaki, H., 127
Klaus, G. G. B., 130
Klausner, R. D., 92
Klein, C., 2
Klein, C., 173
Klein, J., 62
Klein, W. H., 205, 217
Klymkowsky, M. W., 161
Knapik, E. W., 265
Knoll, J. M. H., 164
Knook, D. L., 34
Koller, B. H., 38, 39, 40
Kontogianni, E. H., 28
Kooh, P. J., 187
Koopman, P., 235
Korn, E. D., 98
Kornberg, T. B., 244
Koster, C. H., 118
Koury, M. J., 129
Kraft, B., 2
Krall, M., 56
Krantz, D. E., 185
Kruisbeck, A. M., 199
Krumlauf, R., 267
Kunishada, T., 47
Kyle, J. W., 35

L

Labeit, S., 86
Lacombe, C., 140
Lahm, H., 146
Lalande, M., 164
Lam, P., 89

Lamb, C. J., 94
Landmesser, L., 193
Lane, D. P., 271
Langley, K. E., 147, 149
Leder, P., 145, 146, 294, 295
Lehrach, H., 86
Lemischka, I. R., 134
Lengyel, J. A., 75
Lennarz, W. J., 205
Leslie, I., 147
Levi, E., 147
Levine, M., 252, 273
Li, E., 44
Li, S., 30
Liang, X., 94
Lin, C., 147
Lindquist, S., 11
Lock, L. F., 5
Lohman, P. H. M., 34
Loomis, N. F., 172
Lorimore, S., 136
Loring, J. M., 45
Lovell-Badge, R., 235, 237
Lu, H. S., 147, 149
Lumsden, A. G. S., 229
Lyman, S. D., 143, 148
Lynch, T. J., 98

M

Maas, R. L., 294, 295
Maeda, N. 39
Maguire, J. E., 92
Manley, J. L., 273
Mann, R. S., 247
Manstein, D. J., 4
March, C. J., 143, 148
Marks, A. R., 49
Marrack, P., 40
Martin, F. H., 147, 149
Martin, G. R., 261
Martin, U., 148
Martin-Blanco, E., 244
Mathew, C. G. P., 163
McCarthy, S. A., 92
McConkey, D. J., 152, 239
McConnell, G. K., 54
McGillivray, B., 236
McKearn, J. P., 134
McKee, B. D., 110
McKenna, M., 25
McKeown, M., 223
McMahon, A. P., 42, 113
McNiece, I. K., 147, 149
McRobbie, S. T., 271

Medina, K. L., 240
Melchers, F., 210
Melton, D. A., 159, 160, 290
Mendiaz, E. A., 147, 149
Mercer, J. A., 61
Merriam, J. R., 75
Meyer, B. J., 220, 221
Miyake, K., 240
Mizuno, K., 62
Mochizuki, D. Y., 148
Mock, B., 56
Moerman, D. G., 103
Monte, P., 25
Montpetit, I. C., 217
Moon, R. T., 113
Morata, G., 248, 249
Morris, C. F., 147
Moschel, R. C., 137
Mukai, S., 163
Muller, C. M., 228
Mullins, L. J., 57
Munsterberg, A., 235
Muramatsu, J., 127
Muramatsu, M., 83
Murdock, D. C., 149
Murphy, P., 263
Muskavitch, M. A. T., 187
Mutter, G., 254
Myers, R. M., 53
Mynell, L. A., 161

N

Nadal-Ginard, B., 49
Nagoshi, R. N., 224
Nakano, H., 117
Nakano, Y., 278
Nakayama, M., 195
Nakayama, T., 92, 197
Nauber, U., 280
Neckelmann, N., 103
Nguyen, T., 217
Nicholls, R. D., 164
Nieuwkoop, P. D., 289
Nishikawa, S., 47
Nishikawa, S.-I., 47
Nocka, K., 141, 146, 147
Noegel, A. A., 170
Noguchi, S., 101
Noncente-McGrath, C., 204
Nonoyama, S., 195
Nornes, H. O., 260, 265
Norris, M. L., 162
Nose, A., 194
Nusbaum, C., 220

Nusslein-Volhard, C., 275

O

O'Brien, A., 56
Odaka, C., 127
Ogawa, M., 47
Ohkuma, Y., 245
Okamoto, K., 83
Okamura, H., 47
Okazawa, H., 83
O'Keefe, S. J., 138
Okino, K. H., 147, 149
Okuda, A., 83
O'Leary, D. D. M., 229
Olson, E. N., 78
Olson, M. V., 65
Olson, M., 59
O'Neill, H. C., 151
Ono, S., 240
Orrenius, S., 152, 239
Osmond, D. G., 125
Owen, J. J. T., 131
Ozato, K., 84

P

Pachnis, V., 67
Page, D. C., 236
Page, D., 55
Palacios, R., 241
Palmer, M. S., 234, 237
Palmiter, R. D., 233
Pankratz, M. J., 280
Papp, A., 6
Park, L. S., 148
Park, Y.-H., 125
Parker, V. P., 147
Patapoutian, A., 75
Patel, A. C., 147
Patel, N. H., 244
Paton, K. E., 165
Paules, R. S., 137
Pelkonen, J., 241
Perry, M., 78
Petersen, R., 163
Pevny, L., 67
Phillips, R. A., 165
Pignoni, F., 76
Pluck, A., 79
Podgorski, G. J., 69
Poirier, F., 84
Ponder, B. A. J., 163
Ponder, M. A., 163

Poole, S. J., 244
Pope, J. A., 147
Potter, M., 56
Poutska, A., 86
Pragnell, I. B., 136
Pravtcheva, D., 262
Price, E. R., 53
Pulak, R., 6

R

Raff, R. A., 201
Rapaport, J. M., 163
Rastan, S., 55
Rauch, C., 143, 148
Raulet, D. H., 45
Rawson, E. J., 30
Ray, P., 141
Rebay, I, 186, 187
Regan, C. L., 187
Reinitz, J., 252
Reith, A. D., 141
Renucci, A., 259
Riddle, D. L., 106
Rigby, P. W. J., 84
Rinchik, E. M., 54
Rio, D. C., 17, 19
Robertson, E. J., 37, 41
Rodgrigues-Tébar, A., 121
Roeder, R. G., 245
Rolink, A., 210
Rosen, L. L., 60
Rosenfeld, M. G., 30
Rosner, M., H., 84
Roth, S., 275
Rothstein, R., 67
Rotman, M., 2
Rottapel, R., 141
Ruddle, F. W., 32
Ruppert, S., 85
Rushlow, C. A., 273
Rutishauser, U., 193
Ruvkun, G., 243

S

Sachdev, R. K., 147
Sachdev, R. K., 149
Sajjadi, F., 44
Sakai, M., 83
Sakamoto, H., 224
Samaridis, J., 241
Samelson, L. E., 197
Sargent, T. D., 81

Satyagal, V. N., 149
Savion, N., 126
Scales, J. B., 78
Schedl, T., 179
Schimenti, J. C., 60
Schmid, J., 94
Schnabel, H., 207
Schnabel, R., 207
Schöler, H. R., 82, 85
Schurr, E., 56
Schwartzberg, P. L., 41
Scott, M. P., 277
Scottgale, T. N., 186
Seifert, E., 280
Seldin, M. F., 60
Seydoux, G., 7, 179
Shakes, D. C., 155
Sharpe, P. T., 262
Sheng, M., 37
Shibai, H., 117
Shimada, K., 117
Shimura, Y., 224
Shiohara, T., 195
Shoham, J., 126
Shultz, L. D., 47
Siebel, C. W., 19
Silan, C. M., 52, 61
Silver, L. M., 60
Simister, N. E., 45
Simmons, D. M., 30
Sinclair, A. H., 234, 237
Singer, A., 92, 197
Singer, D. S., 92
Siracusa, L. D., 51, 61
Siu, C., 169
Sive, H. L., 286
Skaer, H., 111
Skamene, E., 56
Sly, W. S., 35
Smith, C. A., 131, 132
Smith, J. C., 119
Smith, K. A., 147, 149
Smith, M. J., 237
Smithies, O., 38, 39, 40
Snape, A. M., 81
Snoek, G. T., 118
Snow, P. M., 190
Sokol, S., 159
Soll, D. R., 2
Sosnowski, B. A., 223
Spence, A. M., 222
Spradling, A. C., 22
Spudich, J. A., 4, 97
Staudt, L. M., 84

Stein, D., 275
Steingrimsson, E., 75
Stephenson, D. A., 54, 57
Stern, C. D., 209, 227
Stern, M. J., 181
Steward, R., 274
Stewart, R. M., 114
Strasser, A., 210
Strome, S., 155
Struhl, G., 156, 248
Studer, S., 241
Subramani, S., 44
Sudo, T., 47
Suggs, S. V., 147, 149
Surani, M. A., 162
Suyemitsu, T., 101
Suzuki, N., 82, 85
Swanson, L. W., 30
Symes, K., 119

T

Tadakuma, T., 127
Takeichi, M., 194
Takeishi, T., 149
Tan, C. H. T., 34
Tanb, J. C., 141
Tang, J., 193
Tannahill, D., 160
Taylor, A., 278
Taylor, D. R., 132
Teale, J. M., 90
Thiery, J., 196
Thomas, J. H., 181, 218
Thomas, K. R., 43
Thompson, D., 22
Timmermans, J. P. M., 289
Timmons, P. M., 84
Ting, J., 147
Tonegawa, Y., 101
Traktman, P., 141
Traynor, D., 271
Tsuji, K., 194
Tung, W., 149

U

Ueno, N., 117
Upadhyaya, M., 163
Urguia, N., 248

V

Valdizan, M. C., 217

Van Ness, K., 148
van den Eijnden-Van Raaij, A., J., M., 118
van Nimmen, K., 118
van Zoelent, E. J. J., 118
Vande Woude, G. F., 137
Veillette, A., 199
Vigano, M. A., 84
Vigneron, M., 261
Vijg, J., 34
Villeneuve, A. M., 221
Vivian, N., 235
Vogler, C., 35
Vogt, T. F., 294, 295
Voncek, V., 62

W

Walton, D., 163
Ward, S., 155
Waring, D. A., 177
Waterston, R. H., 73, 103
Weintraub, H., 286
Wellner, D., 146
Wen, D., 147
Wessel, G. M., 205
Wharton, R. P., 156
Whittle, J. R. S., 278
Wieland, I., 63
Wigler, M., 63
Wignall, J. M., 143
Wilcox, M., 75
Wilkie, N. M., 136
Wilkinson, D. G., 87, 267
Williams, D. A., 149
Williams, D. E., 143, 148
Williams, G. T., 131, 132
Williams, J. G., 271
Williams, K. L., 4
Williams, L. R., 149

Williams, P. J., 39
Wilson, J. H., 44
Winoto, A., 214
Winston, R. M. L., 28
Wisenman, J., 143
Wolfes, H., 138
Wolinsky, E., 10
Wolpe, S. D., 136
Woodard, C., 25
Woodland, H. R., 283
Woychik, R. P., 294, 295
Wray, G. A., 201
Wright, C. V. E., 291
Wright, E. G., 136
Wypych, J., 147, 149

X

Xu, T., 186, 187

Y

Yandell, D. W., 163
Yang, Q., 102
Yata, J., 195
Yisraeli, J. K., 159
Yoshida, H., 47
Yuschenkoff, V. N., 149

Z

Zaleska-Rutczynska, Z., 62
Zeller, R., 294
Zheng, H., 44
Zhu, X., 165
Zijlstra, M., 44, 45
Zipursky, S. L., 185
Zsebo, K. M., 147, 149
Zuniga-Pflucker, J. C., 199
Zwarthoff, E. C., 34

SUBJECT INDEX

A

Activin A, 117—118
Adenosine phosphoribosyltransferase (APRT), 51—52
Adhesion proteins, 71
Adult spinal cord, differential expression of *Hox 3.1* protein in, 257
Altered transcripts, correlation of, 294
Amphibian embryos, 285
Amplified cell lines, gene targeting in, 44
Anterior spinal cord, malformations of, 291—291
Anteroposterior axis, progressive determination during formation of, 286—289
Anteroposterior limb pattern formation, 177—179, 294
Anthocidaris crassispina, 101
Antisense oligonucleotides, 139—140
Apical ectodermal ridge differentiation, 294
Apoptosis, 131—132
APRT, see Adenosine phosphoribosyltransferase
Avian epiblast, 210
Axonal growth, directional inhibition of, 227

B

B cell lymphomas, 130—131
Backcross offspring, initial characterization of, 58
Bacteriophage lambda shuttle vectors, 34—35
BDNF, see Brain-derived neurotrophic factor
Blastomeres, changed identity of, 208
Body plan formation, 245
Bone marrow, stromal cells in, 240
Bone marrow cells, transplants of, 47—49
Bone marrow cultures, 240—241
Bone remodeling, capacity for, 48
Brachyury gene, 86—87
Brain-derived neurotrophic factor (BDNF), 121—122
5′-Bromodeoxyuridine, 111

C

c-abl, homologous recombination of, 41
Cadherin cell adhesion molecules, 194—195
Caenorhabditis elegans, 155—156
  gain-of-function mutations of, 7—9
  informational suppression in, 6—7
  inherited neurodegeneration in, 10—11
  RNA leader sequence in, 9—10
Calcium ionophores, 127—129, 153c-*kit* defects, 141
C. elegans, 243—244
  circumferential migrations in, 71—72
  egg-laying system, 181—185
  embryo, early determination in, 207—208
  genetic analysis of defecation in, 218—220
  myosin activity in, 103—104
  potential inductive interaction in, 179—181
  thick filament assembly, 104—106
  transformation of, 73—75
CD4+CD8+ thymocytes, intrathymic signaling in, 197—198
CD4−8+ cytolytic cells, 45—47
Cell adhesion, 185
Cell cycle control proteins, 222—223
Cell division, patterns of, 108
Cell interactions, 169—200
Cell lineage, 201—214
Cell-adhesion proteins, 186
Cell-cell adhesion molecules, genetic analysis of, 191
Cement gland, 286
Central nervous system, anteroposterior transformation in, 289
Cervical vertebrae, variations of, 254—257
C-factor, 174—175
Chaoptin, 186
Chick somites, isolation of glycoprotein fraction, 227—228
Chick embryo, mesoderm in, 209—210
Chromosomal composition, 221
Chromosome atlas cytological map, 61
Cilia, ontogeny of, 218
c-kit receptor system, 141, 146

c-*kit* transmembrane kinase receptor, 147—148
c-*kit* tyrosine kinase receptor, 149—150
Cloned DNA, 67
c-*mos* oligonucleotides, 138
Colony stimulating factors, 132—133
Contact site A protein, selective elimination of, 170—172
Craniofacial anomalies, 254
Cross-regulatory interactions, 248—249
CS, see Culture supernatant
Culture supernatant (CS), 127
Cyclic nucleotide phosphodiesterase, 69—71
Cytoarchitectural changes, 257
Cytodifferentiation, 215—242
Cytokeratin filament organization, 161—162
Cytokinesis, 4
Cytoplasmic determinants, 155—167

D

D. discoideum development, aggregation stage of, 170—171
D. melanogaster, 247—248
Dauer larva development, 106—107
Dedifferentiation, 2—3
deg-1 mutation, 10
Deletion mutations, 24
Delta, protein products of, 187—190
Designer deletions, 22
Developing excretory system, 260—261
Developmental cell biology, 97—153
Developmental fate, 201—214
Developmental gene expression, 69—95
Developmental genetics, 1—68
Dictyostelium amoebae, 98—99
Dictyostelium discoideum, 169—170
    complementation of myosin null mutants in, 4—6
    complementation of, 4
    cyclic nucleotide phosphodiesterase gene of, 69—71
    development, prestalk-prespore pattern in, 271—273
    erasure mutant, 2—3
    genetic marker in, 3—4
    G protein in, 100—101
    multicellular assemblies of, 272
    myosin-defective mutant, 97—98
    spatial gradient of expression, 216—217

Difference cloning, method for, 63—65
Dilute-short ear deletion complex, 63
DNA
    amplification, 29
    -binding motif, 237—239
    clones, 55, 61
    fragmentation, 127—129
    markers, 62—63
    probes, mapping of, 49
    sequences, internal P element, 17—19
    synthesis, 129
Dorsal morphogen, graded distribution of, 273
Dorsal root ganglia (DRG), 228
Dorsal root ganglion growth cones, 227—228
dorsal protein
    gradient of nuclear localization of, 275—277
    relocalization of, 274—275
Dorsoanterior-ventroposterior (DV) axis, 120
DRG, see Dorsal root ganglia
Drosophila, 264
    anteroposterior body pattern of, 156—159
    BicaudalD protein, 156—159
    chaoptin, 185—186
    chemosensory response in, 25—28
    constructing endpoints in, 22—24
    dorsal-ventral polarity genes of, 274
    embryo, 280
        abdominal segmentation in, 280—281
        early commitment of cells in, 108—109
        Ubx product in, 249—252
    first instar larva of, 248
    genetics, chromosomal deletions in, 22—24
    homeodomain proteins, 245—247
    integrin PS2 α transcripts, 75—76
    melanogaster, 11—14
    mutagenesis of, 15—17
    P-element transposase from, 17—19
    ribosomal RNA genes, 110—111
    segmentation, 244
    selected nuclear transport in, 273
    sex-lethal gene, 224
    targeted gene mutations in, 14—15
Dual axis formation, 114
Duchenne muscular dystrophy, 54
DV, see Dorsoanterior-ventroposterior axis
Dwarf locus mutants, 30—32

E

Early amphibian embryos, mesodermal induction in, 117—118
Early embryogenesis, 78—79, 155—156
EC, see Epithelial cells
Ectoderm, patterning neural signal in, 284—285
EGF, see Epidermal growth factor
Egg-laying system, 182
Embryo, division of, 246
Embryonic induction, 287
Embryonic keratin expression, 81—82
Embryonic spinal cord, differential expression of *Hox 3.1* protein in, 257
Embryonic stem (ES) cells, 37—38, 44—45
Embryonic termini, expression of *tailless*, 76—78
Endogenous releasing activity, 173
engrailed proteins, expression of, 244—245
Enzymatic amplification, 29—30
Epidermal growth factor (EGF), 101—103
Epidermis, basal lamina of, 72
Epithelial cell rearrangement, 205—207
Epithelial cells (EC), 126
Epo, see Erythropoietin
Erythropoietin (Epo), 129—130
ES, see Embryonic stem cells
Eucidaris tribuloides, 206
even-skipped homologue, 258
Experimental embryology, 203
Experimental embryos, 116
Extension sequencing, 9

F

FACS, Fluorescence-activated cell sorting
Fasciclin III, 190—193
Fetal hematopoietic, 134—136
Fetal human bone marrow, B cell ontogeny of, 89—90
Fetal liver, B lymphocyte lineage development, 213
Fetal mouse liver, 212
Flies, behavior of, 25—28
FLP recombinase, 11—14
FLP recombination products, 12
FLP recombination target (FRT), 11—14
Fluorescence-activated cell sorting (FACS), 122—124

Formin-encoding transcripts, disruption of, 294—295
Friend leukemia virus, 129—130
Frog embryo, antero-posterior polarity in, 290—291
FRT, see FLP recombination target
fushi tarazu, 252
Fusion gene product, 95

G

$G\alpha 2$ eletcrophoretic mobility transition, 100
Gene disruption, 170—172
Gene expression, developmental regulation of, 76
Gene expression domains, 259
Gene targeting, 37—38
Genetic imprinting, 164
Genomic difference cloning, flow diagram for, 64
Genomic DNA library, 3—4
Genotype-specific modifier genes, 162
Germ line chimeras, 42
Germ-line transmission, 39—40, 44—45
Germline mutation, 165—167
GHI, see Human Genome Initiative
β-Glucuronidase activity, 94
β-Glucuronidase transgene, 35—36
Glycine, 84
Glycosyl-phosphatidylinositol degrading activity, 173—174
Gonad primordium, 184
Granulocyte-macrophage colony-forming cells, 136—137
Growth-deficiency phenotype, 37
Guanine-rich sequences, 172

H

Haptoglobin, 52
Hematopoietic growth factor, 146
Hematopoietic stem cell factor, 149
Hematopoietic stem cells, characteristics of, 133
Hemopoietic stem cell proliferation, 136—137
Hermaphrodite, diagram of anal muscles in, 219
Heterozygotes, immune function of, 45
High-affinity NGF receptor, 122
hmm protein, cytoplasmic distribution of, 98
HNK-positive cells, lineage mapping of, 211

Homeobox gene, ectopic expression of, 254

Homeobox genes, 243—269

Homeodomain gene products, 247

Homeotic gene expression, 252—254

Homeotic genes, 252

Homologous recombination, difference in frequencies, 44

Homozygous mutants, 45—47

Hox 1.4 gene, 261—262

Hox 3.4 gene, embryonic expression patterns, 262—263

Hox-2 genes, segmental expression of, 267—269

Hox-2.9 gene, isolation of, 261

Hox-3.2 gene, expression pattern of, 260—261

HPRT, see Hypoxanthine phosphoribosyltransferase gene

Human chromosome 19q, myotonic dystrophy region of, 49—50

Human fetal bone marrow, 89—90

Human genome, physical mapping of, 59

Human Genome Initiative (HGI), 67

Human genome project, 59

Human implantation embryos, 28—29

Human YAC libraries, strategy for screening, 66

Hypoxanthine phosphoribosyltransferase (HPRT) gene, 39—40

I

Identical genomes, specific differences between, 65

Immature thymocytes, 152—153, 199, 239—240

Imprinting, 155—167

Informational suppression, 6—7

Integral membrane glycoproteins, 194

Integral membrane protein, 173

Integrated shuttle vectors, rescue of, 34—35

Interleukin 1, 239—240

Interleukin-2, expression of, 152

Interspecific backcross linkage map, 51—52

Intramuscular nerve branching, 193—194

J

Jump response assay, 26—28

K

kit ligand, 145—146

knirps gene products, 280—283

Kruppel gene products, 280—283

Kruppel requirement, 280

L

Lactotrophs, absence of in mature mouse, 30—31

Limb deformity gene, 294

Limb deformity locus, 296

Limb pattern formation, 259—260

limb deformity alleles, 294—295

limb deformity gene, 295—296

lin-12 protein, 8

Lineage specificity, 214

Lineage-specific gene expression, 214

Local polymorphisms, 60—61

Low-affinity NGF receptor, 122

Lsh-Ity-Bcg disease resistance locus, 56—57

luciferase gene, 216

Lytechinus pictus, 201

M

Macrophage colony stimulating factor gene, 47—49

Malformations, anterior expression boundaries of, 256

Malignant hyperthermia, 49

Malpighian tubule cell development, 113

Malpighian tubule development, 111—113

Mammalian brain, target control of, 229—230

Mammalian development, genes essential for, 54—55

Mast cell growth factor (MGF), 143—144

Maternal controls, 155—167

Maternal effect lethal mutations, 207

Maternal mRNA, 80—81, 159—160

Mature immune system, 135

MC, see Mesenchymal cells

Meiotic crossovers, increased frequencies of, 54

Meiotic recombination mapping, 24—25

Membrane-spanning domains, 278—279

Mesenchymal cells (MC), 126

Mesenchymal cell types, 202

Mesoderm formation, 86—87
Mesoderm-inducing factor (MIF), 119—120
Metallothionein-1, 52
MGF, see Mast cell growth factor
MHC class I proteins, 40—41
β_2-Microglobin locus, inactivation of, 38—39
Mid-gastrula-stage embryos, immunofluorescence studies in, 204—205
MIF, see Mesoderm-inducing factor
MIS, see Müllerian inhibiting substance
Missense mutation, 141—142
Molecular complementation, 3—4
Molecular genetic linkage map, 48, 61—62
Molecular mapping, 54—55
Monoclonal antibodies, 209, 240—241
Morphogenesis, 271—296
Mos gene product, 138
Mouse albino-deletion complex, 54—55
Mouse chimeras, 40—41
Mouse chromosome 2
  genetic linkage map of, 62
  molecular genetic linkage map of, 61—62
Mouse chromosome 4, genetic linkage map of, 51
Mouse chromosome 10, linkage map of, 144
Mouse egg, 138—140
Mouse embryonic cells, 83—84
Mouse fetal liver, B cell development in, 210—213
Mouse hindbrain, homeobox-containing gene in, 263—265
Mouse thymidylate synthase gene, 4
Mouse X chromosome, 54, 57—59
Mouse X-inactivation center, 55—56
Mouse Y chromosome, sex-determining region of, 235—236
MPSVII, see Mucopolysaccharidosis type VII
mRNA expression, 140
Mucopolysaccharidosis type VII (MPSVII), 35—36
Multiplex gene regulation, 32—34
Murine bone marrow, 196—197
Murine *int*-1, targeted disruption of, 43
Murine mucopolysaccharidosis storage disease, 36
Murine mutation osteopetrosis, 47—49

Mus domesticus, 51
Mus spretus, 51, 60—61
Muscle gene expression, 80
Muscular dystrophy, 49
Mutant gene, male transmission of, 47
Mutant strains, 5
Mutations, parental origin of, 163
Müllerian inhibiting substance (MIS), 233—234
Myosin genes, proper expression of, 73—75
Myosin isoforms, distribution of, 99
Myosin, 105
Myotonic dystrophy, 49

N

N locus, mutations in, 189
N-CAM, 193
N-CAM, see Neural cell adhesion molecule
National Research Council Committee, 59
Nematode body wall, 72
Nematode, sex differentiation in, 223
Neoplasms, development of, 41
Nerve branching pattern, development of, 192
Nerve growth factor (NGF), 121—122
Nerve growth factor receptor, 121—122
Neural cell adhesion molecule (N-CAM), 104
Neural expression, 284—285
Neural induction, need for mesoderm in, 284
Neural tube, longitudinal organization of, 266
Neurogenesis, analysis of in mammalian neuroepithelium, 231—233
Neuronal control of defecation, genetic pathway for, 220
Neuronal differentiation, stages of, 124
NGHF, see Nerve growth factor
Nonexpressed genes, targeting of, 37—38
Normal cell lines, gene targeting in, 44
Northern blot analysis, use of Oct-4 probe in, 85—86
Notch
  -*Delta* interactions, 188
  intracellular domain of, 187
  locus, 186—187
  protein products of, 187—190

Nuclear transport, 276

O

Observed phenotypes, model to explain, 180
Oct factor, germline-specific expression of, 82—83
Octamer binding protein, 82—83
Ocular dominance plasticity, 228—229
Ocular-dominance plasticity, 229
Olfactory epithelium, 232
Olfactory mutant, 24—25
Olfactory receptor neurons, 231
Olfactory T maze, 25
Oligodeoxyribonucleotide competitors, 81—82
Oogenesis, 137—138
Osteosarcoma susceptibility, 52
Osteosarcoma, 163, 166

P

P elements, deletion of DNA between, 23
P-element-mediated mutagenesis, 15
Parental alleles, modification of, 164
patched gene, 277—278
Pattern formation, 271—296
Pax2 murine neurogenesis, 265—266
PCD, see Programmed cell death
PCR, see Polymerase chain reaction
Peptide growth factors, 290—291
PFGE, see Pulsed-field gel electrophoresis
Phenotypic suppression, 249
Phenylalanine ammonia-lyase-β-glucuronidase, 94—95
Phorbol esters, 127—129
Phorbol esters, DNA fragmentation by, 129
Pluteus larva, mRNA in, 103
PMC, see Primary mesenchyme cells
Polymerase chain reaction (PCR), 15—17, 104
Polysialic acid (PSA), 193—194
POU domain, 85—86
POU-domain transcription factor, 31—32, 84—85
Preimplantation mouse embryos, identification of sex of, 29—30
Prespore zone gradient, model for, 217
Primary axonal outgrowth, 230
Primary mesenchyme cells (PMCs), 175—176
Primary thymic epithelial cells, 242

Programmed cell death (PCD), 152
Proline, 84
PSA, see Polysialic acid
Pseudogene-1, 52
Pulsed-field gel electrophoresis (PFGE), 53—54

R

RA, see Retinoic acid
Radiation hybrid mapping, 53—54
Rat hemopoietic precursor cells, 196—197
Receptor affinity reagent, 145
Receptor protein tyrosine kinases, 142
Recombinants, screening of, 58
Regulator genes, 63
Regulatory relationships, summary of, 253
Resting B cells, 90—92
Restriction fragment length polymorphisms (RFLP), 53—54
Retinoblastoma gene, mutations of, 163
Retinoblastoma, 163
Retinoic acid (RA), 83, 289
RFLP, see Restriction fragment length polymorphisms
RNA splicing, regulation of sex-specific, 224—227
RNA synthesis, 129
RNA, sex-specific alternative splicing of, 223—224

S

Saccharomyces cerevisiae, 11—14, 67—68
SCF genomic sequences, 149
SCF, see Stem cell factor
Sea urchin, gastrulation of, 205—207
Sea urchin embryo, 102—103
  cell interactions in, 175—176
  endoderm of, 205
  fates of mesomeres in, 201—204
  lineage-specific protein of, 204—205
Sea urchin Spe3 protein, localization of, 217—218
Secondary mesenchyme cells (SMCs), 175—176
Segment patterns, generation of, 278
Segmental patterning, 277—278
Segmentation, 263—265
Selective neural degeneration, 11
Sequence-tagged sites (STS), 59
Serine, 84
Sex myoblasts, 181

Sex-determination genes, 222
Shuttle vectors, 35
SMC, see Secondary mesenchyme cells
Somatic gonadal cells, 182
Somatic repression, 19—22
Somatotrophs, absence of in mature
    mouse, 30—31
Somites, MyoD expression in forming,
    79—80
Sporogenesis, 4
Stage-specific genes, 33—34
Staphylococcus aureus, 132
steel locus, 143
Stem cell factor (SCF), 149
Stem cell factor DNAs, 147
STS, see Sequence-tagged sites
Subtractive hybridization, 63—65
Superantigens, 132
Sympathoadrenal lineage, 123—124

T

t complex responder (Tcr), 60
t complex responder genetic locus, 60
T-cell localization, 151
Targeted gene disruption, 41—42
TCR expression, post-translational
    regulation of, 92—94
Tcr, see *t complex responder*
Terminal deoxynucelotidyl transferase,
    125—126
Terminal domains, activation of terminal-
    specific genes in, 76
Testis-determining factor, 237
Threonine, 84
Thymic epithelial cell line, supernatant
    from, 197
Thymic epithelial cells, 195—196, 241—
    242
Thymic epithelium, 196—197
Thymic stromal cell subpopulations,
    126—127
Thymocytes, DNA fragmentation of, 128
Thymus glands, 239
Thymus, 151
Thyrotrophs, absence of in mature
    mouse, 30—31
Tissue-specific splicing, 20
Transgene expression, epigenetic control
    of, 162—163
Two-tiered multiplex regulatory system,
    33
Tyrosine kinase receptor, 148
Tyrosine phosphorylation pathways,
    activation of, 199—200

Tyrosine phosphorylation, 198

U

Ultrabithorax proteins, 247—248

V

Vertebral abnormalities, 255
Vertebrate hindbrain, 267
V_H gene families, 91—92
Von Recklinghausen neurofibromatosis,
    163—164
Vulval induction, 183

W

white spotting mutations, 150
Wnt-1 proto-oncogene, 42—43

X

Xenopus
    alteration of cell identity, 81
    bFGF, 119—120
    blastula, 290
    cell line, 118—119
    development, 160—161
    embryonic axis, dorsal development of,
        114—116
    embryos, 113—114
        homeobox gene in, 291—293
        neural induction in, 285—286
    homeobox gene, 284—285
    β globin gene, 82
    neural induction in, 283
    mesoderm-inducing factor, 117
    mid-blastula embryos, 80
    oocyte, 161—162
XX animals, 220—221

Y

Y chromosome repeat, enzymatic
    amplification, 29—30
Yeast, homologous recombination in,
    12—13
Yeast artificial chromosome
    libraries, 65—67
    transfer of, 67—68

Z

ZFY, 234—235
Zfy gene expression, 235